The Scientific Counter-Revolution

Bloomsbury Studies in the Aristotelian Tradition

General Editor:
Marco Sgarbi, Università Ca' Foscari, Italy

Editorial Board:
Klaus Corcilius *(University of California, Berkeley, USA)*; Daniel Garber *(Princeton University, USA)*; Oliver Leaman *(University of Kentucky, USA)*; Anna Marmodoro *(University of Oxford, UK)*; Craig Martin *(Oakland University, USA)*; Carlo Natali *(Università Ca' Foscari, Italy)*; Riccardo Pozzo *(Consiglio Nazionale delle Ricerche, Rome, Italy)*; Renée Raphael *(University of California, Irvine, USA)*; Victor M. Salas *(Sacred Heart Major Seminary, USA)*; Leen Spruit *(Radboud University Nijmegen, The Netherlands)*.

Aristotle's influence throughout the history of philosophical thought has been immense and in recent years the study of Aristotelian philosophy has enjoyed a revival. However, Aristotelianism remains an incredibly polysemous concept, encapsulating many, often conflicting, definitions. *Bloomsbury Studies in the Aristotelian Tradition* responds to this need to define Aristotelianism and give rise to a clear characterization.

Investigating the influence and reception of Aristotle's thought from classical antiquity to contemporary philosophy from a wide range of perspectives, this series aims to reconstruct how philosophers have become acquainted with the tradition. The books in this series go beyond simply ascertaining that there are Aristotelian doctrines within the works of various thinkers in the history of philosophy, but seek to understand how they have received and elaborated Aristotle's thought, developing concepts into ideas that have become independent of him.

Bloomsbury Studies in the Aristotelian Tradition promotes new approaches to Aristotelian philosophy and its history. Giving special attention to the use of interdisciplinary methods and insights, books in this series will appeal to scholars working in the fields of philosophy, history and cultural studies.

Available titles:
The Aftermath of Syllogism, edited by Marco Sgarbi, Matteo Cosci
A Political Philosophy of Conservatism, by Ferenc Hörcher

The Reception of Aristotle's Poetics in the Italian Renaissance and Beyond,
by Bryan Brazeau
Early Modern Aristotelianism and the Making of Philosophical Disciplines,
by Danilo Facca
Phantasia in Aristotle's Ethics, by Jakob Leth Fink
Virtue Ethics and Contemporary Aristotelianism, edited by Andrius Bielskis,
Eleni Leontsini, Kelvin Knight

The Scientific Counter-Revolution

The Jesuits and the Invention of Modern Science

Michael John Gorman

BLOOMSBURY ACADEMIC
LONDON • NEW YORK • OXFORD • NEW DELHI • SYDNEY

BLOOMSBURY ACADEMIC
Bloomsbury Publishing Plc
50 Bedford Square, London, WC1B 3DP, UK
1385 Broadway, New York, NY 10018, USA
29 Earlsfort Terrace, Dublin 2, Ireland

BLOOMSBURY, BLOOMSBURY ACADEMIC and the Diana logo are trademarks of Bloomsbury Publishing Plc

First published in Great Britain 2020
This paperback edition published in 2022

Copyright © Michael John Gorman, 2020

Michael John Gorman has asserted his right under the Copyright, Designs and Patents Act, 1988, to be identified as Author of this work.

For legal purposes the Acknowledgements on pp.x–xii constitute an extension of this copyright page.

Cover design by Catherine Wood
Cover image: Christoph Grienberger, Hydrostatic experiments, from De ijs quae vehuntur in aquis recited by Giovanni Bardi, 23 June 1614, APUG Fondo Curia 2052. Pen and ink with wash. Scanned by the author with permission of the Pontifical Gregorian University

All rights reserved. No part of this publication may be reproduced or transmitted in any form or by any means, electronic or mechanical, including photocopying, recording, or any information storage or retrieval system, without prior permission in writing from the publishers.

Bloomsbury Publishing Plc does not have any control over, or responsibility for, any third-party websites referred to or in this book. All internet addresses given in this book were correct at the time of going to press. The author and publisher regret any inconvenience caused if addresses have changed or sites have ceased to exist, but can accept no responsibility for any such changes.

A catalogue record for this book is available from the British Library.

Library of Congress Control Number: 2020941216.

ISBN: HB: 978-1-3500-9195-5
PB: 978-1-3502-1143-8
ePDF: 978-1-3500-9196-2
eBook: 978-1-3500-9197-9

Series: Bloomsbury Studies in the Aristotelian Tradition

Typeset by RefineCatch Limited, Bungay, Suffolk

To find out more about our authors and books visit www.bloomsbury.com and sign up for our newsletters.

For my parents, Marianne and Michael Gorman

... for Jesuits never content themselves with the Theory in anything, but straight proceed to practise ...

John Donne, Ignatius his conclave (1611)

It is amazing what you can accomplish if you do not care who gets the credit

Anonymous

Contents

Acknowledgements	x
Resources for the Book	xiii
List of Figures	xiv
List of Abbreviations	xviii
Introduction: What Scientific Counter-Revolution?	1
1 Establishing Mathematical Authority: The Politics of Christoph Clavius	9
2 Mathematics and Modesty: The Problemata of Christoph Grienberger	41
3 Discipline, Authority and Jesuit Censorship: From the Galileo Trial to the *Ordinatio Pro Studiis Superioribus*	85
4 The Jesuits and the Vacuum Debate	125
5 The Angel and the Compass: Athansius Kircher's Geographical Project	167
6 Between the Demonic and the Miraculous: Athanasius Kircher and the Baroque Culture of Machines	197
7 From 'The Eyes Of All' to 'Usefull Quarries in Philosophy and Good Literature': The Changing Reputation of the Jesuit *Mathematicus*	249
Index	267

Acknowledgements

The creation of this book has been a long and fascinating journey over more than two decades, during which I have accumulated many debts. I first became fascinated with Jesuit involvement in early modern science on reading Pietro Redondi's book *Galileo Heretic*, and soon afterwards went to the European University Institute in Florence to work on my doctoral thesis, *The Scientific Counter-Revolution: Mathematics, Experimentalism and Natural Philosophy in Jesuit Culture (1999)* on which much of this book is based. I would like to thank my PhD supervisor, Dominique Julia, my external supervisor, Paolo Galluzzi and my doctoral jury members Mario Biagioli, John Brewer and Simon Schaffer (Jury Chair) for all their support and advice.

I was very fortunate to gain access to archives of the Pontifical Gregorian University in Rome thanks to the then archivist Fr Marcel Chappin, and worked extensively in Rome, mining those archives, Jesuit administrative archives (ARSI) and the rich libraries of that city. During this time, I also benefited greatly from an Erasmus at the École des Hautes Études en Sciences Sociales in Paris supported by Roger Chartier and then moved to London for a fellowship at the Warburg Institute. I then had the memorable experience of investigating Jesuit missionary science in Brazil, in Rio de Janeiro and Salvador da Bahia, thanks to Carlos Ziller-Camenietzki and the Museu de Astronomia e Ciências Afins (MAST). I was able to continue working on Jesuit natural philosophy and Athanasius Kircher's correspondence at the Dibner Institute in MIT, and then thanks to the support of Paula Findlen, at Stanford University, a veritable hotbed of Kircherians. It has been enjoyable, after several years of doing radically different things (perhaps somehow indirectly inspired by the transdisciplinary public exhibitions of seventeenth-century Jesuits), to blow off the dust, and look at this fascinating period through new eyes.

I would also like to thank the following people for their generous advice, suggestions and feedback along the long journey (in alphabetical order): Jean-Robert Armogathe, Ugo Baldini, Jim Bennett, Marco Beretta, Ann Blair, Pascal Brioist, Lawrence Brockliss, Massimo Bucciantini, Jed Buchwald, Charles Burnett, Filippo Camerota, Antonio Clericuzio, the late Alistair Crombie, Kirti Chaudhuri, Joseph Connors, Peter Dear, Sven Dupré, William Eamon, Mordechai Feingold, Joshua Foer, David Freedberg, Anthony Grafton, Michael Hunter,

Michel-Pierre Lerner, Luce Giard, John Glassie, Susanna Gomez-Lopez, Steve J. Harris, Marcus Hellyer, Rob Iliffe, Jill Kraye, Christopher Ligota, Noel Malcolm, Aliocha Maldavsky, Alexander Marr, Massimo Mazzotti, Mara Miniati, Paul Nelles, Pietro Redondi, Daniel Roche, Guido Rebecchini, Eileen Reeves, Craig Rodine, Antonella Romano, Alain Ségonds, William R. Shea, Daniel Stolzenberg, Gerhard Strasser, René Taylor, Albert van Helden, Robert Westman and last, but certainly not least, Nick Wilding.

I owe a tremendous debt to the former archivist of the Pontifical Gregorian University, Fr. Marcel Chappin, SJ, for making the rich and relatively unexplored archives of that institution available to me and allowing Nick Wilding and me to digitize the Kircher correspondence and related materials. I am also deeply grateful to the archivists of the Jesuit archives in Rome, Fr Mario Zanardi, SJ, and Fr J. De Cock, SJ for their help with the archives. For extremely generous hospitality in Rome, Vienna, and Paris while conducting archival research, I thank (respectively) the Zaccagnini family, the late Georg and Alice Eisler and Desirée Hayter. For help with photographic materials, I am grateful to Franca Principe and Elizabeth McGrath. I would also like to express my enormous gratitude to the staff of the École Française de Rome, the Institute of Historical Research in London, the Warburg Institute, the Herzog August Bibliothek in Wolfenbüttel and the Museo Galileo in Florence.

Stanford University Libraries deserve particular thanks, for assistance and permission to reproduce images from Kircher's works in their collections. Paula Findlen has played a key role in supporting research on seventeenth-century Jesuit culture, especially around Athanasius Kircher, in Stanford, through a host of publications, digital projects and conferences. I am also grateful to Paula and to Stanford Libraries for commissioning Caroline Bouguereau and me to recreate Athanasius Kircher's magnetic clock for their 2001 Kircher exhibition. Lawrence Weschler, David Wilson, Joshua Foer and John Glassie have been a source of wonderfully stimulating and wide-ranging conversations about Kircher over the years. I am very grateful to Marco Sgarbi for persuading me that it made sense to publish this as a book, and to Becky Holland and her colleagues at Bloomsbury for all their help. I thank my wife Lenka for her endless patience and encouragement, and my children Maia, Thaddeus and Benjamin for being a refreshing distraction from Jesuit machinations.

Some of the chapters of this book have appeared in whole or in part in other publications previously:

Chapter 3 appeared as 'Mathematics and Modesty in the Society of Jesus: The Problems of Christoph Grienberger', in *The New Science and Jesuit Science:*

Seventeenth Century Perspectives, edited by Mordechai Feingold, Dordrecht: Kluwer, 2003, 1–120. Parts of Chapter 4 were included in 'A Matter of Faith? Christoph Scheiner, Jesuit censorship and the Trial of Galileo', *Perspectives on Science*. 1996; 4(3): 283–320. Chapter 6 appeared, in modified form, as 'The Angel and the Compass: Athanasius Kircher's Magnetic Geography' in *Athanasius Kircher, The Last Man Who Knew Everything*, edited by Paula Findlen, New York and London: Routledge, 2004, 239–259. A shortened version of Chapter 7 appeared as 'Between the Demonic and the Miraculous: Athanasius Kircher and the Baroque Culture of Machines' in *The Great Art of Knowing: The Baroque Encyclopedia of Athanasius Kircher*, edited by Daniel Stolzenberg, Stanford: Stanford University Libraries, 2001, and parts of Chapter 8 appeared as 'From "The Eyes of All" to "Usefull Quarries in philosophy and good literature": Consuming Jesuit Science, 1600–1665', in *The Jesuits: Culture, Sciences and the Arts, 1540–1773*, ed. John W. O'Malley, SJ, Gauvin Alexander Bailey, Steven J. Harris, and T. Frank Kennedy, SJ, Toronto: University of Toronto Press, 1999, pp. 170–189. I am grateful for permission to reproduce relevant material here.

Resources for the Book

For those who are interested in further reading and resources connected to the scholarship in this book please visit the book's webpage at Bloomsbury.com where you will find an Online Resources tab which includes:

Further Reading

Principal Manuscript Sources

Select Bibliography

Figures

1.1 Portrait of Christoph Clavius, 1606, by Francesco Villamena, etching, Metropolitan Museum, New York, The Elisha Whittelsey Collection, The Elisha Whittelsey Fund, 1951, public domain. https://www.metmuseum.org/art/collection/search/342372   29

2.1 Christoph Grienberger's enhanced version of Mario Bettini's scenographic instrument, from Mario Bettini, *Apiaria Universae Philosophiae Mathematicae*, Bononiae: Io. Baptistae Ferronij; 1645, Apiarium V, Caput VI, p. 45. Engraving. Public domain. Available from: https://reader.digitale-sammlungen.de/de/fs1/object/display/bsb11199743_00001.html   46

2.2 A schematic version of Grienberger's enhanced version of Mario Bettini's scenographic instrument showing detail of *cursores*, from Mario Bettini, *Apiaria Universae Philosophiae Mathematicae*, Bononiae: Io. Baptistae Ferronij; 1645, Apiarium V, Caput VI, p. 45. Engraving. Public domain. Available from: https://reader.digitale-sammlungen.de/de/fs1/object/display/bsb11199743_00001.html   47

2.3 Christoph Scheiner's pantograph, from C. Scheiner, *Pantographice seu Ars delineandi res quaslibet per parallelogrammum lineare seu cauum, mechanicum mobile*, Romae; 1631. Engraving. [Stanford University Libraries, Rare Books, TJ181.9.S34 1631]   48

2.4 Daniel Widman, Various methods for observing sunspots, from Christoph Scheiner, *Rosa ursina, sive, Sol, ex admirando facularum et macularum suarum phaenomeno varius*, Bracciani: Apud Andream Phaeum Typographum Ducalem; 1630. Etching. [Stanford University Libraries: Barchas Collection QB525.S28F]   50

2.5 Christoph Grienberger's 'Heliotropic Telescope' or 'Telescopic Heliotrope', from Christoph Scheiner, *Rosa ursina, sive, Sol, ex admirando facularum et macularum suarum phaenomeno varius*, Bracciani: Apud Andream Phaeum Typographum Ducalem; 1630, p. 349. Engraving. [Stanford University Libraries: Barchas Collection QB525.S28F]   51

2.6 The Conversion of Kingdoms and Provinces by the Society, from *Imago Primi Saeculi Societatis Iesu A Provincia Flandro-Belgica eiusdem Societatis Repraesentata.* Antwerp: Balthasar Moretus; 1640, p. 321. Available from https://reader.digitale-sammlungen.de/resolve/display/bsb11054470.html 58

2.7 Christoph Grienberger, Device for raising a golden earth by the force of one talent, from Grienberger, *Problema, Terra Auream, Talenti potentia movere.* APUG Fondo Curia 2052. Pen and ink. Scanned by the author, with permission of the Pontifical Gregorian University. 59

2.8 Earth-moving machine, from Kaspar Schott, *Magia universalis,* Bamberg: J. M. Schönwetter, 1677. Public domain. Available online from https://reader.digitale-sammlungen.de/resolve/display/bsb10908666.html 60

2.9 Christoph Grienberger, Hydrostatic experiments, from *De ijs quae vehuntur in aquis* recited by Giovanni Bardi, 23 June 1614, APUG Fondo Curia 2052. Pen and ink with wash. Scanned by the author, with permission of the Pontifical Gregorian University. 66

4.1 Otto von Guericke's *antlia pneumatica,* from Kaspar Schott, *Mechanica hydraulico-pneumatica,* Würzburg, 1657 [available from https://digital.staatsbibliothek-berlin.de/werkansicht?PPN=PPN817182462&PHYSID=PHYS_0565&DMDID=DMDLOG_0018&view=picture-download or Stanford University Libraries Rare Books QC143.S3 1657] 145

5.1 A magnetic Habsburg eagle, from Kircher, *Magnes, sive de arte magnetica* (1643 edn.), frontispiece, Courtesy of Special Collections, Stanford University Libraries. 172

5.2 Jan van der Straet (Stradanus), *The longitudes of the globe discovered by the declination of the magnet from the pole,* from the series *Nova Reperta,* circa 1600. [available from Stanford University Libraries – Rumsey Map Center]. 174

5.3 Table of magnetic declinations, from Kircher, *Magnes, sive de arte magnetica,* 1643, p. 401, Courtesy of Special Collections, Stanford University Libraries. [https://www.e-rara.ch/zut/content/pageview/4319054] 178

5.4 Letter from Giovanni Battista Zupi to Kircher, Naples, 21 January 1640, APUG 567, f. 286r. Courtesy of Athanasius Kircher Correspondence Project, Stanford University Libraries. 179

5.5 The Catholic Horoscope of the Society of Jesus, from Athanasius Kircher, *Ars Magna lucis et umbrae*, Romae: Ludovico Grignani; 1646, facing p. 553, Courtesy of Special Collections, Stanford University Libraries. 183

5.6 Frontispiece showing Astrea, goddess of justice, as a winged angel, from Giambattista Riccioli, *Almagestum Novum*, 1651, Courtesy of Special Collections, Stanford University Libraries. 187

5.7 Frontispiece of Athanasius Kircher, *Ars magna lucis et umbrae*, 1646, Courtesy of Special Collections, Stanford University Libraries. 189

6.1 Frontispiece depicting the *Museum Kircherianum*, from G. de Sepibus, *Romanii Collegii Musaeum Celeberrimum cuius magnae antiquariae rei* ... Amsterdam: Ex Officina Janssonio-Waesbergiana, 1678, engraving. Courtesy of Special Collections, Stanford University Libraries. 198

6.2 Speaking tubes connected to statues, from Athanasius Kircher, *Musurgia universalis*, vol. 2, p. 303, Courtesy of Special Collections, Stanford University Libraries. 204

6.3 Multimammary goddess, from Kaspar Schott, *Mechanica Hydraulico-Pneumatica*, p. 255, Courtesy of Special Collections, Stanford University Libraries. 210

6.4 The magnetic anemoscope that Kircher built in Malta, from *Magnes, sive de arte magnetica* (1643 edn.), p. 322, Courtesy of Special Collections, Stanford University Libraries. 211

6.5 Vomiting fountain, from Kaspar Schott, *Mechanica Hydraulico-Pneumatica*, p. 210, Courtesy of Special Collections, Stanford University Libraries. 216

6.6 Various hydraulic machines, from Kaspar Schott, *Mechanica Hydraulico-Pneumatica*, p. 181, Courtesy of Special Collections, Stanford University Libraries. 217

6.7 The sunflower clock, from Kircher, *Magnes, sive De Arte Magnetica* (1643 ed.), p. 644, Courtesy of Special Collections, Stanford University Libraries. 219

6.8 The catoptric theatre, from Kircher, *Ars Magna Lucis et Umbrae* (1671 ed.), p. 776, Courtesy of Special Collections, Stanford University Libraries. 220

6.9 Kircher's reconstruction of the legendary sphere of Archimedes, from Kircher, *Magnes, sive de Arte Magnetica* (1643 ed.) p. 305, Courtesy of Special Collections, Stanford University Libraries. 226

6.10	The spagyrical furnace of the Collegio Romano, from Kircher, *Mundus Subterraneus* (1665 edn.) vol. 2, p. 392, Courtesy of Special Collections, Stanford University Libraries.	235
7.1	A surviving example of Kircher's *Organum Mathematicum*, conserved in the Museo Galileo, Florence (photograph: Franca Principe)	251

Abbreviations

APUG	Archivio della Pontificia Università Gregoriana, Rome
ARSI	Archivum Romanum Societatis Iesu, Rome
Birch	Thomas Birch, The History of the Royal Society of London, London; 1756–7
BL	British Library, London
BN	Bibliothèque Nationale, Paris
BNCF	Biblioteca Nazionale Centrale di Firenze
BNR	Biblioteca Nazionale 'Vittorio Emmanuele II', Rome
Boyle, Works	The works of the honourable Robert Boyle, ed. Thomas Birch, London: J & F Rivington, 1772 (2nd edition)
CC	Christoph Clavius: Corrispondenza, ed. by Ugo Baldini and Pier Daniele Napolitani. Pisa: Università di Pisa, Dipartimento di Matematica, Sezione di Didattica e Storia della Matematica; 1992
FG	Fondo Gesuitico
HAB	Herzog August Bibliothek, Wolfenbüttel
MP	Monumenta Paedagogica Societatis Iesu, Nova editio penitus retractata, ed. Ladislaus Lukács, Rome, Institutum Historicum Societatis Iesu, 1965–
OC	The Correspondence of Henry Oldenburg, ed. and transl. by A. Rupert Hall and Marie Boas Hall, Madison, Milwaukee, and London, 1965–
OG	Le Opere di Galileo Galilei, Edizione Nazionale a cura di A. Favaro (1890–1909)
Sommervogel	Augustin de Backer, Aloys de Backer and Auguste Carayon, Bibliothèque de la Compagnie de Jésus, Nouvelle éd. par Carlos Sommervogel, 12 vols., Brussels: O. Schepens, Paris: A. Picard (1890–1932), Repr. Louvain (1960)

Introduction: What Scientific Counter-Revolution?

What does it really mean to speak of the 'invention of modern science'? Surely to suggest that the Jesuits played a role in the 'invention of modern science' because of certain practices that they developed for the investigation of nature, is tantamount to saying that the Jesuits 'invented audio-books', simply because they used to enjoy listening to edifying texts being read aloud to them while they ate their meals in the Refectory of the *Collegio Romano*, the flagship Jesuit educational institution in Rome.

Not so – as we have moved away from the celebration of heroic individual discoveries at the centre of a narrative of the so-called 'Scientific Revolution' of the seventeenth century, we have gained new appreciation of the emergence of certain types of socially-embedded practices for natural investigation.[1] From the application of mathematics to the study of natural phenomena, to the use of collective networks to accumulate and compare observations, from the use of experimental methods in natural philosophy, and the associated challenges in establishing the credibility of experimental results, to the emergence of an early form of 'peer review' through the operation of the Jesuit book censorship system, these practices provide an alternative narrative of the emergence of modern science as a socially-embedded web of behaviours, processes and institutions rather than a series of charismatic discoveries. This book chronicles not a glorious revolution but what we might describe as a 'scientific counter-revolution', the messy and troublesome establishment of the invisible operating systems and institutional bureaucracies – what Bruno Latour has called 'centres of calculation' or Paula Findlen has recently termed 'Empires of Knowledge' – that transformed scientific practice, from arguments from universally accepted experiences to contrived experimentation and mathematical natural philosophy.[2] The role of the Jesuits in the establishment of such institutional practices was at first celebrated, with the world looking to the Jesuit networks of missionary correspondents as 'the eyes of the world', then reviled, as the Jesuits were discredited as doctrinally-suspect 'jugglers', and finally all but effaced from history, through a Protestant historiographical tradition that enjoyed painting the Catholic Church as the villain of the standard narrative of the Scientific Revolution, in which the Jesuits were relegated to bit parts in relation to their clashes with Galileo, Pascal and other 'heroes' of the new science.

A significant raft of studies of Jesuit natural philosophy and mathematics in the seventeenth century has helped to document the important role of the Jesuit order in establishing such practices. Several studies have suggested that the 'expert book censorship' practices established by the Jesuit *Revisores* established practices that informed the emerging peer review practices of the earliest scientific journals.[3] Scholars such as Ugo Baldini and Antonella Romano have provided extensive documentation of Jesuit mathematical practice and education especially in the Italian and French contexts based on primary archival sources. Marcus Hellyer has provided a rich account of Jesuit physics in the German lands.[4] Mark Waddell has examined Jesuit natural philosophy in relation to natural magic and the revelation of nature's secrets. Athanasius Kircher, once consigned to the dustbin of history, has been reappraised, critiqued, eulogised and celebrated in a number of new collections, conferences and monographs, including Daniel Stolzenberg's important study of his Egyptology and Paula Findlen's studies situating Kircher and his museum in the history of collecting. Mordechai Feingold has provided important new scholarship on Jesuit natural philosophy, and edited collections relating to Jesuit science, and most recently Jesuit philosophy has been the focus of a new collected volume edited by Cristiano Casalini.[5]

In the context of this significant revival of interest in Jesuit science in the sixteenth and seventeenth centuries, the present book, derived in large part from my 1999 doctoral thesis and subsequent publications, presents Jesuit science 'in action' at the peak of the Counter-reformation, and explores the complex institutional context of the work of key figures, from college to correspondence networks. The focus of the present book is largely on mathematics and physics/natural philosophy, but this is not intended in any way to diminish the equally fascinating contributions of the Jesuits to early modern botany, natural history and other areas of study.

The mid-seventeenth century satire of the Jesuit order, *Monarchia solipsorum*, or the 'Monarchy of the Solipsists', began its biting attack on the institutions of the order with a depiction of the paradoxes inherent the geography of the Jesuit 'kingdom':

At the outset it would be worth teaching the Reader about the skies and place of the Kingdom, the expanse of its dominions and the confines of its people. But no mortal who has penetrated there has accomplished this simple task, so I will not attempt it. For although, like Homer's Ulysses, I have seen many of the towns and the customs of the men of that Kingdom, I have been unable to observe anything certain about the position of the heavens. For the constitution of their universe is very different from ours, as are the number and various names of the poles. For them the Moon is more often in the vertical than the Sun at midday. They have no differences of place or position, so that what is above is simultaneously to the right and what is to the left is simultaneously below. There is even no fixed or single centre. The manner of all of this diversity depends on the will of the Monarch.[6]

The passage cited alludes humorously to three of the aspects of the Society of Jesus that would have been most readily familiar to a broad seventeenth century readership – its global expanse, the extreme mobility of its members and its emphasis on obedience to a central authority – the 'Monarch', or General, in Rome. It is now clear that the Society of Jesus provided an institutional home for an enormous variety of scientific practices during the sixteenth and seventeenth centuries. The types of activities in which Jesuit natural philosophers and mathematicians engaged were crucially conditioned by the changing local contexts in which they carried out their work. Their activities were nonetheless also conditioned by the changing non-local geography of the Society of Jesus as a whole, alluded to by the author the *Monarchy of the Solipsists*. It is precisely this tension between specific local contexts of natural investigation and a centralised global bureaucratic structure allowing the mobility of trained people and letters on natural subjects between distant sites that is uniquely characteristic of natural investigation by members of the Jesuit order during the early modern period. In the Jesuit collegiate network, the *Collegio Romano* had a special place. The efforts of Christoph Clavius and his successors established its status as a European centre of mathematical expertise. As well as being the flagship didactic establishment of the Jesuit order, the *Collegio Romano* functioned as a clearing-house for people, letters, natural curiosities and instruments.

In the first chapter of the present book, I examine the origins of the distributed Jesuit mathematical community, largely through the political and pedagogical efforts of Christoph Clavius. Clavius argued for the importance of mathematical learning within the order both as an essential concomitant to the study of natural philosophy and theology and as a means for entering into relationships with the mathematically curious aristocracy that had emerged in sixteenth-century

Europe. The private mathematical academy run by Clavius in his *cubiculum* in the *Collegio Romano* to train future mathematics teachers for the other Jesuit provinces rapidly became a centre for advanced mathematical research, fuelled by the correspondence carried out by Clavius and his successors with mathematicians and natural philosophers both inside and outside the Jesuit order.

As Paolo Galluzzi, Nick Jardine and Peter Dear have pointed out,[7] the late sixteenth and early seventeenth centuries witnessed a transformation in the status of the mathematical disciplines which was closely related to the activities of the Jesuit 'school' of mathematics founded by Clavius.

Chapter 2 examines the institutional context of mathematics in the local context of the *Collegio Romano*, focussing on the career of Clavius's successor, Christoph Grienberger, a Jesuit mathematician who worked almost exclusively within Jesuit-controlled institutions. I argue that the strategies of self-abnegation deployed by Grienberger, availing of every opportunity to remove his name from texts written with his pen and optical and astronomical instruments designed by him and built with his own hands, can reveal much about what it was to be both a Jesuit and a skilled mathematical practitioner in the early seventeenth century. Where Galileo found a legitimatory resource for certain types of mathematical practice in the colourful world of the Medici court in Florence, his contemporary Grienberger found his Archimedean point for the upward leverage of the status of mathematics deep within the complex bureaucratic structure of the Jesuit order. Grienberger developed the public role of mathematical presentations within the College, a role that had been suggested by Clavius, and my interpretation of his career is based on manuscript drafts of these presentations which I had the good fortune to discover in the archives of the Pontifical Gregorian University in Rome.

The following chapter considers the constraints placed on the identity and authority of the Jesuit mathematical practitioner by disciplinary structures of the order, particularly as manifested in the changing structures of censorship during the disciplinary crisis of the 1640s and 1650s. The origins of the disciplinary crisis, previously suggested in the work of Claudio Costantini and Ugo Baldini,[8] are traced to an attempted internal reform of organization of the order led by Melchior Inchofer. Different Jesuit mathematical practitioners responded to the increased emphasis on adherence to Aristotle in matters of natural philosophy in different ways, and this chapter focuses on the problematic clash between the corporate identity of the Jesuit mathematical practitioner and his participation in polemics concerning the natural world.

One of the most violent natural philosophical debates of the mid-seventeenth century surrounded the experiment performed by Evangelista Torricelli, and later carried out, with significant variations by Valeriano Magni, Blaise Pascal and others, to demonstrate the existence of the vacuum. Chapter 4 explores the Jesuit response to the vacuum experiments, which raise questions about the expert status of Jesuit mathematical practitioners and natural philosophers during the period of disciplinary enforcement discussed in the previous chapter. The violence of the Jesuit polemic with Valeriano Magni is related to his attempt to liberate the Charles university in Prague from the Aristotelian natural philosophy taught by the Jesuits. The Jesuit responses to the later vacuum experiments of von Guericke demonstrate simultaneously an enforced adherence to Aristotelian plenism, a desire to participate actively in experimental disputes, and a highly collectivist approach to the validation or invalidation of experiments. This chapter considers the capacity of Jesuit colleges to function as split laboratories in the seventeenth century, allowing local and non-local knowledge to be combined to back up positions adopted in a dispute.

Chapter 5 considers the role of Jesuit correspondence networks in the scientific output of the order. The correspondence network of Athanasius Kircher, building on a tradition founded by Clavius and Grienberger, and centred in the *Collegio Romano*, allowed the accumulation of information sent to Kircher from distantly stationed correspondents both inside and outside the Society. The case of the measurement of magnetic variation at different points in the globe, as an 'extrinsic' solution to the problem of the measurement of longitude at sea, provides a rich case study for the discussion of this particular institutional aspect of Jesuit scientific practice.

The disciplinary enforcement of the 1640s and 1650s was accompanied by a rising number of publications by Jesuits, particularly Athanasius Kircher and his disciples, relating to experimental magic, which drew their context from the courtly function of Jesuit experimentalism. works avoided the disciplinary problems associated with deviant natural philosophy after the 1651 *Ordinatio pro studiis superioribus* by claiming to relate to art, or technique, rather than *physica*. Chapter 6 looks at Athanasius Kircher and the Baroque culture of ingenious machines and mechanical marvels designed to deceive the senses.

The 'showiness' associated with Jesuit experimentation, particularly by members of the early Royal Society and their correspondents constituted grounds for questioning the credibility of Jesuit reports about experiments. By 1670, the *Collegio Romano* could no longer claim the degree of authority over the natural world that it had attained during its corroboration of Galileo's celestial observations in 1610. A variety of institutional factors constrained experimental novelties to be presented by many Jesuits in the form of monstrosities or curiosities. Mathematics and experimentation were increasingly insulated in this new situation from the domain of natural philosophy. The final chapter examines the discrediting of Jesuit natural philosophy, and the process whereby early scientific societies such as the Royal Society emulated Jesuit information-gathering networks, while discrediting Jesuit interpretations.

While arguing for a series of transformations in the way mathematically trained Jesuits might conduct their careers and engage in debates concerning matters of natural philosophy during the seventeenth century, the present book also aims to express a deep continuity between the period in which Christoph Clavius was the senior mathematician of the *Collegio Romano*, and the period in which that position was occupied by Athanasius Kircher. Despite the manifest differences between the careers, opinions and literary output of Clavius, Grienberger and Kircher, they laid claim to a common space – the mathematical *cubiculum*, or 'mathematical museum' of the *Collegio Romano*, where they accumulated instruments, manuscripts, letters on mathematical subjects and natural phenomena and other topics and taught private disciples the mathematical disciplines to an advanced level. They participated in a local collegiate culture that was simultaneously global, occupying a vantage point that had no obvious equivalent in early modern Europe, enabling them to take advantage of the enormous range of the Jesuit missionary network to collect the results of observations and experiments performed at remote stations.

Rather than attempt a chronological history of Jesuit scientific practice, I describe here a series of illuminating episodes, in the hope of illustrating the complexity of the relationship between the local and the global in Jesuit scientific practice. Much historiography of science has laid particular emphasis on the importance of minute local studies in attempts to understand the resolution of scientific disputes and the micro-politics of the laboratory. If the Jesuit example can teach us anything, however, it is that the place of scientific practice can be a centralised global network of trained correspondents just as much as an enclosed laboratory.

Notes

1   For discussion of the problematic nature of the concept of 'Scientific Revolution', see for example Steven Shapin, *The Scientific Revolution*, Chicago: University of Chicago Press, 1996, 1–14, and Peter Harrison, *Was there a Scientific Revolution?* European

Review, Vol. 15, No. 4, 445–457. More recently David Wootton has made a polemical revisionist argument to restate the importance of the scientific revolution of the seventeenth century in his book *The Invention of Science: A New History of the Scientific Revolution*, New York: Harper, 2015.

2 For the original use of 'centres of calculation' see Latour's classic *Science in Action*, Cambridge: Harvard University Press, 1987 esp 232–247. For new scholarship on early modern information-gathering networks, see Findlen, ed., *Empires of Knowledge: Scientific Networks in the Early Modern World*, Oxford: Routledge, 2018

3 For the prehistory of academic peer review see Mario Biagioli, 'From Book Censorship to Academic Peer Review', *Emergences: Journal for the Study of Media & Composite Cultures,* 2002, Vol. 12 Issue 1, 11–45, which analyses the censorship practices of the early *Académie Royale des Sciences* and the Royal Society of London, who were granted the privilege to publish their own works, and traces the transition from book censorship to article peer review. There is evidence that Jesuit book censorship practices, discussed in Chapters 3 and 4 below, instigated a form of qualitative 'peer review' even prior to the scientific academies. See also Ugo Baldini, *Uniformitas et soliditas doctrinae. Le censure librorum e opinionum*, in Baldini, *Legem impone subactis. Studi su filosofia e scienza dei gesuiti in Italia, 1540–1632*. Rome: Bulzoni; 1992, 75–119, Daniel Stolzenberg, 'Utility, Edification and Superstition: Jesuit Censorship and Athanasius Kircher's *Oedipus Aegyptiacus*' in *The Jesuits II: Cultures, Sciences and the Arts, 1540–1773*, edited by John W. O'Malley, S.J., Gauvin Alexander Bailey, Steven J. Harris and T. Frank Kennedy S.J., Toronto and London: University of Toronto Press, 336–354, M.J. Gorman, *A Matter of Faith? Christoph Scheiner, Jesuit censorship and the Trial of Galileo.* Perspectives on Science. 1996; 4(3): 283–320 and Christoph Sander, 'Uniformitas et soliditas doctrinae. History, Topics and Impact of Jesuit Censorship in Philosophy (1550–1599)', in *Jesuit philosophy on the eve of modernity*, edited by Cristiano Casalini, Boston: Brill, 2019.

4 See Marcus Hellyer, *Catholic Physics: Jesuit Natural Philosophy in Early Modern Germany*, Notre Dame, Indiana: University of Notre Dame Press, 2005.

5 See especially Baldini, *Legem impone subactis*, Antonella Romano, *La Contre-Réforme mathématique. Constitution et diffusion d'une culture mathématique jésuite à la Renaissance (1540–1640)*, Rome: École française de Rome, 1999, Mark A. Waddell, *Jesuit Science and the End of Nature's Secrets*, Farnham: Ashgate, 2015, Daniel Stolzenberg, *Egyptian Oedipus; Athanasius Kircher and the Secrets of Antiquity*, Chicago: University of Chicago Press, 2013, *Athanasius Kircher, The Last Man Who Knew Everything*, edited by Paula Findlen, New York and London: Routledge, 2004, *Jesuit Science and the Republic of Letters*, edited by Mordechai Feingold, Cambridge, MA., London: The MIT Press, 2003, *The New Science and Jesuit Science: Seventeenth Century Perspectives*, edited by Mordechai Feingold, Dordrecht: Kluwer, 2003, *The Jesuits: Culture, Sciences and the Arts, 1540–1773*, edited by John W. O'Malley et al.,

Toronto: University of Toronto Press, 1999, *Jesuits II : cultures, sciences, and the arts, 1540–1773*, edited by John W. O'Malley et al., Toronto: University of Toronto Press, c2006, and *Jesuit philosophy on the eve of modernity*, edited by Cristiano Casalini, Boston: Brill, 2019.

6   *Lucii Cornelii Europaei Monarchia Solipsorum, Ad Virum Clarissimum Leonem Allatium*, Venice, 1645.

7   Paolo Galluzzi, *Il 'Platonismo' del tardo Cinquecento e la filosofia di Galileo*. in Paola Zambelli, ed., *Ricerche sulla cultura dell'Italia Moderna*. Bari: Laterza; 1973: 37–79, Nicholas Jardine, *The forging of modern realism: Kepler and Clavius against the sceptics*. Studies in history and philosophy of science. 1979; X: 141–173, Peter Dear, *Jesuit mathematical science and the reconstitution of experience in the early seventeenth century*. Studies in History and Philosophy of Science. 1987 Jun; 18(2): 133–175.

8   Claudio Costantini, *Baliani e i Gesuiti*. Florence: Giunti Barbèra; 1969, Ugo Baldini, *Legem impone subactis. Studi su filosofia e scienza dei gesuiti in Italia, 1540–1632*. Rome: Bulzoni; 1992.

1

Establishing Mathematical Authority: The Politics of Christoph Clavius

Prior to any detailed discussion of the work carried out by the globally distributed community of Jesuit mathematical practitioners in seventeenth-century Europe, it seems worth asking just how such a community ever came to exist in the first place. How, we might ask, did its members become equipped with the mathematical instruments, books and training that allowed them to enter into astronomical disputes, to perform geometrical demonstrations and to correspond with each other on topics of mathematical and philosophical import? How did mathematical practices come to be granted a sufficient amount of social and cognitive status within the political structure of each Jesuit college to nurture their continued existence within the Jesuit collegiate network?

To attempt to answer these questions, the present chapter will look at Christoph Clavius's project for establishing the Jesuit intellectual flagship in Rome, the *Collegio Romano* as a centre of mathematical skill and authority, with a view to granting special training to gifted Jesuit mathematical practitioners before redistributing them to the different provinces of the order.[1] In attempting to create an authoritative, distributed community of Jesuit mathematicians, Clavius faced a number of obstacles both inside and outside the order, which he was only able to overcome by enlisting powerful political support. Clavius's defence of the newly promulgated Gregorian calendar against the attacks of Michael Maestlin, Joseph Scaliger, François Viète and others played an important part in securing Jesuit mathematical authority in a wider European context.

As well as being indispensable for mastery of the other arts and sciences and for the administration of civic affairs, mathematical knowledge was presented by Clavius to his Jesuit superiors as an antidote to conversational embarrassment. Noblemen were interested in mathematical problems, and it would bring disgrace to the Jesuit order if its members were unable to discourse intelligently on mathematical subjects in the company of princes. The dinner-table was a politically charged space during the Counter-reformation. The training of Jesuits as rhetoricians and humanists formed part of an attempt to reconquer this space, which had been encroached upon by Luther's *Tischreden* and by the myriad works of the Protestant humanists. Clavius inscribed mathematical training firmly within this project.

Unlike his Jesuit predecessor Balthasar Torres and many other sixteenth-century mathematicians of a practical bent, Clavius embraced print culture. Apart from his own printed works, he was closely involved in the production of the mathematical part of enormous *Bibliotheca Selecta* produced by the Jesuit diplomat and scholar Antonio Possevino. Possevino's work was intended to be a sanitised reworking of the Protestant Conrad Gesner's *Bibliotheca Universalis*, as part of a larger vision of global evangelization through print, centralised correspondence and coordinated pastoral work. While Possevino intended to establish a centre for the training of missionaries in Rome, a project that eventually led to the foundation of the Roman *Collegium de Propaganda Fide*, Clavius established a centre for mathematical training, his mathematical academy.[2] His purposes were closer to Possevino's than one might initially expect. The distribution throughout the Jesuit provinces of mathematical experts trained in the Roman centre, like the distribution of Jesuits skilled in *eloquentia*, Greek and Hebrew, would bring great glory to the Jesuit order and recover some of the souls to the Catholic church that had been lost through the alluring erudition of the great Protestant humanists of the sixteenth century.

Jesuit humanism and the Counter-reformation

At intermittent moments between 1562 and 1565, the Majorcan Jesuit Jerónimo Nadal took time off from the Council of Trent to compose a dialogue. Originating as a response to an attack on the Jesuits penned by Melanchthon's pupil and friend Martin Chemnitz, the dialogue was staged as an ecumenical meeting of three travellers, whose largely pacific encounter was made plausible by the peace of Augsburg of 1555, finally granting official tolerance to religious diversity in in German lands. The participants in Nadal's apologetic *mis-en-scène* were a Lutheran (Philippicus), a Catholic ill-disposed towards the Jesuits (Libanius), and Philalethes, a 'friend and past pupil of the Society, who learned about the nature our Society's organization from the Jesuits in Cologne'.[3] Two of the projected four parts of the Dialogue were completed by Nadal. In the first,

Philalethes explains the early origins of the Society to Libanius and Philippicus. In the second part, he elaborates on the organizational structure of the order. These were to be followed by two 'negative' parts, attacking, respectively, Chemnitz and his heretical associates, and the Dominican theologian Melchor Cano, a vigorous opponent of the early Society and the root of many of its initial troubles with the Inquisition.[4]

The defence of the Jesuits mounted by Philalethes in Nadal's dialogue is threatened intermittently by the undisciplined wrath of Philippicus. Irritated by the learned Greek and Hebrew citations with which Philalethes sprinkles his discourse, Philippicus, transparently a mask for Melanchthon and his followers, threatens to destabilise the very conventions governing dialogue form:

> What shall we do, Libanius? What are we waiting for? We will not hear any more from this man. Let us prepare our blades for combat.[5]

Despite these moments of dramatic tension, the dialogue between the three travellers continues for long enough to allow Nadal's mouthpiece Philalethes to expound upon the various aspects of the Jesuit ministry in some detail. The discussion of the Jesuit educational ministry, composed by an insider deeply involved in its development, draws heavily on the relevant sections of the recently published *Constitutions* of the order, and prefigures the *Ratio studiorum*, which Nadal was to be intimately involved with at the early stages.[6] Nadal defends the involvement of the Jesuits in pedagogy in purely apostolic terms:

> [T]hey [i.e. the Jesuits] judge that teaching the youth pertains to the ministry of the word of God; their only reason for opening the schools was so that with this hook they might draw students of literature to piety[7]

Shortly afterwards, after an official visit of inspection to the Jesuit college of Cologne, Nadal elaborated that:

> The Society would never have undertaken the task of giving lessons in colleges, if it did not also understand that by so doing it was also giving a moral training [...] So for us lessons and scholarly exercises are a sort of hook with which we fish for souls.[8]

As Gabriel Codina Mir, has observed, 'on reading some of the statements made by Nadal, it is easy to believe that the Colleges of the Jesuits were only conceived in order to combat Protestantism, at least in Germany and in the countries affected by the Reformation, and that the study of letters was only envisaged with the aim of fighting the Protestants with their own weapons. As *Belles-lettres* were in vogue, the Jesuits occupied themselves industriously with them, but always kept the apostolic goal which they had set for themselves clearly in view'.[9]

The apostolic ends of Jesuit education were reflected clearly in the pyramidal disciplinary structure which characterised teaching in Jesuit colleges, as has been emphasised in a number of studies.[10] This structure, in which theology ruled over philosophy, which in turn ruled over mathematics and the lower disciplines, was inherited from the *modus parisiensis* – the educational structure of the University of Paris where the first Jesuits had received their education – praised by Nadal as 'the most exact and the most fruitful'.[11]

While the basic structure of the Jesuit *cursus* was provided by Paris, the apostolate of the order led to certain important departures. In particular, if a student was perceived to have particular intellectual gifts in a single direction, these would be nurtured by the Jesuit preceptors. Nadal gives a characteristically concise resumé:

> PHILALETHES [The Jesuits] observe the natural faculties and propensities of minds, and if someone is seen to have sufficient talent for a particular discipline, he is ordered to devote all of his studies to that discipline towards which he is most inclined [...] In this way they hope that in the future they will train outstanding practitioners and teachers of the different arts; but they don't wish anyone to be ignorant of those things that are necessary or even useful for helping souls
>
> LIBANIUS Such devotion to studies might be seen to tend towards curiosity or arrogance.
>
> PHILALETHES It doesn't, Libanius, it doesn't; but towards necessity and utility, as I have said.[12]

Nadal's text allows us to see certain elements of the Jesuit educational project in a nakedness that the complex later debates surrounding the *Ratio Studiorum*, as different Jesuit colleges presented their cases for modifications of the Jesuit cursus, would disguise. While the details of the *Ratio* would undergo numerous modifications in the light of reports of the local problems facing teachers, as documented by Ladislaus Lukács in his remarkable edition of the *Monumenta Paedagogica*,[13] the overtly apostolic goals of the Jesuit educational enterprise described by Nadal at this early moment would rapidly become tacit and submerged in the immense bureaucratic structure engendered by Ignatius and his prolific secretary Juan de Polanco.[14] The efforts of Christoph Clavius ensured that mathematical practices were perceived as converging with these goals.

Mathematics and humanism

The sixteenth century witnessed a dramatic transformation in the perceived status of the mathematical disciplines and their practitioners in the Italian peninsula. As Mario Biagioli has observed, after Charles VIII's sweeping invasion of Italy in 1494 and 1495, 'the cannon-syndrome and the introduction of the bastion forced the *milites*, the professional warriors of aristocratic origins, to begin to rely less on their horses and more on Euclid for their survival as a distinct social group.'[15]

Elsewhere in Europe, the social role of the mathematical practitioner and the prestige accorded to mathematical practices were undergoing related changes in the sixteenth century. In Philip II's Spain, military and navigational concerns, a renewed interest in Vitruvian architecture and a fervour for astrological prediction, only slightly dampened by the efforts of the Spanish Inquisition, combined to give increasing political importance to astronomy, mechanics and cosmography, epitomised in the careers of Philip's architect, Juan de Herrera, and his charismatic cosmographer Giovan-Battista Gesio.[16] Philip's foundation of the chair in the art of navigation and cosmography at the *Casa de la Contratación*, the training centre for pilots in Seville, in 1552, marked an important moment in this process, as did the creation of Juan de Herrera's mathematical academy in Madrid, which availed of a rich supply of instruments and a lavish library of ancient and recent works on mathematics, astrology and alchemy.[17] Navigational and mercantile concerns similarly governed the creation of Gresham college in London,[18] while in the German lands the liminal and ambiguous personage of the astrologer was becoming an ever more stable figure at court, a feat of social mobility that was both confirmed and reproached by the Faust legends and plays that enjoyed such popularity during this period.[19] Musical fountains, anamorphoses, automata, perpetual-motion machines and other feats of mathematical magic were objects of fascination to the late sixteenth-century European court, epitomised in Rudolphine Prague.[20]

Into this transforming European cultural landscape arrived a new hybrid creature – the Jesuit *mathematicus*. Cleric and geometer, humanist and astronomer, court-confessor and experimenter, like all hybrids the Jesuit mathematician posed a threat to the *status quo* of the environments in which he made his presence felt, both inside and outside the Society of Jesus.

In 1582, the Jesuit General Everard Mercurian invited the professors of the Collegio Romano to give their opinions on the ways in which the different disciplines should be taught in Jesuit colleges. Christoph Clavius, who had occupied the official post of mathematics professor at the college since 1567,[21] responded with a detailed report entitled *The way in which the mathematical disciplines can be promoted in the Society*.[22] The report emphasised the increased importance of mathematical competence for inserting Jesuits in courtly *conversazione* in late sixteenth-century Europe:

> [Mathematics] will also bring a great ornament to the Society when noblemen understand that ours [i.e. Jesuits] are not ignorant of mathematics, for it is discussed most frequently in their conversations and meetings. For this reason, ours would incur great shame and disgrace if they were to remain silent in gatherings of this kind. This has been related most frequently by those people who were embarrassed in this way.[23]

There is a clear convergence between Clavius's remarks and Nadal's angling metaphor for the Jesuit educational apostolate. It would, however, be misleading to dismiss Jesuit mathematical practice as a response to purely external stimuli. In the Gregorian reform of the calendar, the late sixteenth century witnessed an unprecedented alignment between the business of the post-Tridentine Catholic church and mathematical expertise. As one of the chief elaborators of the new calendar, and its most vociferous defender in print, Clavius presented the Jesuit *mathematicus* as the mathematical voice of the papacy.[24]

Calendar reform and mathematical authority

> And yet not onely for this is our Clavius to bee honoured, but for the great paines also which hee tooke in the Gregorian Calender, by which both the peace of the Church, & Civill businesses have beene egregiously troubled: nor hath heaven it selfe escaped his violence, but hath ever since obeied his appointments: so that S. Stephen, John Baptist, & all the rest, which have bin commanded to worke miracles at certain appointed daies, where their Reliques are preserved, do not now attend till the day come, as they were accustomed, but are awaked ten daies sooner, and constrained by him to come downe from heaven to do that businesse.
>
> John Donne, *Ignatius his conclave (1611)*, ed. T.S. Healy, Oxford: Clarendon Press; 1969, pp. 17–19.

The papal bull '*inter gravissimas*' of 24 February 1582, promulgating the Gregorian calendar, began: 'Among the most serious tasks, last perhaps but not least of those which in our pastoral duty we must attend to, is to complete with the help of God what the Council of Trent has reserved to the Apostolic See'.[25] The Council of Trent had left it to the papacy to complete the reform of the mass book and breviary, which incorporated correction of the calendar. While for Donne and other Protestant divines the resulting calendar may have been the object of ridicule and resistance, it is difficult to overstate the degree of legitimation which calendar reform conferred on astronomical, and hence mathematical, practices within the post-Tridentine church, despite their unpalatable contemporary associations with judicial astrology and fatalism. A suggestive illustration of the point is Francesco Ingoli's report to the Cardinals of the Congregation of the Index of 2 April 1618, which prefaces his list of corrections to Copernicus's *De Revolutionibus* with the entreaty that:

> [T]he aforementioned books of Copernicus must at all costs be conserved and supported for the use of the Christian Republic. For [measurements] of time, very much needed by the Christian people both for the celebration of divine solemnities and for the carrying out of business, derive from the calculations of Astronomers especially of the sun and the moon and the precession of the equinoxes, as is clear from the corrections carried out to the year during the happy reign of Gregory XIII.[26]

Although calendar reform raised the status of mathematical practices at the heart of the Roman curia, opposition to the new calendar was manifested immediately in the Lutheran stronghold of the University of Tübingen. Jakobus Heerbrandus cited the prophesy in the book of Daniel, '*putabit se posse mutare tempora*' (Dan. 7, 25), as evidence that calendar reform was the handiwork of the Antichrist. Another Tübingen professor, Lucas Ossiander, unmasked the Gregorian reform of the calendar as an illegitimate attempt by the papacy to establish a monopoly for the distribution of standards, under threat of penalties, on German soil. The Reichsconstitution and a number of Reichstags, Ossiander reminded his readers, had explicitly forbidden the existence of such monopolies.[27] The most technical and sustained attack, however, came from Kepler's teacher Michael Maestlin, mathematics professor in Heidelberg, in his *Außführlicher und Gründlicher Bericht*.[28]

Maestlin took issue with the new calendar on all levels. One of his more fundamental objections was that the imminence of doomsday made calendar reform superfluous, and a source of unnecessary confusion. Indeed, the attempted papal reform of the calendar was itself a symptom of the approach of the apocalypse. At a less radical level, he claimed that the attempted promulgation of the new calendar by Gregory XIII was an invasion of the evangelical freedom of Protestant Germany that followed the Peace of Augsburg. Additionally, the connection between religious rites and time-calculations was purely conventional for Maestlin, who followed Luther in suggesting the stipulation of a fixed date for Easter. The dates of religious festivities could not, he argued, be made articles of religious faith.[29]

As well as undermining the <u>authority</u> of the papacy to reform the Julian calendar, Maestlin questioned the <u>expertise</u> of the pope's mathematicians. His only *ad hominem* attack was reserved for Clavius, involved in calendar reform 'nit alleyn als ein Clericus, sonder auch als ein Mathematicus' ('not just as a cleric, but also as a mathematician'). In the 1581 edition of Clavius's *Commentary on the Sphere of Sacrobosco* Clavius had announced the impending reform of the calendar.[30] However, Maestlin continues,

> In the whole of the same book he mentions no new observation or new tables, but doubtless he would have made this proclamation or announcement more worthily if the author had had such [observations and tables] and used them. How can he make a certain calculation of the year and month by the Sun and Moon when he does not know how late or soon the Sun and Moon complete their courses? For this reason, this first astronomical basis of the Pope's calendar is utterly disorderly, and anything might follow from it.[31]

Such an accusation clearly posed a threat to the credibility of Clavius as a mathematician representing, simultaneously, the papacy and the Society of Jesus. Clavius did not pick up the gauntlet immediately, but allowed the task of defending the calendar to be carried out by someone more used to battling with heretics, the Mantuan Jesuit Antonio Possevino, who inserted a section *On the emendation of the Year and Easter* into his 1587 work *Muscovy, and other Works on the state of this century against the enemies of the Catholic church*.[32] The section, reviewed by Clavius prior to publication, gives short shrift to 'the "Examination" of a certain heretic, Michael Maestlin, who writes that he is a mathematician from Tübingen'. Criticising Maestlin's discussion of the vernal equinox on technical grounds, Possevino concludes that 'not only does [his Examination] display his incompetence [*imperitia*] and vanity, but it even [...] provides further confirmation of the Gregorian emendation'.[33]

In response, Maestlin taunted Clavius further with a *Defensio alterius sui examinis* published in 1588, which noted that the only Roman answer to his 1583 Außführlicher *und Gründlicher Bericht* was Possevino's *Moscovia*, and insinuated that Rome was unable to provide a stronger rebuttal. This, at last, brought Clavius directly into the fray, and in the letter of dedication to Emperor

Rudolph II of his *Novi Calendarii Romani Apologia, Adversus Michaelem Maestlinum Gaeppingensem, in Tubingensi Academia Mathematicum*[34] he drew on the anti-heretical rhetoric of his Mantuan friend Possevino[35]:

> Such is either the natural vice or the depraved perversion of the studies of certain men, MOST INVINCIBLE CAESAR, that just as there is nothing so false and absurd that it will not have its defenders, and there is nothing so true and praiseworthy that it will escape from all of the calumnies of quibblers. This might be demonstrated [...] by the Roman Calendar lately liberated from errors [...] through the authority of Pope Gregory XIII, with the immense approval of the other princes and schools of the Catholic world, and then of your Holy Majesty, and finally either published or accepted in almost all nations as a result of great study. Although this was done rightly and correctly, and produced in accordance with the method and use that the church of God has always held in celebrating the holy day of Easter, and the other feasts that are termed moveable, Michael Maestlin, a mathematician from Tübingen, has contrived to oppose this most excellent and esteemed Calendar and aims to dissolve the concordance of the Catholic church by nefarious fraud, as he is a man infected with the stain of Ubiquitarian heresy.[36]

In the body of the *Apologia*, Clavius elaborated on Maestlin's heretical tendencies,[37] castigated his pride and arrogance,[38] and came to the defence of Possevino, the target of Maestlin's most recent work, while dealing with his technical objections patiently and at great length. All of Maestlin's works appeared on Sixtus V's Index of 1590, representing a further fortification of the Roman defence.[39] The calendar polemic raged on, nonetheless, with further volleys at Clavius coming from Joseph Scaliger. Reflecting bitterly on Clavius's comportment at a later moment, Scaliger neatly summarised the new public persona forged by Clavius for the Jesuit mathematician during this period: '*Est Germanus, un esprit lourd & patient, & tales esse debent Mathematici; praeclarum ingenium non potest esse magnus Mathematicus*' ('He is a German, a heavy and patient mind, and mathematicians must be like this. An outstandingly ingenious person cannot be a great mathematician').[40] During the protracted polemic, the reputation of the Society of Jesus and the propagation of the new calendar became inextricably linked. In 1609, Clavius's editor, Johann Reinhard Ziegler, made this clear when wrote to his master about the most recent opponent of the Gregorian calendar Georg Germann that

> Even this fly is worth driving away [...] This is important both for the good name of the Society and for the reputation [*existimatio*] of the calendar[41]

Clavius and the Bibliotheca Selecta

Clavius's battle with Maestlin over calendar reform was related to Antonio Possevino's plan for Catholic 'world-evanglization' through arms and print culture, a relationship that was to have enormous significance for the status of mathematical practices in the Jesuit order.[42] During the 1580s, as we have seen, the interests of Possevino and Clavius converged significantly over the issue of the calendar, and the brief collision of these two contrary Jesuit career trajectories – the political man turned scholar and the mathematician turned politician – coincided with the crucial years for the development of Jesuit educational policy. In 1586, after almost 30,000 miles of travel as papal nuncio and legate,[43] Antonio Possevino was sent by the Jesuit General Claudio Aquaviva to the Jesuit college in Padua, in an attempt to remedy his perceived detachment from the internal life of the Society of Jesus, which he had joined relatively late in life.[44] While Possevino had been very close to Aquaviva's predecessor Everard Mercurian, working for some time as his personal secretary, the new General apparently feared him as something of a loose cannon. In Padua, relieved from teaching duties in the Jesuit college, Possevino read voraciously. His biographer recounts a turning-point in this studious exile:

> While in Padua, *he says in a letter to one of his friends*, I was penetrated with pain on seeing that the *Bibliotheca* of a certain Gesner was filled with an infinity of books equally dangerous for the Faith and morals. I wondered if I couldn't engage my friends, both within the Society and outside, to work each according to his talent to collect that which could allow one to become competent in each faculty, after having purged it of all of the errors which might have slipped in, and to make a Library from this collection, that one could consult fruitfully and without danger. I didn't flatter myself that I could carry out such a great plan alone. With this in mind, I cast my eyes on Fr. Francesco Turriano, who had special knowledge of the Church Fathers, on Fr. Clavius, who is excellently skilled in mathematics, and on some lay-people who were perfectly versed in Civil and Canon Law. I attempted to persuade them to undertake a part of the Work each, which I judged to be one of the most important for the glory of God and the service of the Church.[45]

Conrad Gesner's *Bibliotheca Universalis*,[46] the source of Possevino's discomfort, had enjoyed immense success in Catholic and Protestant territories alike after its publication in Zurich in 1545–8, despite its appearance on the Venetian Index of 1554 and the Roman Indices of 1559 and 1564.[47] Possevino's *Bibliotheca Selecta*, composed as a sanitized, Catholic answer to Gesner, took its lead from the disciplinary divisions adopted by the earlier work.[48] The mathematical disciplines, within which Gesner had included a lengthy section *De Divinatione cum licita tum illicita, & magia*, little short of anathema to the Roman censors,[49] were no exception to Possevino's purgative enterprise. In opening Book XV of the *Bibliotheca Selecta*, which deals with mathematics, Possevino explicitly acknowledges Clavius's assistance in selecting a mathematical bibliography more suitable for a Catholic readership than Gesner's indiscriminate agglomeration of potentially dangerous texts:

> In this matter, among others, the judgement, and the excellence (as it may truly be said) of Christoph Clavius, Mathematician of our Society, were of enormous help to me[50]

As Adriano Prosperi has shown, Possevino's project went beyond a merely virtual library. He functioned as something of a walking-library at times. On a visit to Montemagno, as Prosperi recounts, Possevino met schoolmasters, preachers and parish-priests in a church without doors which allowed animals to enter freely. Opening his chests, he distributed books to everyone according to his particular needs.[51]

Clavius's relationship with Possevino, formed during the calendar polemic,[52] gave mathematics a prominent place in the latter's project for fomenting piety through the judicious deployment of print culture. The long section on mathematics in the *Bibliotheca Selecta* extolls the necessity, dignity and utility of mathematics in terms highly redolent of the prefaces to Clavius's own published works. While enlisting Benito Pereira's attack on illicit magic to exclude judicial astrology and the other forms of divination that inhabited Gesner's *Bibliotheca Universalis*,[53] Possevino placed sanitised mathematical practices firmly within the domain of the pious Catholic scholar. The heroic stories of the mathematically-enabled military exploits of Archimedes and Proclus were linked in his account with the technical wonders of Renaissance Italy and the use of astronomical knowledge to excite admiration for the divine opus.[54] The mathematician and post-Tridentine theologian, in Possevino's description, were close allies, not just because of the Gregorian reform of the calendar, but also because of the widespread mentions in Scripture of stars, orbs, measures, the architecture of the temple of Solomon, and 'of more than six-hundred other things'.[55] Given a seal of authority by a warm prefatory letter from Clement VIII, Possevino's *Bibliotheca Selecta*, despite undergoing a lengthy period of censorship prior to its publication, situated mathematics within a Jesuit-led strategy of spiritual recovery in late sixteenth-century Europe.

Shortly after the appearance of Possevino's *Bibliotheca Selecta*, Clavius presented his own plan for world-evangelisation through polite letters, probably on the occasion of the Fifth General Congregation of the Society (1593–4) in Rome. The text, which remained unpublished for reasons that will shortly become apparent, was entitled *Discourse of a very close friend of the Society of Jesus on the method and way in which the Society can improve the opinion which men have of it, to the greater honour of God and the profit of souls, and by which the estimation of all heretics in literary matters (upon which they greatly depend) can be most rapidly and easily destroyed*.[56]

In his *Discourse*, Clavius summarised the centrality of erudition to the success of the global Jesuit apostolate:

> The great reputation that the Society of Jesus generally holds amongst foreigners in distant parts of the world has come from [its contributions to] the Universe of Letters. For, although, as we believe, it has possessed singular moral uprightness, this has only been clearly perceived by those who are very close to it, and most people, ignorant of this, have regarded it as no different in this respect to the other religious orders. However the praise of its most elegant erudition, in comparison with the other barbarous orders, is such that even one of the enemies will perceive a great unanimity. It is by this one thing alone that [the Society] has acquired so much authority amongst many men living very far away that people who were hesitant in faith have retained the Catholic one, and people who have entered into heresy have revoked it, persuaded solely by their belief that so many men of such prodigious learning who are so unanimously agreed cannot be ignorant of the truth.[57]

Learning and unanimity played a fundamental role in ensuring the superior credibility of the Jesuit line on matters of faith with respect to the other religious orders. However, the Ignatian emphasis on *mediocritas*, manifested in the *Constitutions* and in the organization of studies in the *Ratio Studiorum*, still undergoing modification at this point, threatened to undermine this privileged position:

> This opinion of the erudition of the Society and its high reputation must therefore be safeguarded and be amplified in every possible way, as some would attempt to diminish it for the following reason: there are in the Society many men who are moderately learned, but none who are exceptionally so. For besides philosophy and scholastic theology, which the heretics deprecate, calling one a hodge-podge of opinions and the other Thomist and Scotist sophistries, in the rest of the so-named polite arts and in the knowledge of various languages they claim that the Society is inferior to them. They would demonstrate this

with the examples of Tremellius in Hebrew, Wolf in Greek, Sturm in oratory and Melanchthon, most excellently versed in every kind of history. They judge that there are no equals in the Society against whom these men could be set.[58]

In mathematics, oratory, Greek and Hebrew, Clavius contended, the Jesuit order lacked specialists who could compete with heretical laymen for the attention of Catholic princes. Clavius's diagnosis of the reasons for this state of affairs provides a rich illustration of the conditions of Jesuit training:

> The cause of this situation is manifest. For, since members of the laity often concentrate on a single subject, it is necessary by this approach, that they stand out with the course of time as extremely learned. In the Society, however, all follow the same method and way of studies. For after a short, but sufficient introduction to *Literae Humaniores*, they devote their constant attention to three years of philosophy and four of scholastic theology, so during their career their eyes neither can, nor should, be deflected towards anything else. When, however, these years are over everyone is sent to work immediately, pressed by incessant affairs, with the result that there is no spare time afterwards to allow them to touch any part of another science in any depth whatsoever. Thus it happens that although most or almost all men of the Society attain a moderate knowledge of many things, very few attain excellent knowledge of even one. I mean to except theology. I also mean to except some men who studied certain subjects to an outstanding level before they joined the Society.[59]

The future reputation of the Society, and hence its ability to perform its work in the vineyard of the Lord, demanded the cultivation of excellence:

> [T]here is no one who does not perceive how much it is central to every objective of the Society to have some men who are most outstandingly erudite in these minor studies of mathematics, rhetoric, and language [...] who would spread the eminent reputation of the Society far and wide, unite the love of noble youths, curb the bragging of the heretics in these arts, and institute a tradition of excellence in all those disciplines in the Society.[60]

The public gaze, so abhorred by Ignatius Loyola, was to be reconquered hypnotically by Jesuit intellectual celebrities in Clavius's project:

> I believe it to be so ordained by nature that eminence in any subject, even of the least importance, causes the eyes of everyone to converge on oneself. This was the cause of that veneration of ancient kings towards remarkable painters and sculptors. It is for this reason that in these times many Catholics have surrendered

their sons by the reputation of more excellent erudition to be instructed and lost to heretics; the noble King of Scots his son to the poet Buchanan, noble Frenchmen their sons to Petrus Ramus, and now the Germans to Hieronymus Wolf, an impious heretic excellently versed in Greek Letters. [...]

It is only the Society [of Jesus] that can pursue [eminence] both very quickly and very easily. For it possesses diverse and most beautiful minds of youths, it has free time, it has masters, it has the authority to direct its subjects to whichever kind of study that suits them the most. Thus, with no effort and no extra expense, in a very brief time in all of these matters that I have mentioned, in eloquence, mathematics and discoursing in Greek and Hebrew, the Society of Jesus can have brilliant and most eminent men. When they are distributed in various nations and kingdoms like sparkling gems these will be a source of great fear to all enemies, and an incredible incitement to make young people flock to us from all the parts of the world, to the great honour of the Society[61]

Clavius proposed to implement his project through the establishment of four Jesuit academies, one for eloquence, one for Greek, one for Hebrew and one for mathematics. Each would be located in one of the great colleges of the order, Rome, Coimbra, Milan and Paris, and would contain ten men, chosen from ten different provinces. Following their completion of the philosophy course, the chosen academicians would undergo four years of private training in which they would be permitted 'to pursue their own excursions independently in the works of various authors'.[62] The teacher would be present as a *moderator* of the academy, ensuring that his gifted pupils were not led astray. Once the alumni of these academies were redistributed to the different Jesuit provinces, Clavius hoped,

> by the praise of [their] most excellent erudition, snatched away from the impious, and, as it were, gathered under the banner of the Society, the youth of the world will be drawn to it, the booty of the heretics will be recovered and an infinite multitude of souls will be acquired for Christ our Lord and the Doctors[63]

The Ignatian *Constitutiones* had stipulated the necessity of knowledge of the mathematical disciplines for members of the Jesuit order only 'in so far as they are suitable to the end which is before us'.[64] While his role as defender of the Gregorian calendar was establishing the reputation of the Jesuit mathematician in the wider *orbis christianus*, Clavius's internal campaign for mathematics sought to convince his Jesuit superiors that a high degree of carefully cultivated mathematical expertise, like oratorical skill, might be entirely consonant with the higher goals of the Society.

Mobile mathematicians: Clavius's disciples

Although the fate of Clavius's three other projected academies is uncertain, Ugo Baldini has shown that a private level of mathematical teaching appears to have existed in the Collegio Romano since the time of Balthasar Torres, long before Clavius's dramatic presentation of the apostolic role of mathematical expertise.[65] Given the vast production of mathematical works by Jesuits during the seventeenth century,[66] the existence of a private level of tuition can hardly come as a surprise. The public mathematics course provided in Jesuit colleges during the second year of the philosophy course[67] was simply too basic to allow Jesuits to reach a level of mathematical competence sufficient to compose original mathematical works or construct instruments, unless they had received a mathematical apprenticeship prior to entering the Society. The alumni of the informal *academia mathematica* became mathematics teachers in the different provinces of the Society, or used mathematics in other ways to further the apostolic goals of the order. Matteo Ricci was merely the most famous of Clavius's private pupils. His translation into Chinese of the first six books of Clavius's *Euclidis Elementorum*[68] and his use of cartographic and astronomical skills, which combined with his mnemonic abilities and dress-sense to ingratiate him into the late Ming court,[69] can be directly related to his period of private mathematical tuition with Clavius in the Collegio Romano between 1575–7.[70] From China, Ricci wrote to Clavius and his fellow academician Giulio Fuligatti to discuss sundials and globes that he had made with the help of Clavius's printed works to display to his Chinese callers.[71] The pattern of recruitment from the Provinces, apprenticeship in Rome, redistribution to the periphery and correspondence with the master was to become standard. Clavius's first recorded private academician, the Scottish Jesuit John Hay, subsequently resided in Vilnius, Bordeaux, Paris, Tournon, Louvain, Liège and Pont-à-Mousson.[72] James Bosgrave, an English Jesuit, left Clavius's tutelage to live in Olomouc, Vilnius, Braunsberg, Poznan and Kalisz.[73] Other early disciples left the Collegio Romano for Cluj in Transylvania, Vienna, Prague, Lisbon, Coimbra, Douai, Louvain, Macao and almost anywhere else in the globe where the Jesuits had a foothold. Rapidly Clavius's mathematically trained offspring began their own private 'academies' in these outposts, training disciples who were unable to make the journey to Rome. These second-generation disciples, however, still recognised Clavius as their mathematical ancestor, with his printed works acquiring the status of devotional objects. Johann Falckestein wrote to Clavius '*ut Oraculum consulam*' about a new astrolabe, and added that:

> Besides the fact that your commentaries were like teachers [*Magistrorum instar*] to me, I had Your Reverence's pupil the Scotsman John Hay as my master[74]

The geographical displacements of pupils of the Roman academy run by Clavius and subsequently continued under Christoph Grienberger (himself a second-generation disciple[75]) and Athanasius Kircher opened up new channels along which letters, books, instruments and mathematical problems could flow to and from Rome. Although Rome retained special status as a centre of authority, other important geographical centres emerged around the principal courts of Catholic Europe. Lisbon, as the point of departure for Jesuit missions to India, China, Japan and Brazil, became particularly significant. The *Indepetae* of the Jesuit archives in Rome are full of letters from mathematically trained Jesuits requesting to be sent to the Indies to use their abilities for the saving of souls. Grienberger left Clavius in Rome temporarily in 1599 to supervise the training of Jesuit mathematicians for the missions in Lisbon directly. One of his pupils, Giovanni Antonio Rubino, wrote from Chandrapur, the seat of the Rajah of Vijayanagar, to Clavius in 1609:

> I am in the great Kingdom of Bisnagà, attempting to procure the conversion of these souls, but for the moment *clausa est ianua*, we are waiting for the Lord to open it, so that many souls will be saved from going miserably to hell. The Brahmans, who are the *literati* of this kingdom, are very given to the cognition of the movements and conjunctions of the planets and stars, and in particular of 27 of them by which they govern and rule themselves. Your Reverence will be amazed at how they predict the hour and minute of eclipses of the sun and the moon, without knowing the way in which eclipses occur. I have attempted many times to make them state the way in which they derive the conjunctions of the planets, but I was never able to get them to declare it, and they don't wish to teach the things they know to others, except in secret to their relatives[76]

Rubino added that 'There is nothing that I desire more than Your Reverence's *Astrolabium*, which is not to be found in the whole of the Indies',[77] demonstrating the importance of the transmission of Clavius's published works in the contests of mathematical expertise that were to become a trademark of Jesuit missions to the Orient. The mathematicians who stayed in Europe, often frustrated would-be missionaries, sometimes placed their mathematical activities within an apostolic context. Paul Guldin reminded readers of his *Centrobaryca* that to save a single human soul was more important than any mathematical problem, including the squaring of the circle.[78] Despite Guldin's distinction, the labours of the school of Clavius were arguably based on a firmly held belief that circle-squaring (which Clavius held to be possible[79]) and the saving of souls were entirely congruent forms of activity. Clavius's earlier suggestions to his Jesuit superiors about the importance of mathematical conversation to forming powerful links between the Jesuits and European aristocrats were confirmed in practice. Bernardino Salino wrote to Clavius from Genoa in 1595 to describe such an encounter:

> A few days after I arrived in Genoa from Corsica, as Marchese Pietro Francesco Malaspina had heard from one of ours [i.e. a Jesuit] that there was a father in our college who understood a little mathematics, he came immediately to find me in the College, and wished to talk with me about many things. He asked me to teach him some new demonstration, if I had any.[80]

Later Malaspina became the '*Moecenas [. . .] munificentissimus*' of works written by another of Clavius's disciples, Giuseppe Biancani.[81] Biancani and Malaspina had '*stretta conversatione*' in Piacenza in the Winter of 1602–3, discussing the conics of Apollonius of Perga, and the wealthy Marchese, delighted by Biancani's demonstrations and inventions, such as a sundial that could tell the time in the shade, later took theology courses in the Jesuit college of S. Rocco in Parma where Biancani taught.[82]

As Clavius had predicted, mathematical works could play an important role in ensuring local Jesuit patronage in Europe. Johann Reinhard Ziegler wrote to Clavius in 1607 to say that

> [T]here is a certain Catholic Count in Aquisgrana who is extremely studious of mathematical matters, and has even wanted to transcribe with his own hand that which I have taught in the various parts of this science in classes. He is a friend and great patron [*fautor*] of the Society and, especially of the College of Aquisgrana, and if you would dedicate some little work to him in the future it would be worthwhile[83]

Where munificent, mathematically literate patrons were not locally available, lack of books and mathematical instruments often made the replication of the Roman academy in the Provinces difficult. Bernardo Salino wrote to Clavius to give a very clear picture of the situation in Genoa, after Malaspina's interest and financial support had been transferred to the more mathematically sophisticated province of the Veneto.[84] His letter is revealing of the immense difficulties involved in exporting the Clavian model, so it is, I think, worth quoting at some length:

> I think your Reverence will already know that I teach mathematics here in Genoa,[85] and, as this science has never been taught in this place before, I have

found the College to be completely unfurnished with both the books and the instruments necessary for this profession. And as, until know, I have had to teach Cases of Conscience, I have not been able to take care of making much provision of books or of instruments, having no time to attend to these studies apart from that required by the lessons. Now the Rector of the College, seeing that I couldn't attend well to both the one and the other to satisfy the desire of these Gentlemen, who, in addition to the lesson, often want to chat with me and ask various things relating to this science, has obtained with the Fr Provincial that I should be relieved of the burden of Cases, but as he has found no one for the moment who was capable of lecturing in them, he has begun to teach them himself in my place. Thus, desiring to attend more diligently to the teaching that has been imposed on me, I have acquired some few books by way of Milan, but some of those that I included in my list could not be found. For this reason, necessity has forced me to have recourse to Your Reverence, to make provision via Rome for all of the things that I require. I have all of your works apart from Theodosius *de Sphericis* with the table of sines[86] and the Gregorian Calendar,[87] and two *Compendii de Horologiis*.[88] I would also like Apollonius of Perga *De Conicis Elementis*,[89] the *Tavole perpetue* of Magini,[90] and some new ephimerides if they can be found. I have borrowed Magini's from a gentleman,[91] and even if no others can be found I can still take these. I'd like the book of a Dane who has made experiments [*esperienze*] of the celestial motions,[92] the book of a German gentleman who saves the motions of the planets with new hypotheses,[93] and some curious books about Mechanics and practical geometry. As for the instruments, I need, first, a compass for drawing mathematical figures on the large board in the classroom, a celestial and terrestrial globe, of medium size. Because they are very insistent that I should teach the recognition of the stars, and the art of measuring distances, heights and depths, I need some instruments for these purposes, a quadrant and an astrolabe. If these instruments are secondhand, I don't mind, as it might cost too much to make new ones. For the money to purchase these things, you need only ask the *Procurator Generalis*, who knows where to procure it. Your Reverence need only ensure that the Fr Rector orders someone to buy these things, and look to see if they are of good quality, before sending them to me when you have a convenient moment.[94]

Three months later, without an anwer from Clavius, Salino reiterated that '*de libri ne ho pochissimi, e de instromenti nissuno penitus*'. The college was financially crippled by the costs of building work and land-acquisitions, so Salino had to be content with a few books and some wooden instruments that he had made himself.[95] From Poland, Simon Kaczorononski wrote that 'we suffer from a terrible penury of mathematical books, so Your Reverence will not be surprised if we do not give sufficient satisfaction with our minds'.⁹⁶ With the same letter, Kaczorononski sent an astronomical quadrant that he had made 'in furtive hours' using Clavius's *Astrolabium*. He pleads for his crude efforts to be corrected by the master: 'If something might be added or made in a better way, you would teach one who desires to learn'⁹⁷:

> In the division of the horizon, I have erred. And I have placed the stars in the parallels of the same, not in their places according to longitude, but I know this now.⁹⁸

The correspondence of Clavius is filled with examples of disciples seeking correction of demonstrations, instruments and opinions from Rome. They looked to Clavius to hunt out the *'anguis latens in herba'* – the serpent in the grass, or concealed paralogism, in new attempts to square the circle. They asked Clavius's colleague in Rome Juan Bautista Villalpando to send metrological assistance through the post – a *'piede antico Romano, in legno o altra materia salda'*, and an *'oncia antica agiustata'*,⁹⁹ to correct unreliable local measures. Most of all, apprentice Jesuit mathematicians in the Provinces looked to Clavius and his colleagues in Rome for painless discipline. On 15th August 1593, the Sicilian mathematician Nicolò Calandrino wrote to Clavius from Reggio Calabria:

> It seems to me that our Sicilians are *in rebus mathematicis* like a sciotheric sundial, on which the lines have been drawn correctly, placed in the proper place and site, which has no gnomon to tell the time. Your Reverence was to be that gnomon.¹⁰⁰

When Clavius, or his successor Grienberger, did grace a college with their presence, most notably the case in Naples, the replication of the Roman mathematical academy was achieved most easily.¹⁰¹ Not all confessions that made their way to Clavius were of a mathematical nature. One of the closest mathematical collaborators of Clavius, Grienberger and Villalpando, Marino Ghetaldi,¹⁰² left Rome hurriedly after involvement in a dirty deed:

> I never thought I'd have to leave Rome without saying a word to my friends, but unthinkable things intervened. Your Reverence knows that I am a stranger to affrays [*costioni*], but believe me that I was driven to do what I have done. Nonetheless, even in that rage I never wanted to kill him, though he had given me good reason, but just meant to teach him a lesson. But, because one doesn't measure blows, I did more than I meant to do.¹⁰³

Once he had reached the safety of Ragusa, Ghetaldi's morally questionable action did not exclude him from continued epistolary commerce with Clavius and

Grienberger, with whom he continued to exchange mathematical demonstrations in the years after the homicide.[104]

Conclusion – The circulation of Clavius's image

The precise nature of teaching, including mathematical teaching, in Jesuit colleges could vary dramatically from place to place, depending on local needs and competition from other educational institutions, as a number of recent local studies have demonstrated.[105] One could not, thus, construct a simple 'family-tree' of Jesuit mathematicians having Clavius as its only root.[106] Nonetheless Clavius's project for raising the status of the mathematical disciplines was at base a social project of distributing trained practitioners, instruments and books throughout the Jesuit empire.

In May 1611 Johann Reinhard Ziegler wrote to Paul Guldin to discuss the publication of the first volume of Clavius's *Opera Omnia*.

> I have written the dedication in the *persona* of Clavius. If I wrote anything unworthy of him or other than he desired, I ask for indulgence. I wished to obey, not to offend. I have not yet presented the first volume to his Reverence the Bishop of Bamberg.[107] However I shall do so soon, and it will be elegantly bound. To honour Fr. Clavius, he is taking care to ornament the front of the work with an engraved title page at his expense. He even wishes for the likeness of Fr. Clavius that has been circulating in Germany to be reprinted. If it is not a good representation, be patient, as the book itself will certainly express the mind.[108]

The likeness (*effigies*) of Clavius to which Ziegler referred was the engraving carried out by Franciscus Villamena in 1606 [**See Figure 1.1**], which was rapidly copied by other engravers. During sixteenth-century Europe (particularly in the German lands) woodcuts of famous men had become stock-in-trade of the *colporteur*, along with the more traditional images of saints, miracles and *canards* describing monstrous births. Dürer's engravings of Erasmus and Melanchthon, both carried out in 1526, are early examples of this privilege being extended to scholars for a more elite audience. The title page of the first volume of Clavius's *Opera Omnia* incorporates a rough copy of Villamena's engraving carried out by Johannes Leypolt, in which Clavius has been artificially aged by means of a grizzly beard. The elderly mathematician, pictured in his Roman *cubiculum*, is surrounded by the tools of his trade – a pair of compasses, a quadrant, an

Establishing Mathematical Authority

Figure 1.1 Portrait of Christoph Clavius, 1606, by Francesco Villamena

armillary sphere, an astrolabe and a number of books, containing geometrical diagrams. Whereas the authenticity of the likeness presented in the original engraving was guaranteed by a papal privilege and the 'authority of the Superiors', the book itself was sufficient warranty for the integrity of Leypolt's copy. 'God gave me knowledge of the course of the year and the positions of the stars'[109] the title page announced, citing the book of Wisdom, and the remaining elements – the opposite figures of *Astronomia* and *Geometria*, presided over by the Virgin Mary, and the medieval saints Heinrich, founder of the bishopric of Bamberg, and his notoriously chaste wife, Kunigund, are indicative of the world in which Clavius's mathematical authority was being proclaimed. The publication of the *Opera*, which consciously emulated the successful Mainz edition of Suarez's *Metaphysicae Disputationes*,[110] was a laborious and thankless task for Ziegler, who wrote to Clavius constantly asking for advice and copper-plates, but only received meaningful collaboration from Odo Maelcote and Paul Guldin in Rome. '*Totus sum in delineandis figuris*' ('I am all [immersed] in drawing the figures'), he wrote to Clavius in October 1609. 'I am now writing my fifth letter to Your Reverence,' he wrote three months later, 'and I have not received a single answer. But I do not despair. I am the willing servant of Your Reverence, from love of the Society and of Mathematics'.[111] The circulation of Clavius's authority and reputation, epitomised by the publication of his *Opera mathematica*, was enabled by the circulation of his obedient and self-effacing disciples.

Notes

1. On Clavius, see especially James M. Lattis, *Between Copernicus and Galileo: Christoph Clavius and the collapse of Ptolemaic cosmology*, Chicago and London: University of Chicago Press; 1994, and Romano, 1999, pp. 85–178.
2. On Clavius's mathematical academy, see Ugo Baldini, The Academy of Mathematics of the Collegio Romano from 1553 to 1612, in Mordechai Feingold, ed., *Jesuit Science and the Republic of Letters*, Cambridge, MA., London: The MIT Press, 2003, 47–98.
3. J. Nadal, Dialogus I (1562–1563), in *P. Hieronymi Nadal Commentarii de Instituto Societatis Iesu*, ed. Michael Nicolau, S.J. (= *Epistolae et Monumenta P. Hieronymi Nadal*, Tomus V) Romae: apud Monumenta Historica Societatis Iesu, 1962, pp. 524–600, on p. 536.
4. On Cano's opposition to the Jesuits, see O'Malley, *The First Jesuits*, Cambridge, Massachusetts: Harvard University Press; 1993, pp. 292–3.
5. J. Nadal, Dialogus II (1562–1565), in *P. Hieronymi Nadal Commentarii de Instituto Societatis Iesu*, ed. Michael Nicolau, S.J. (= *Epistolae et Monumenta P. Hieronymi*

Nadal, Tomus V) Romae: apud Monumenta Historica Societatis Iesu, 1962, p. 545, O'Malley, *The First Jesuits*, cit., pp. 200-242.

6   See L. Lukáks, *De Prima Societatis ratione studiorum Sancto Francisco Borgia Praeposito Generali constituta*, AHSI 27 (1958) 209-232.

7   Nadal, *Dialogus II*, cit., p. 666.

8   J. Nadal, *Exhortatio Coloniensis 6a* (1567), in *P. Hieronymi Nadal Commentarii de Instituto Societatis Iesu*, ed. Michael Nicolau, S.J. (= *Epistolae et Monumenta P. Hieronymi Nadal*, Tomus V) Romae: apud Monumenta Historica Societatis Iesu, 1962, p. 832, n. 21.

9   Gabriel Codina Mir, *Au Sources de la Pédagogie des Jésuites. Le 'Modus Parisiensis'*, Rome: Institutum Historicum Societatis Iesu; 1968, p. 283

10 See especially Ugo Baldini, *Legem impone subactis. Studi su filosofia e scienza dei gesuiti in Italia, 1540-1632*, Rome: Bulzoni; 1992: pp. 19-73. Baldini argues convincingly that the organizing function of the apostolic goal of Jesuit philosophical and mathematical activities was of far greater importance to the development of these activities than specific conflicts between philosophical positions and Scriptural truth, such as the case of heliocentrism, emphasised in many past discussions of Jesuit science in the seventeenth century.

11 'PHILI. - Quam rationem docendi tenent isti iesuitae? / PHILA. - Parisiorum Academiae, quae exactissima visa est ac fructuosissima.'Nadal, Dialogus II, cit., p.738. See also O'Malley, *The First Jesuits*, cit., pp. 200-242.

12 Nadal, Dialogus II, cit., p. 737.

13 MP V (*Ratio atque Institutio Studiorum Societatis Iesu*), MP VI (*Collectanea de Ratione Studiorum Societatis Iesu (1582-1587)*), MP VII (*Collectanea de Ratione Studiorum Societatis Iesu (1588-1616)*)

14 On Polanco, see O'Malley, cit., pp. 10-11. On the relationship between bureaucratic expertise and the organization of the early Society see Dominique Bertrand, *La politique de Saint Ignace de Loyola: L'analyse sociale*, Paris: Editions du Cerf; 1985.

15 Mario Biagioli, *The Social Status of Italian Mathematicians*. History of Science. 1989; 27(Part 1 Number 75): pp. 41-95. On mathematics and humanism in sixteenth-century Italy, see especially Paul Lawrence Rose, *The Italian Renaissance of Mathematics: Studies on Humanists and Mathematicians from Petrarch to Galileo*. Geneva: Librarie Droz; 1975.

16 On Gesio, see David C. Goodman, *Power and Penury: Government, Technology and Science in Philip II's Spain,* Cambridge: Cambridge University Press; 1988, passim.

17 On the mathematical academy, see M.I. Vicente Maroto and M. Esteban Piñeiro, eds., *Aspectos de la Ciencia Aplicada en la España del siglo de oro*, León: Consejería de Cultura y Bienestar Social; 1991, pp. 69-134. On Herrera's library and instruments, see Luis Cervera Vera, *Inventario de los bienes de Juan de Herrera*, Albatros: Valencia; 1977.

18 See J. A Bennett, *The Mechanics' Philosophy and the Mechanical Philosophy*, History of Science, 1986; pp. 24, 1-28, E. G. R. Taylor, *The Mathematical Practitioners of Tudor an Stuart England*, Cambridge; 1954.

19 A survey of the German scene, which discusses the vogue for Faust plays, is William Clark, *The scientific revolution in the German nations* in R.S. Porter and M. Teich, eds. *The scientific revolution in national context*, Cambridge: Cambridge University Press; 1992: 90–114. On the status of mathematics and astronomy, see Robert S. Westman, *The Astronomer's role in the sixteenth century: A preliminary study*. History of science, 1980; 18: pp.105–147, idem., *The Melanchthon Circle, Rheticus and the Wittenberg interpretation of the Copernican theory*. Isis. 1975; 66: pp.165–193, idem, *Humanism and scientific roles in the Sixteenth century* in Rudolf Schmitz and Fritz Krafft (eds.), *Humanismus und Naturwissenschaft*. Boppard: Harald Boldt; 1980: 83–99.

20 For a rich study of the culture of the Rudolphine court in Prague, see R. J. W. Evans, *Rudolph II and his world: A study in intellectual history, 1576-1612*, Oxford: Clarendon Press, 1973. On illusionism in Prague see Thomas DaCosta Kaufmann, *The Mastery of Nature: Aspects of Art, Science, and Humanism in the Renaissance*, Princeton, NJ: Princeton University Press; 1993. On optical devices, see Jurgis Baltrusaitis, *Anamorphoses ou magie artificielle des effets merveilleux*, Paris: Olivier Perrin; 1969.

21 For biographical details, see James M. Lattis, *Between Copernicus and Galileo: Christoph Clavius and the collapse of Ptolemaic cosmology*, Chicago and London: University of Chicago Press; 1994, pp. 12–29, CC I.1, pp. 33–58, Ugo Baldini, *Christoph Clavius and the scientific scene in Rome*. in G.V. Coyne and O. Pedersen, eds., *Gregorian Reform of the Calendar: Proceedings of the Vatican Conference to Celebrate its 400th Anniversary 1582-1982*; 1982; Vatican Observatory. Città del Vaticano; 1983: 137–169.

22 Christoph Clavius, *Modus quo disciplinae mathematicae in scholis Societatis possent promoveri*, in MP VII, pp. 115–117.

23 Clavius, *Modus quo disciplinae mathematicae in scholis Societatis possent promoveri*, cit., p. 116.

24 On Clavius's involvement in the development of the Gregorian calendar, see especially Ugo Baldini, *Christoph Clavius and the scientific scene in Rome*, in G.V. Coyne and O. Pedersen, eds., *Gregorian Reform of the Calendar: Proceedings of the Vatican Conference to Celebrate its 400th Anniversary 1582-1982*, Città del Vaticano: Vatican Observatory, 1983, pp. 137–169.

25 *Bullarium Romanum*, Vol. 8, pp. 386–390, cited in August Ziggelaar, *The papal bull of 1582 promulgating a reform of the calendar*. in G.V. Coyne and O. Pedersen, eds., *Gregorian Reform of the Calendar: Proceedings of the Vatican Conference to Celebrate its 400th Anniversary 1582-1982*, Città del Vaticano: Specola Vaticana; 1983: 201–239.

26 BAV Barb. Lat. 3151 ff. 58r–61v (published in Massimo Bucciantini, *Contro Galileo: Alle origini dell'Affaire*, Florence: Olschki; 1995, pp. 207–9), on 58r.
27 Ferdinand Kaltenbrunner, *Die Polemik über die Gregorianische Kalenderreform*, Vienna: In Commission bei Karl Gerold's Sohn; 1877.
28 Michael Maestlin, *Außführlicher und Gründlicher Bericht* (Heidelberg: Jakob Müller; 1583).
29 Maestlin, *Außführlicher und Gründlicher Bericht*. On the Clavius-Maestlin controversy, see F. Kaltenbrunner, *Die Polemix über die Gregorianische Kalenderreform*, Wien, 1877, pp. 32–6, 48–50, 57–62, 63–9 and CC 2.2, pp. 80–8 (letter 46 note 2).
30 *Christophori Clavii Bambergensis ex Societate Iesu in Sphaeram Ioannis de Sacro Bosco Commentarius Nunc iterum ab ipso Auctore recognitus, et multis ac varijs locis locupletatus*, Romae: Ex Officina Dominici Basae; 1581.
31 Maestlin, *Außführlicher und Gründlicher Bericht*, cit., p. 159.
32 Antonio Possevino, *Moscovia, et, alia Opera, de statu huius seculi, adversus Catholicae Ecclesiae hostes*, [Coloniae, Agrippine]: In officina Birckmannica, sumptibus Arnoldi Mylij; 1587, Sectio IV, pp. 206–223, *De anni et Paschae Emendatione, Sectio IIII, Olim a Nicaenis Patribus facta, ac nunc ab Ecclesia Catholica ad pristinam normam ac rationem revocata: Quae ab haereticis nescientibus quid loquantur reprehensa satis ostendit, quanti Chytraeus, & reliqui veritatem, & decreta Synodorum probatarum, & purae, quam vocant, Ecclesiae faciant*.
33 Possevino, *Moscovia*, cit., p. 223.
34 Christoph Clavius, *Novi Calendarii Romani Apologia, Adversus Michaelem Maestlinum Gaeppingensem, in Tubingensi Academia Mathematicum Tribus Libris Explicata*. Romae: Apud Sanctium, & Soc.; 1588.
35 See, in particular, Antonio Possevino, *Atheismi Lutheri, Melanchthonis, Calvini, Bezae, Ubiquetariorum, Anabaptistarum, Picardorum, Puritanorum, Arianorum, & aliorum nostri temporis haereticorum*. Vilnae: Apud Ioannem Velicensem; 1586. On Ubiquitarians, see ff. 32v–39r.
36 Clavius to Rudolph II, dedicatory letter to *Novi Calendarii Romani Apologia, Adversus Michaelem Maeslinum Goeppingensem, inTubingensi Academia Mathematicum, tribus libris explicata*, ... Romae, Apud Sanctium, et Soc. 1588 (sig. a2), Rome, 18 October 1588, published in CC II.1, pp. 130–133. Ubiquitarianism, as the name suggests, is a term used to refer to the doctrine adopted by some Lutherans that the body of Christ is in some sense everywhere, rather than localised in the Eucharist.
37 Clavius, *Apologia*, cit., p. 316. Clavius pointed out Maestlin's heresy again in the dedicatory letter, to Cardinal Francisco Toledo, of his reply to Scaliger, *Iosephi Scaligeri Elenchus, et Castigatio Calendarij Gregoriani a Christophoro Clavio Bambergensi Societatis Iesu castigata*, Romae: Apud Aloysium Zannettum; 1595, p. 3–7 (also in CC III.1, pp. 118–9), on p. 4.

38 Clavius, *Apologia*, cit., p. 323 and ibid., pp. 313–4.
39 F. H. Reusch, *Der Index der verbotenen Bücher: ein Beitrag zur Kircher- und Literaturgeschichte*, Bonn: Verlag von Max Cohen & Sohn; 1883–5, pp. 504, 566.
40 *Scaligerana. Editio altera. ad verum exemplar restituta, & innumeris iisque foedissimis mendis, quibus prior illa passim scatebat, diligentissime purgata*. Coloniae Agrippinae: Apud Gerbrandum Scagen; 1667, p. 51 (s.v. 'Clavius').
41 Johann Reinhard Ziegler to Clavius, Mainz, 16 January 1609, in CC VI.1, 128–9
42 For a useful summary of Possevino's enterprise, see John Patrick Donnelly, *Antonio Possevino's plan for world evangelization*. The Catholic Historical Review. 1988; 74: 179–198.
43 The figure is provided by Donnelly, *Antonio Possevino's plan*.
44 Donnelly, *Antonio Possevino's plan*, CC I.2, pp. 80–1, Giuseppe Castellani, *La vocazione alla Compagnia di Gesù del P. Antonio Possevino da una relazione inedita del medesimo*, Archivum Historicum Societatis Iesu, 14 (1945–6), 102–124.
45 Jean Dorigny, *La vie du Père Antoine Possevin de la Compagnie de Jésus*, Paris: Chez Etienne Ganeau, 1712.: pp. 500–1.
46 *Bibliotheca Universalis, sive Catalogus omnium scriptorum locupletissimus, in tribus linguis, Latina, Graeca, & Hebraica*, Tiguri, apud Christophorum Froschoverum, 1545. (Tom. 1), *Pandectarum sive Partitionum univeralium*, Tiguri, apud Christophorum Froschoverum, 1548 (Tom. 2)
47 J.M. de Bujanda, ed., *Index des Livres Interdits*, Québec: Centre d'Études de la Renaissance, Éditions de l'Université de Sherbrooke, Librairie Droz, 1990, vol. VIII (Index de Rome),: p. 396, A. Moreni, La *Bibliotheca Universalis* di Konrad Gesner e gli Indici dei libri proibiti, in *La Bibliofilia*, LXXXVIII (1986), pp. 131–150
48 On the relationship between the two works, see Alfredo Serrai, *Storia della Bibliografia. IV. Cataloghi a stampa. Bibliografie teologiche. Bibliografie filosofiche. Antonio Possevino*. A cura di Maria Grazia Ceccarelli, Roma: Bulzoni; 1993, pp. 713–760, and Helmut Zedelmaier, *Bibliotheca Universalis und Bibliotheca Selecta: Das Problem der Ordnung des gelehrten Wissens in der frühen Neuzeit*, Köln, Weimar, Wien: Böhlau Verlag; 1992, especially pp. 128–150 which provides a systematic comparison. On the *Bibliotheca Selecta*, see also Albano Biondi, *La Bibliotheca Selecta di Antonio Possevino. Un progetto di egemonia culturale* in Gian Paolo Brizzi, ed., *La 'Ratio Studiorum': Modelli culturali e pratiche dei Gesuiti in Italia tra Cinque e Seicento*, Rome: Bulzoni; 1981: 43–75.
49 Gesner, *Bibliotheca Universalis*, cit., Tom. 2 pp. 73–76: Arithmetica, pp. 77–80: Geometria pp. 81–86: De musica, pp. 87–94: De astronomia, pp. 95–98: De Astrologia, pp. 99–106: De Divinatione cum licita tum illicita, & magia, pp. 107–116: De Geographia.
50 'Quam ad rem, praeter alios, Christophori Clavij Mathematici Societatis nostrae, iudicium, &, quae vere dici potest, praestantia, magno mihi auxilio fuit', Possevino, ref.

51 Adriano Prosperi, *Tribunali della coscienza. Inquisitori, confessori, missionari*, Milan: Einaudi; 1996, pp. 617–8, citing ARSI Ven. 105, II, f. 368r.
52 On Possevino's relationship with Clavius, see also CC II.2, p. 44 (note 2): 'Tra 1584 e 1588 le relazioni di Clavio con Possevino furono più intense di quelle documentate nelle quattro lettere collegabili all'affaire Lathos [i.e. J. Latho's objections to the Gregorian calendar, which Possevino sent to Rome]; esse conpresero una consulenza data da Clavio per la parte dedicata al calendario nelle Notae divini verbi [...] In seguito Clavio sarà consulente per il libro XV della Bibliotheca selecta, dedicato alle scienze matematiche. [...] i principali momenti di collaborazione tra i due coincisero con periodi di soggiorno di P. a Roma'.
53 Possevino, *Bibliotheca Selecta*, cit., Tom. 2, pp. 202–206, esp. p. 205.
54 Possevino, *Bibliotheca Selecta*, cit., Tom. 2, pp. 177–8
55 Possevino, *Bibliotheca Selecta*, cit., Tom. II, Lib. XV, p. 177.
56 Christoph Clavius, *Discursus cuiusdam amicissimi Societatis Iesu de modo et via qua Societas ad maiorem Dei honorem et animarum profectum augere hominum de se opinionem, omnemque haereticorum in literis aestimationem, qua illi multum nituntur, convellere brevissime et facillime possit*, (c. 1594), ARSI Stud. 3, ff. 485–487 (Clavius autograph), published in MP VII, pp. 119–122.
57 Clavius, *Discursus*, ed. cit., p. 119.
58 ibid.
59 Ibid.
60 Ibid.
61 Ibid
62 Ibid
63 Ibid.
64 'Tractabitur [...] logica, physica, metaphysica, moralis scientia et etiam mathematicae, quatenus tamen ad finem nobis propositum conveniunt' MP I, p. 283
65 On the mathematical academy of the Collegio Romano, see Baldini's description in CC I.1, 68–89, which describes its activities up to 1610–11, when Clavius's old age and illness forced him to cede its direction fully to Grienberger. Baldini (CC I.1, p. 69) cites the suggestion made by Torres between 1557 and 1560: 'Y si alguno discipulos, los mas ingeniosos y aptos a la mathematica, pareciere ser cosa conveniente que oyan mas que esto, para ser mas sufficientes, se les podrà leer las fiestas del año una lectiòn familiar en camara el tercer año, en la qual se les declaren sphaerica Theodosii et Menelai et Maurolici, y una introduction de tablas, o almanach perpetuo, con algun quadrante o anulo o radio', published in MP II, pp. 433–5.
66 For a quantitative analysis of Jesuit printed works on mathematics during this period, based on Sommervogel, see Steven J. Harris, *Apostolic Spirituality and the Jesuit scientific tradition*. Science in Context. 1989 Mar; 3(1): pp. 29–65.

67 See MP V, pp. 109–110, 177, 236, 284–5, 402, MP VII 109–115, Giuseppe Cosentino, *Le mathematiche nella 'Ratio Studiorum' della Compagnia di Gesú*. Miscellanea Storica Ligure. 1970; II(2): 171–213, idem., *L'insegnamento delle mathematiche nei collegi Gesuitici nell'Italia settentrionale. Nota introduttiva*. Physis. 1971; 13: pp.205–217, A. C. Crombie, *Mathematics and Platonism in the Sixteenth-century Italian Universities and in Jesuit Educational Policy* in Y. Maeyama and W.G. Daltzer (eds.), *Prismata: Naturwissenschaftsgeschichtliche Studien (Festschrift für Willy Hartner)*, Wiesbaden: Franz Steiner; 1977: pp.63–94.

68 Christoph Clavius, *Euclidis Elementorum libri XV*, Romae: Apud Vincentium Accoltum, 1574.

69 On Ricci's translation of the *Elements*, mentioned in almost all treatments of the Jesuit presence in China, see in particular Pasquale D'Elia,*Presentazione della prima traduzione cinese di Euclide*, Monumenta Serica, 1956; XV-1: 161–202, Peter Engelfriet, *The Chinese Euclid and its European Context*, in Catherine Jami and Hubert Delahaye, eds., *L'Europe en Chine: Interactions Scientifiques, Religieuses et Culturelles aux XVIIe et XVIIIe siècles*, Paris: Collège de France, 1993, pp. 241–252. On his mnemonic techniques, see Jonathan D. Spence, *The memory palace of Matteo Ricci*, London: Faber and Faber; 1985. On his dress-sense, see Willard J. Peterson, *What to wear? Observation and participation by Jesuit missionaries in late Ming society*, in Stuart B. Schwartz, ed. *Implicit Understandings: Observing, Reporting, and reflecting on the encounters between Europeans and other peoples in the early modern era*, Cambridge: Cambridge University Press; 1994: 403–421.

70 See CC I.1, p. 85, note 62.

71 Ricci to Giulio Fuligatti and Clavius, Nanjuang, 12 October 1596, in *Opere storiche del P. Matteo Ricci S.J.*, ed. P. Tacchi Venturi, Macerata; 1911–13, Vol. II, pp. 213–8, also published in CC III.1, 175–81.

72 CC I.2, 59–60

73 CC I.2, 20–21

74 'Nam praeterquam quod eius commentarii fuere mihi Magistororum instar, Magistrum habui V. Rae. discipulum P. Io. Hayum Scotum', Johann Falckestein to Clavius, Chambéry, 28 May 1594, CC.III.1, 82–3.

75 Grienberger was first taught mathematics in Prague by Paul Pistorius, who was an academician in the Collegio Romano from 1577–8. CC I.1, p. 85.

76 Rubino to Clavius, Chandragiri, 25 October 1609, CC VI.1, 142–3, also in P. Tacchi Venturi, *Alcuni lettere del P. Antonio Rubino D.C.D.G.*, Torino, 1901.

77 '[N]iuno cosa desidero più, che l'Astrolabio di VR, il quale non si ritrova in tutte l'Indie', ibid.

78 Paul Guldin, *De centro gravitatis, liber tertius, de fructu et usu centri gravitatis* Viennae: Formis Matthaei Cosmerovij in Aula Coloniensi, 1641, p. 209: 'unamque animam perditam Conditori suo restituere pluris esse iudicavi, quam omnia Mathematica

Inventa, ipsamque Circuli Quadraturam', cit. in Duhr, *Geschichte der Jesuiten in den Ländern deutscher Zunge*, Freiburg: Herdersche Verlagshandlung; 1913, Vol. II. 2, pp. 433.

79 'Ego sane nullo modo dubitare possum de possibilitate quadraturae circuli', Clavius to Johann Hartmann Beyer, Rome, 19 December 1609, CC VI.1, p. 146.

80 Salino to Clavius, Genova, 19 July 1595, CC III.1, 120–134

81 G. Biancani, *Aristotelis loca Mathematica*, Bononiae: Apud Bartholomaeum Cochium, 1615, idem., *Sphaera mundi, seu Cosmographia*, Bononiae: Typis Sebastiani Bonomij, 1620. The description of Malaspina is from the dedicatory letter of *Aristotelis loca mathematica* (see CC V.2, pp. 36–7, note 2).

82 Biancani to Clavius, Bologna, 27 May 1603, CC V.1, 79–81, and CC V.2, pp. 36–7 note 2, which cites two letters from Biancani to Guldin (January, March 1615, in the Universitätsbibliothek, Graz, ms. 159, letters 58. 59) mentioning Biancani's conversations with Malaspina.

83 Ziegler to Clavius, Mainz, 9 November 1607, CC VI.1, 65–8, on p. 67.

84 On the Jesuit mathematical school in the Veneto, see Baldini, *Legem impone subactis*, cit., pp. 347–465.

85 Salino had already written several letters, apparently unanswered, to Clavius in this capacity.

86 C. Clavius, *Theodosii Tripolitae Sphaericorum libri III. A Christophoro Clavio Bambergensi Societatis Iesu perspicuis demonstrationibus, ac scholiis illustrati*, Romae: Ex Typographia Dominici Basae, 1586.

87 idem, *Romani Calendarii a Gregorio XIII. P.M. restituti explicatio S.D.N. clementis VIII. iussu edita*, Romae: Apud Aloysium Zannettum, 1603.

88 idem, *Horologiorum nova descriptio*, Romae: Apud Aloysium Zannettum, 1599, and *Compendium brevissimum describendorum horologiorum Horizontalium ac Declinantium*, Romae: Apud Aloysium Zannettum, 1603.

89 F. Commandino, *Apollonii Pergaei conicorum libri quattuor*, Pistorii: ex nova typographia Stephani Gatti, 1596^2.

90 G. A. Magini, *Tabulae primi mobilis quae directionum vulgo dicunt*, Venetiis: Apud Damianum Zenarium, 164.

91 Probably G. A. Magini, *Ephemerides coelestium motuum . . . Ab Anno Domini 1598 usque ad Annum 1610*, Venetiis, Apud Damianum Zenarium, 1599.

92 Presumably Tycho Brahe, *Astronomiae instauratae progymnasmata*, Prague, 1602.

93 Presumably J. Kepler, *Mysterium Cosmographicum*, Tubingae: Georgius Gruppenbachius, 1596.

94 Bernardo Salino to Clavius, Genova, 14 January 1605, CC V.1, pp. 143–4.

95 Salino to Clavius, Genova, 19 April 1605, CC V.1, 155–6.

96 'Maximam penuriam librorum mathematicorum patimur, unde non miretur V.R. si non ingenio nostro, ut volumus, satisfacimus', S. Kaczorononski to Clavius, Kalisz, 27 August 1606, CC VI.1, 42–44

97 'Mitto V.R. quadrantem meum, horis furtivis factum ex Astrolabio V.ae R.ae ut si quid addi possit, aut melius fieri, discere cupientem doceat', ibid. p. 43

98 'In partitione horizontis erravi, et stellas in parallelis ipsarum longitudinem posui, sed hoc iam scio', ibid., p. 44.

99 Mark Welser to J.B. Villalpando, Augsburg, 18 October 1602, CC V.1, pp. 44–5.

100 'Parmi che i nostri Siciliani siano in rebus Math<ematic>is come saria un'horologio Sciotherico il quale tutto ben lineato e posto nel debito luogo e sito non ha gnomone che mostri l'hore. V.Rev. havea d'esser questo gnomone', Nicolò Calandrino to Christoph Clavius, Reggio Calabria, 15 August 1593, in CC II.1, pp. 27–30, on p. 27.

101 The financial resources available to Jesuit mathematicians in Naples were clearly on a different scale to those in other cities. In 1606, Giovanni Giacomo Staserio could ask Clavius for thirty copies of the new edition of his Sphaera, writing 'et si vuole, li farò dare costì li danari da adesso; credo non passaranno alla stampa cinque giulii l'una' (Staserio to Clavius, Naples, 13 January 1606, CC VI.1, p. 15).
On the academy in Naples, see Romano Gatto, *Tra Scienza e immaginazione. Le matematiche presso il collegio gesuitico napoletano (1552–1670 ca.),* Florence: Olschki; 1994, especially pp. 59–120. On the academy founded by Grienberger in Lisbon, see Grienberger to Clavius, Lisboa, 24 March 1601, in CC IV.1, pp. 136–9. On the academy in Palermo, see Grienberger to Clavius, Palermo; 21 April 21 1609 in CC VI.I, pp. 136–9. On the situation in Paris, see Chastellier to Clavius, Paris, 4 December 1594, CC III.1, 98–111, on p. 99. More often than a fully-fledged academy, it seems that Jesuit mathematics professors were permitted to have one or two 'private' pupils, who would also assist them in preparing printed works.

102 On Ghetaldi's involvement with the mathematicians of the Collegio Romano, see especially P. D. Napolitani, *La Geometrizzazione della realtà fisica: il peso specifico in Ghetaldi e in Galileo.* Bolletino di Storia delle Scienze Matematiche. 1988; VIII: pp.139–237.

103 Ghetaldi to Clavius, Venice, 21 June 1603, CC V.1, p. 82

104 For example Ghetaldi to Clavius, Ragusa, 6 June 1604, CC V.1, pp. 101–2, Ghetaldi to Clavius and Grienberger, Ragusa, 20 February 1608, CC VI.1, p. 79, Ghetaldi to Clavius, Ragusa, 20 May 1608, CC VI.1, p. 84, Ghetaldi to Clavius, Ragusa, 13 September 1608, CC VI.1 pp. 89–90.

105 For France, see especially Romano 1999. For Germany see Hellyer 2005.

106 This picture is made more complex by a number of other factors – the recruitment of members of other religious orders into Clavius's circle (e.g. Gulio Fuligatti's efforts to teach the Observant Franciscan Bonaventura da Cingoli to make sundials), the importance of other teachers (the founder of the mathematical school in the Veneto, Giuseppe Biancani, to give one example, learned mathematics both outside of the order and with Marc Antonio de' Dominis before coming into

contact with Clavius. See Bonaventura da Cingoli (OFM Obs.) to Clavius, Recanati, 16 August 1601, CC IV.1, 152–3: 'se bene non la cognosco per vista, tuttavia gli porto aff<etio>ne grande per le sue rare virtù', Lorenzo Terzo (Rector of the Jesuit College of Padua) to Clavius, Padua, 28 February 1598 (following Biancani to Clavius, Padua, 28 February 1598 CC IV.1, pp. 34–37): 'questo [i.e. Biancani] ha imparato parte al secolo, e parte dal già P. Marcantonio de Dominis, hora eletto di Segna'. Despite the difficulties, and necessary simplifications, involved in constructing a genealogical tree of Clavius's second and third generation 'disciples', the type of prosopographical method employed by Antonella Romano in tracing the teaching activities of Clavius's pupils in France might bear much fruit if applied to the other provinces of the order. See Romano, 1999, esp. pp. 287–352, 551–613.

107 Johann Gottfried von Aschhausen, a pupil at the Jesuit colleges of Würzburg, Pont-à-Mousson and Mainz, made Prince Bishop of Bamberg in 1609.

108 Johann Reinhard Ziegler to Paul Guldin, Mainz, 14 May 1611, Universitätsbibliothek, Graz, ms. 159, letter 2, quoted in CC VI.2, p.9.

109 'Dedit mihi Deus ut sciam anni cursus et stellarum dispositiones' (Wisdom, 7), title-page of Clavius, *Opera mathematica*, vol. 1.

110 'De forma editionis, illud placuit, ut fieret in folio ea quantitate, qua alias hic recusa metaphysica R. Patris Franc. Suarez', Ziegler to Clavius, Mainz, 27 June 1608, CC VI.1, pp. 86–7, on p. 86, referring to Suarez, *Metaphysicarum disputationum*, Moguntiae: Balthasar Lippius; 1606.

111 'Iam quintam ad R.V. scribo epistolam, nec quidquam responsi accipio. Tamen non despero. Ego R.V. servus sum volontarius, amore Societatis et Matheseos', Ziegler to Clavius, Mainz, 1 December 1609, CC VI.1, p. 144.

2

Mathematics and Modesty: The Problemata of Christoph Grienberger[1]

CENODOXUS: *Wakeful and easeless are my days and nights, consumed in careful studies*

SELF-LOVE: *But time cannot consume what all men's praises render immortal.*

CENODOXUS: *Yet how easily such honours can be gained. My life's whole purpose is therefore this: by glorious deeds to ensure that I and all my glory never perish. This die I've cast.*

Jakob Bidermann, *Cenodoxus*, I. iii, transl. D. G. Dyer and C. Longrigg, Edinburgh, 1975, p. 47

Modesty

In 1609 Jakob Bidermann's 'Comico-Tragedy' *Cenodoxus, or the Doctor of Paris* was performed on the stage of the Jesuit college in Munich. The play, first produced seven years earlier in Augsburg, deals with the story of a Parisian scholar who, despite maintaining an ascetic public demeanour, privately prided himself on his unparalleled erudition. In Bidermann's graphic account, based loosely around the legend of St Bruno, the eleventh-century founder of the Carthusian order, Cenodoxus, recast as a Renaissance humanist, is finally condemned to eternal torment for the sin of κενοδοξία or vaingloriousness.[2] The Munich production of the play provoked a memorable reaction, described in the preface to the first collected edition of Bidermann's dramatic works.[3] At first the audience laughed at the opening comic scenes, but as the play progressed the mood gradually changed to one of astonishment and horror as the spectators realised the enormity of the sins portrayed and became aware of the power of hell. By the end of the play, the terrified members of the audience were contemplating their own sins in stunned silence. The impact of the play was immediate. Fourteen members of the audience went into retreat to perform the *Spiritual Exercises* of St Ignatius, just as in the play Bruno retreated into the wilderness to found his monastery and lead a life of spiritual contemplation. The actor who played Cenodoxus himself then joined a Jesuit novitiate, and passed the rest of his life in the religious modesty of the Society of Jesus.[4]

It is difficult to find a more poignant example of the way the Jesuit order in general, and the Jesuit spiritual teachings embodied in the *Spiritual Exercises* in particular, were perceived amongst the ruling elites of early modern Europe as constituting a powerful antidote to pride, *superbia*, or vaingloriousness. Ignatius himself, following Gregory the Great and Thomas Aquinas, frequently emphasised the interdependence of modesty and obedience in his writings, arguing that disobedience, the ultimate enemy to the social fabric of the Jesuit order that he had founded, was an inevitable consequence of vaingloriousness.[5] The *Rules of the Society of Jesus*, first published in 1582 as a guide to the different functions and modes of social behaviour of Jesuits, contained a series of *Rules on Modesty* due to Ignatius. These rules, originally composed around 1555,[6] and well entrenched by the 1580s, really amounted to rules of bodily deportment. Members of the Society, in order to display modesty, humility and religious maturity, had to keep their heads pointing straight forward, with their necks inclined slightly downward. Eyes were to be kept down, especially when talking to others, wrinkling of the nose was to be avoided, walking more quickly than necessary was discouraged, and all gestures were to display humility and move the observer to devotion.[7] Speech too was to display modesty and edification.[8] Biographical writings about eminent Jesuits, taking their lead from Ribadeneyra's widely read biography of Ignatius,[9] laid great emphasis on the qualities of modesty, humility and self-abnegation advocated by the Jesuit *Constitutions* and *Rules*.

Deportment and scientific practice

Before the development of societies and institutions exclusively devoted to scientific pursuits in Europe from the 1660s onwards, and the subsequent emergence of codified and tacit forms of professional ethics specific to such institutions, natural philosophers and mathematicians attempting to make novel claims about the natural world were obliged to look outside science for models of acceptable conduct in the prosecution and presentation of their work. Rather than being obliged to acquiesce into a single model of personhood, scientific practitioners were free to make their own creative synthesis from a smorgasbord of religious and courtly models, to name just two of the more obvious options. Steven Shapin has emphasised the extent to which Robert Boyle drew on the social mores of the English gentleman in order to provide a social basis for credibility in the reporting of scientific observations. In a similar vein, Mario Biagioli has argued that Galileo fashioned himself as a natural philosopher by successfully deploying the vocabulary of Medicean dynastic emblematics.[10]

Whereas the court environment in which Galileo worked for at least part of his life promoted visibility and authorship – the attachments of texts, inventions and observations to a proper-name[11] –, the cultural values promoted in the Jesuit order generally emphasised invisibility and self-abnegation, and denied 'authorship' to all but a relative few, sometimes denoted by the term *scriptor* in the catalogues of the Jesuit houses. Individual glory was, in general, to be shirked in favour of the collective glory of the order. In disciplining their adversaries in theological and philosophical disputes, Jesuit authors made frequent use of terms like *jactantia* and *jactatores*, using the inappropriate deportment of opponents to religious or philosophical orthodoxy to discredit their arguments. The playwright Jakob Bidermann himself, after the successes of his theatrical castigations of *superbia*, was brought to Rome to act as General Revisor for Jesuit literary works, where he had the opportunity to police the humility of a large number of learned Jesuit writers in person for almost twenty years.[12]

Admittedly many Jesuit mathematicians also worked in a courtly environment. Galileo's opponent in the dispute over sunspots, Christoph Scheiner, is a prominent example.[13] Nonetheless, careers such as Scheiner's manifest the deep tensions between the type of deportment suitable to a court and the ready-made, modest 'personality' provided by the Jesuit prescriptive literature and inculcated through the practice of the *Spiritual Exercises*.[14] Precisely for this reason I would like to look more closely in the present chapter at a Jesuit mathematician who worked almost exclusively within Jesuit-controlled institutions. I believe that the strategies of self-abnegation[15] deployed by the Jesuit mathematician Christoph Grienberger, who availed himself of every opportunity to remove his name from texts written with his pen and optical and astronomical instruments designed by him and built with his own hands, can reveal much about what it was to be both a Jesuit and a skilled mathematical practitioner in the early seventeenth century. At the outset, this appears to be a task of some difficulty, as the 'person' that we would like to understand is a person who manifests himself by disappearing – erasing his tracks in the history of science with remarkable dexterity and even managing to avoid an entry in the *Dictionary of Scientific Biography*. However, through the indiscretions of some of his Jesuit colleagues, through his own epistolary confessions to his senior mathematical colleague, Christoph Clavius, and through the existence of a significant number of anonymous manuscripts that I attribute to Grienberger,[16] the the public and private selves of this elusive individual begin to emerge. Where Galileo found a source of legitimation for certain types of mathematical practice in the colourful world of the Medici court in Florence, his exact contemporary Grienberger found his Archimedean point for the upward leverage of the status of mathematics deep within the complex bureaucratic structure of the Jesuit order.

Who was Christoph Grienberger?

Bamberga, Bamberger, Banbergiera, Gamberger, Ghambergier, Granberger, Panberger – the list of names used by his contemporaries to refer to Christoph Grienberger goes on and on.[17] Print has a tendency to fix the orthography of proper names, and Grienberger's name was one that, with the exception of a slim book of star-charts and a set of trigonometric tables,[18] rarely appeared in print during his life. In approaching the question 'Who was Christoph Grienberger?' I do not aim to provide anything like a biography.[19] Instead, I would like to look at how people wrote about Grienberger and how Grienberger wrote about himself. I would like to examine Grienberger's own production in terms of texts and instruments, and his moderation of the productions of others, in his work as a revisor of mathematical works written by Jesuits and in his strategies of engagement in epistolary relationships with natural philosophers and mathematicians outside the Jesuit order.[20]

Christoph Grienberger died on 11 March 1636. Before his death he was in charge of the technical censorship of all mathematical works written by Jesuit authors. Often Grienberger would send detailed calculations and corrections to an author, demanding that they be incorporated before allowing the work to be published. In some cases, as in Gregorius a St Vincent's attempt to square the circle, Grienberger advised the Jesuit General Muzio Vitelleschi to refuse publication altogether, on the grounds that the errors contained in the proofs would damage the reputation of the Society of Jesus.[21] When Grienberger died, he clearly lost control over the mathematical publications of his fellow Jesuit mathematicians. Perhaps more interestingly, he lost control over his own authorial presence, or rather, absence. A case in point is Mario Bettini's *Apiaria*, an encyclopedic collection of mathematical curiosities.[22] The censorship of the book took place in the mid-1630s, but publication was held up, possibly through a lack of a suitable patron.[23] The book finally appeared in 1645, and unlike other works, which merely incorporated Grienberger's corrections unacknowledged, Bettini takes great pains to highlight the contributions of the late *Revisor*, whom he hails at the outset of his book as having the stature of an 'Archimedes of our time', combining 'most ingenious practices and wonderful machinery' with 'very acute theories'.[24] Later in the work, Bettini confessed that 'I have benefited, my Reader, from the mind and industry of the very learned and exceedingly modest man, Grienberger, who, while he would have discovered many marvellous things by himself, preferred to make himself serviceable to other people's inventions and other people's praises'.[25] In his *Aerarium*, published three years later, Bettini included a *Scholion Parergicon* eulogising Grienberger, and continuing to compare him to Archimedes, adding that 'Grienberger has no greater enemy than his own modesty, by which it has come to pass that his ingenious inventions have been neglected, and he will be consigned to oblivion'.[26] Bettini added, echoing the *Apiaria*, that 'It was a remarkable characteristic of [Grienberger] that, following the example of Archimedes, he combined most acute theories with extraordinary practices',[27] and his claims for Grienberger's achievements in designing instruments and machines are closely echoed by other contemporary mathematical authors.[28]

Instruments and invisibility

> And yet Archimedes possessed such a lofty spirit, so profound a soul, and such a wealth of scientific theory, that although his inventions had won for him a name and fame for superhuman sagacity, he would not consent to leave behind him any treatise on this subject.
>
> Plutarch, *Life of Marcellus*, XVII.3–4

In Bettini's *Apiaria*, we see Grienberger's instrumental proficiency forcibly exposed to the public gaze. In composing his corrections to the *Apiaria*, in his role as *Revisor*, Grienberger had noticed that a scenographic instrument described by Bettini could be improved in a way that would make it easier to use and more accurate. The instrument [**See Figures 2.1** and **2.2**], rather similar to Christoph Scheiner's pantograph [**See Figure 2.3**],[29] allowed the user to make accurate drawings from life with little effort and less skill. Grienberger wrote to Bettini in 1635 to describe his modifications:

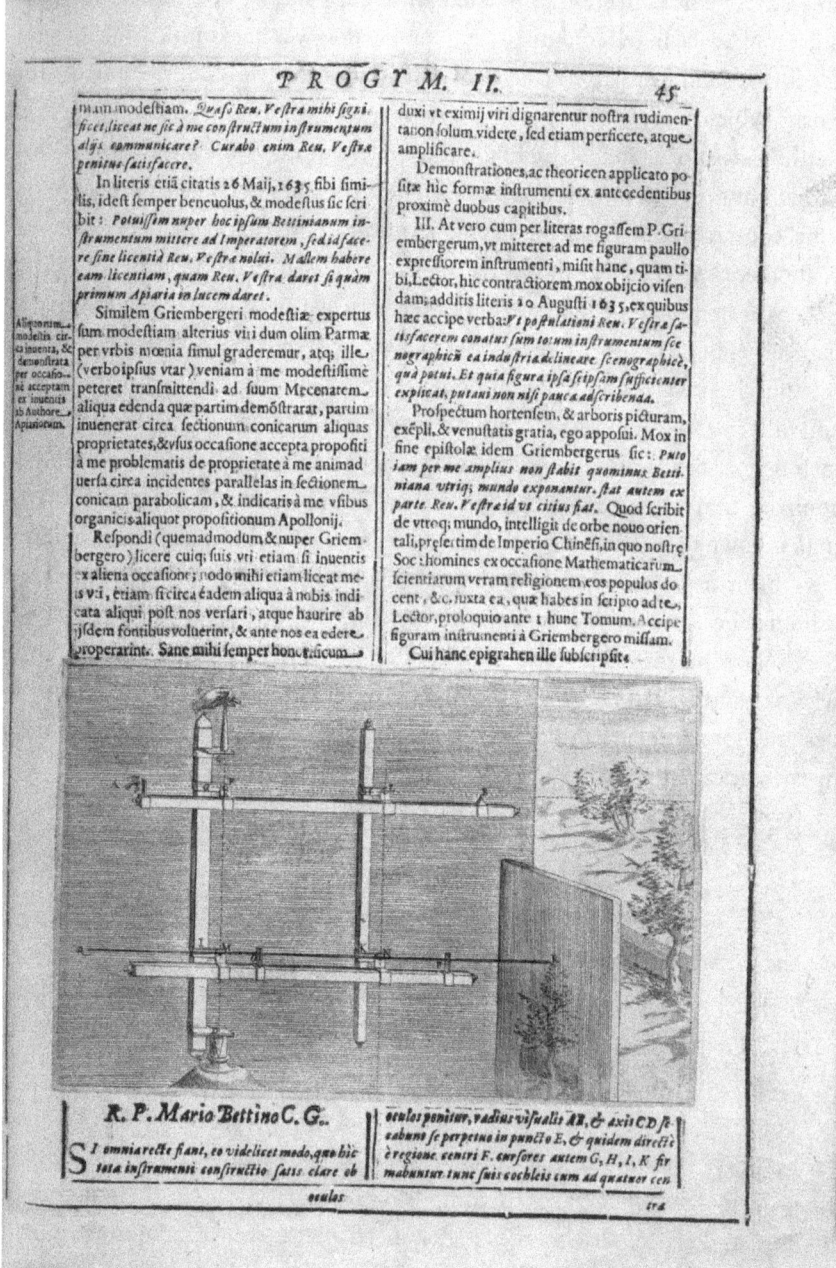

Figure 2.1 Christoph Grienberger's enhanced version of Mario Bettini's scenographic instrument, from Mario Bettini, *Apiaria Universae Philosophiae Mathematicae*, Bononiae: Io. Baptistae Ferronij; 1645

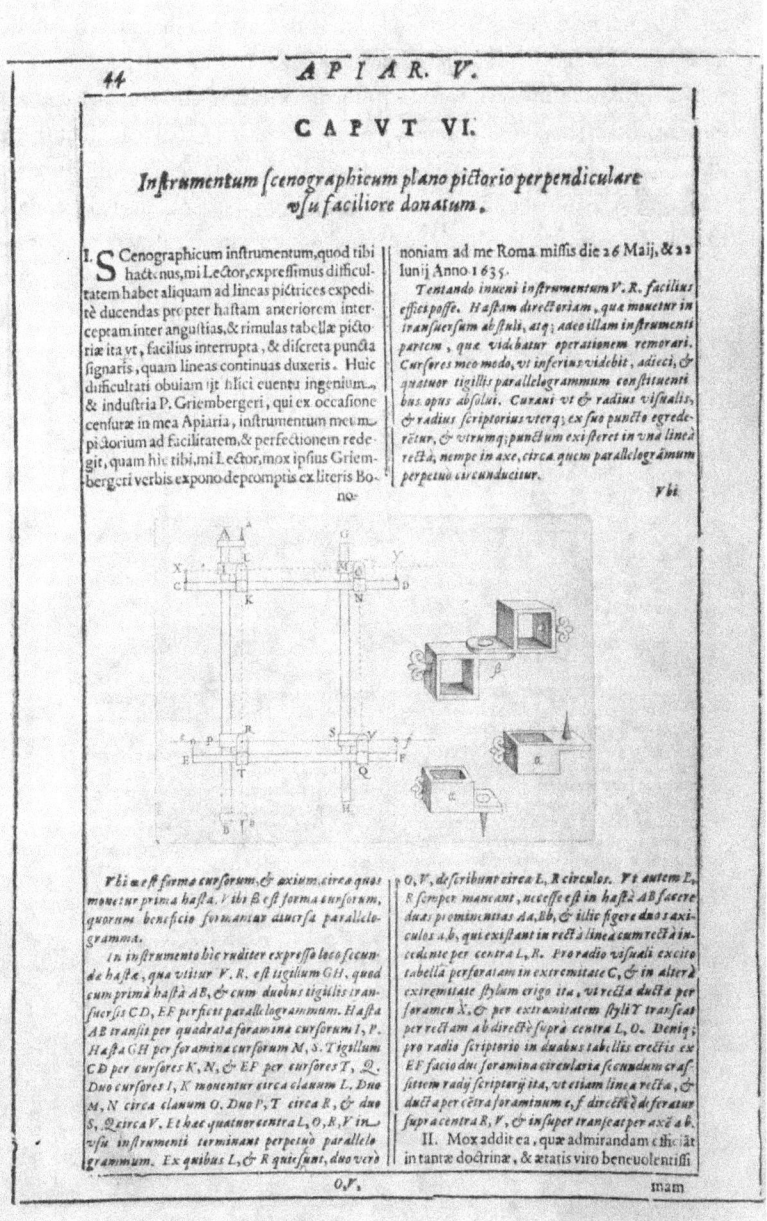

Figure 2.2 A schematic version of Grienberger's enhanced version of Mario Bettini's scenographic instrument showing detail of cursores, from Mario Bettini, *Apiaria Universae Philosophiae Mathematicae*, Bononiae: Io. Baptistae Ferronij; 1645

On experimenting [*tentando*], I discovered that Your Reverence's instrument might be made more easily. I removed the directing rod that moved transversely, until now the part of the instrument that appeared to obstruct its operation. I added *cursores* in my own way, as you will see below, and completed the job by means of four small beams, making a parallelogram. I took care that the line of sight [*radius visualis*] and the line of writing [*radius scriptorius*] would both depart from one of its points, and that both points would exist in a single straight line, namely the axis around which the parallellogram will be rotated continuously.[30]

Figure 2.3 Christoph Scheiner's pantograph, from C. Scheiner, *Pantographice seu Ars delineandi res quaslibet per parallelogrammum lineare seu cauum, mechanicum mobile*, Romae; 1631

In addition to providing a lengthy description of the device, arguably at least as different from Bettini's own rude contraption as Scheiner's pantograph, Grienberger sent Bettini two engravings[31] for inclusion in his book, one showing a schematised form of the instrument accompanied by Grienberger's trademark *cursores*, and the other showing the instrument manipulated by the eyes and hand of an invisible Grienberger (see **Figures 2.1, 2.2**).

Grienberger's pathological modesty is at work here again. Ever keen to divest himself of any vestige of authorship, he writes to Bettini of the modified scenographic instrument that

> I could have sent this Bettinian Instrument to the Emperor recently, but I did not wish to do this without the permission of Your Reverence. I would rather receive that permission which Your Reverence would bestow if [the instrument] were first published in the *Apiaria*.[32]

Another work in which Grienberger's instrumental manipulations in the *Collegio Romano* lie tantalisingly in the shadows is Christoph Scheiner's 1630 book on sunspots, *Rosa Ursina*.[33] The dichotomy between court and Curia that characterised the work of Scheiner and of many other Jesuit astronomers is eloquently expressed by Daniel Widman's etching of the different techniques for observing sunspots [**See Figure 2.4**]. At the top we see Scheiner in the company of various members of the Orsini household, observing the sun on a viewing platform, complete with obelisks, on the banks of the Lago di Bracciano, close to the Orsini Castle, which can be seen in the left background. At the bottom we see Scheiner in duplicate, compasses still in hand, making observations from his room in the Jesuit *Domus Professa* in Rome.[34] The instrument used by Scheiner in the lower vignette is the telescope that he claimed to have used to discover sunspots before Galileo observed them in 1611, and suffered from the disadvantage of being difficult to move from a fixed position, thus making protracted observations over any length of time a very awkward business.

To cope with this problem, Grienberger developed a 'telescopic heliotrope' or 'heliotropic telescope', an instrument [**See Figure 3.5**] which avoided the difficulties of the other device by being simultaneously mounted on two axes around which it could rotate freely to follow the trajectory of the sun, like the sunflower from which it took its name. Again, Grienberger seems to have been responsible for the engraving of this device published by Scheiner.

Again, Grienberger as machine-operator is invisible, in marked contrast to the multiple representations of Scheiner in the previous figure. Scheiner, ever one to emphasise the collective nature of the scientific enterprise,[35] asked

Figure 2.4 Daniel Widman, Various methods for observing sunspots, from Christoph Scheiner, *Rosa Ursina*, Bracciani: Apud Andream Phaeum Typographum Ducalem; 1630

Mathematics and Modesty 51

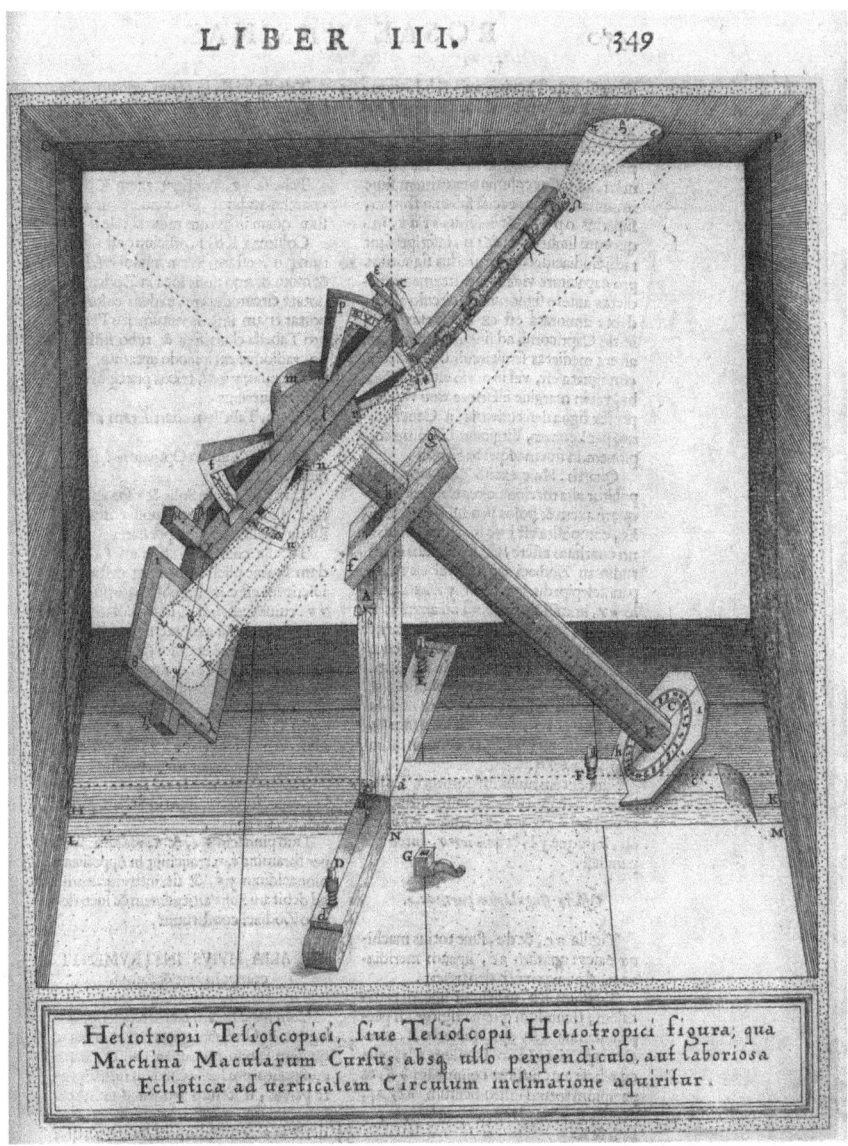

Figure 2.5 Christoph Grienberger's 'Heliotropic Telescope' or 'Telescopic Heliotrope', from Christoph Scheiner, *Rosa ursina*, Bracciani: Apud Andream Phaeum Typographum Ducalem; 1630

Grienberger to provide him with a description of his instrument, but he refused, to Scheiner's surprise:

> And thus this machine is not entangled in as many difficulties as the other one; and additionally [Grienberger's] machine is more convenient, and carried out the work more quickly than that one. For this reason, it will be worthwhile to

write a short explanation of its nature, *since the Architect of the Machine himself seemed to be unwilling to furnish this*: despite having later edified many things with his demonstrations, and hastened and urged me to finish the work, [as well as having] helped me most opportunely with similar services that were virtually necessary to me in such a short space of time.[36]

Undoubtedly the polemic between Scheiner and Galileo was part of the reason for Grienberger's attempt to distance himself from the text of Scheiner's work. The rift between Galileo and the Jesuit mathematicians of the *Collegio Romano* that followed Galileo's attacks on Orazio Grassi's public disputation on the comets of 1618 was a source of much distress to Grienberger,[37] who could not see any reason for this turnaround.[38] In fact, Galileo's gesture seems to have been the result of a cynical, and somewhat shortsighted attempt to cultivate the patronage of Archduke Leopold of Austria, also a patron of Scheiner.[39] Nonetheless, Grienberger's participation in the *Rosa Ursina*, performing observations (not with Scheiner in the *Domus Professa*, but in the *Collegio Romano* only a few hundred yards away) and refining observational instruments is characteristic of the way he chose to present himself in other works. To understand the development of this pattern of effacement of claims to intellectual ownership, I would like to turn to Grienberger's earlier career in the Jesuit order.

Private lessons

On 15 September 1590, Grienberger, then mathematics teacher and student of theology at the Jesuit College in Vienna, wrote the earliest of his surviving letters to Christoph Clavius in Rome. Although Grienberger, who had spent the ten years since he first entered the Jesuit order in Prague and Olomouc,[40] had not yet met Clavius face to face, his letter betrays an unexpected degree of intimacy:

> Why should I not love my teacher? And indeed so much mine that he seems almost to be mine alone. Are you not mine, who are so present to me always, that I began immediately to love you and now for almost the four years for which I have known you have hardly ever placed a foot outside my bedroom?[41]

Grienberger is, of course, cohabiting with Clavius's textual body – his commentaries on Euclid's *Elements* and the *Sphere of Sacrobosco* as well as other works.[42] Nonetheless, a short time after this letter was sent, along with the *demonstratiunculae* on spherical trigonometry that Grienberger, like a good pupil, sent to his virtual master,[43] Grienberger was summoned to Rome so that the two mathematicians could really live under the same roof.⁴⁴ The pattern was to become relatively common – Giuseppe Biancani and Odo van Maelcote were also brought to Rome to assist Clavius (and to be fashioned as mathematicians in his image) after sending unsolicited solutions to celebrated problems or instruments to the famous professor in Rome,⁴⁵ and many others sent demonstrations hopefully.

In 1595 Clavius went to Naples, leaving Grienberger in charge in Rome. Grienberger wrote to Clavius shortly after his departure:

> Now the Mathematical Museum has put on new clothes, nor does it cry out for anything other than the speedy return of its master. In the meantime it will have me as a custodian. On Monday next I will give my old [room] to two others.⁴⁶

The bedroom was a multifunctional space for the Jesuit mathematician. Generally, the rooms of Jesuits were not provided with keys, but, along with the rooms of the Superiors, the Procurator (responsible for the financial affairs of the College), the room of the senior mathematician of the College formed an exception.⁴⁷ The added security of a key meant that the mathematics professor could store valuable mathematical instruments in his domestic space, which was often referred to as a mathematical museum, or *musaeum mathematicum*.⁴⁸ Later, while in Lisbon, Grienberger would tell Clavius of a valuable clock that he had kept for several months in the privacy of his bedroom.⁴⁹ As well as constituting a space for the storage and construction of instruments, the mathematician's bedroom was the focus for the studies carried out by the private mathematical academy of the college.⁵⁰ Printed books currently being used by the academy, manuscripts of mathematical works and, perhaps most crucially, the letters sent to successive professors of mathematics in the *Collegio Romano*, were all stored in this space.⁵¹ Whereas the private papers of a Jesuit were generally destroyed after his death unless deemed to be of particular importance,⁵² the mathematicians of the college enjoyed the security of a place apart, allowing the correspondence and manuscripts accumulated by successive professors to constitute what Athanasius Kircher and his colleagues were later to use as a private mathematical archive.⁵³

Humdrum mathematical culture

During Clavius's absence in Naples, Grienberger kept him informed with regular bulletins on the vicissitudes of college life. These allude to his own research, the work of the private mathematical academy under his guidance and the normal mathematics classes of the College. Grienberger's letters are punctuated by descriptions of humorous events, such as Fabricio Mordente's pompous display of his beautiful, but imprecise, geometrical compasses to the mathematicians of the college[54] and a rather excessive number of ponderous jokes about Clavius's penchant for Neapolitan pastries.[55]

On 12 January 1596 Grienberger told Clavius of a possible addition to his other duties:

> I fear that perhaps I may have to teach privately to a certain Count whose name escapes me. But I hear that he has studied little else, and it appears to me that he is rather young, not to say a boy, so I hope for little profit, even on my side, as *I do not know how to deal with that type of person correctly*.[56]

Shortly afterwards, Grienberger's fears came true, making unfair demands on both his time and his character:

> I do not have much free time, apart from in the mornings. For after lunch all is taken up by the class and the academies, of which there is the domestic one, as you know, and another at the Gate, to which Count St. George, as he's known, comes, a boy with a reasonable mind, together with a certain other [boy] of around the same age, called Orazio, from Perugia, also of good family. Both of these asked the Fr. General if I could lecture them privately. Your Rev. will wonder that I am suitable for this task, as it should really require not a German but a Tuscan, who would be more affable than me. But seeing that it has pleased them thus I hope that they will have patience with me.[57]

Grienberger, unlike his more famous Tuscan contemporary Galileo, was clearly no courtier, and elsewhere diagnosed himself as having a particularly frigid nature, when speculating that Clavius might be prolonging his stay in Naples because Grienberger was occupying his bedroom:

> But is [Clavius] perhaps excluded from his bedroom? On the contrary, it is so ready that it would invite him there freely even against his will. For I will easily find another one that is equally cold, unless perhaps all rooms are cold that are occupied by exceedingly cold [*frigidissimus*] me.[58]

Despite pandering on occasions to a cardinal's desire for a sundial,[59] or to the wishes of young aristocrats to have private tuition, Grienberger's concerns lay more with the well-being of his young disciples in the mathematical academy, bound to him by a common love of mathematics, than with courtly aspirations.

> Unless the Superiors change their plans, I believe that I will be freed from the ordinary domestic academy. [. . .] The other private academy is creeping forward

slowly [in the study] of Clocks. Out of the three pupils, one (Janos Nagy, of course) as he was trying impetuosly to go up two steps at a time four or five days ago, almost suffocated on his catharr. However, Nature won, and made herself a way forcibly, but not without blood, as together with the phlegm he vomited up no small quantity of blood.[60]

The impetuous mathematician clearly pays a price. Grienberger went to visit Nagy in the college infirmary, where he found two other indisposed mathematical practitioners:

> As I was visiting Nagy today, I found Fr. Villalpando and Fr. Mario (the one who saluted you in Naples when you were in your sedan chair) in the same place [i.e. the infirmary]. Reading your letter they rejoiced to hear of your good health, and indeed we sensed some unknown fragrance from your letter, and some unknown pleasant odour, but without a taste.[61]

Although the convalescent mathematicians might have detected the smell of the Neapolitan sweetmeats that Clavius hoarded in his bedroom in Naples, imbibed by his writing paper, Grienberger is suggesting with lumbering jocularity that the elderly mathematician had kept the taste of the pastries for himself.[62]

Public mathematics in the Collegio Romano: the *Problemata*

> What gain is to be had from disputations, which reduce everything to the musicians, party-givers (*festaroli*) and printers? Who cannot see that they are altercations in which, as the ancient poet [Persius] said so well, truth is lost instead of found, and that on these occasions one can only prove one's ready cheekiness and sarcasm? And that by making a great show and expending thousands of conclusions, one ends up without having concluded a single thing?
>
> Federico Cesi, *Del natural desiderio di sapere* (1616)

Shortly after Clavius left Rome for Naples, Grienberger castigated him for suggesting to the Rector that Grienberger might give a public oration to mark the commencement of studies in the *Collegio*:

> I do not know what Your Reverence expected when you promised our Rev. Fr. Rector that I would give an oration [*Praefatio*], for I happened to hear this from him at least twice, in the presence of others. *For you know extremely well that, to me, that has always seemed an extremely difficult task.* Certainly, if they expected an oration, they did not get one, but instead I explained the dimension of the circle from Archimedes so slowly that it could not be completed in half an hour.[63]

Grienberger did not enjoy speaking in public. No great surprise here, but what might appear initially to represent something of a paradox is a statement made in Mario Bettini's *Aerarium*, when concluding his eulogy of Grienberger and 'correcting' Giuseppe Biancani's entry on Grienberger in his Chronology of Illustrious Mathematicians, appended to his 1615 *Aristotelis loca mathematica*.[64] Listing Grienberger's extant manuscript works, Bettini writes:

> There are many optical and mechanical [*machinaria*] experiments present in our Roman College that were once exhibited to the eyes and admiration of princely men visiting that place.[65]

To understand how Grienberger's modesty and distaste for public speaking might be reconciled with his authorship of a large number of experimental problems presented publicly to the applause of princes visiting the Collegio Romano, I would like to consider the emergence of a highly specific genre – the *Problemata*.

As discussed above, the 1586 first edition of the Jesuit *Ratio Studiorum* had proposed that Clavius should give private lessons in mathematics to eight or ten Jesuits, selected from all the different provinces of the order, in order to furnish the provinces with mathematics teachers.[66] The next published edition of the *Ratio* (1591) suggested that in addition to this private *academia*,

> once or twice a month one of the students should recount [*enarret*] an illustrious [*illustre*] mathematical problem in a large gathering of philosophers and theologians, having first been instructed [*edoctus*], as is proper, by the master.[67]

Some of the surviving mathematical problems presented in the *Collegio Romano* have now been published.[68] Although these *Problemata* are generally anonymous, a significant amount of evidence in addition to Bettini's attribution, discussed in the notes accompanying each problem, ranging from references in letters, literary style, internal evidence and Grienberger's distinctive handwriting, points to Grienberger as the author of all of these *Problemata*, which range in date from 1591, the year of Grienberger's arrival in Rome, to 1614. As a ceremonial form of culture, such presentations clearly had much in common with the extravagant public defences of philosophical theses made by aristocratic students in the *Collegio* and so disparaged by Federico Cesi. In the thesis defences at the *Collegio*, studied in detail by Louise Rice,[69] the script read by the student was generally written by one of the professors, although if the theses (or the odes composed for the occasion) were printed, they were accompanied by the student's name. The same practice seems to have been adopted for the mathematical *Problemata*, as suggested by the 'instruction' by the master advocated by the *Ratio Studiorum*. Publication was a rarer matter in the mathematical presentations, but when it happened, it followed the same patterns. The Roman publishers Zannetti and Mascardi, favourites for such philosophical 'vanity publications', were also used for the mathematical problems.[70]

Mixed mathematical themes in the *Problemata*

It should perhaps be stressed that the individual modesty that I ascribe to the behavioural patterns of Christoph Grienberger was utterly different from the prescriptions of cognitive humility with regard to the mysteries of the natural world that characterised much theological discourse of the late sixteenth century, as discussed by Carlo Ginzburg.[71] Indeed, Grienberger's refusal to accept authorial dignities, and his confessions of bodily weakness[72] coexisted with the flow from his pen of a series of claims for the exalted powers and cognitive capacities of mathematicians with respect to the natural world. Such a combination of individual modesty with elevated claims for the power of a collectivity is a feature that can be found elsewhere in Jesuit culture, perhaps reaching its zenith the 1640 *Image of the First Century of the Society of Jesus* published to mark the centenary of the order.[73] As Marc Fumaroli has shown, the anonymous Jesuit compilers of this work excused its rather immodest claims for the achievements of the Society by attributing these achievements indirectly to Jesus, in whose hands the Society that took his name was merely a passive instrument. This relationship was captured emblematically by a device in which the Society of Jesus was the moon, reflecting the light of the Sun, representing Christ.[74] Another emblem in the same book reinforces the idea of the Society of Jesus as a passive, mechanical device, manipulated by Divine Love to raise the earth towards heaven by means of conversion [**See Figure 2.6**]. The device used is similar to one which forms the topic for one of Grienberger's mathematical *Problemata*, dating from 1603. The problem in question, later cited by Paul Guldin in his controversial *Physico-mathematical dissertation on the motion of the earth*,[75] provides a graphic example of the enormous power over nature which Grienberger ascribed to the collectivity of mathematical practitioners [**See Figure 2.7**].

Grienberger's speaker intends to demonstrate to his audience that 'by means of no more than 24 wheels with toothed axes, the Earth's globe, even if it were

Figure 2.6 The conversion of kingdoms and provinces by the Society, from *Imago Primi Saeculi Societatis Iesu A Provincia Flandro Belgica eiusdem Societatis Repraesentata*. Antwerp: Balthasar Moretus; 1640

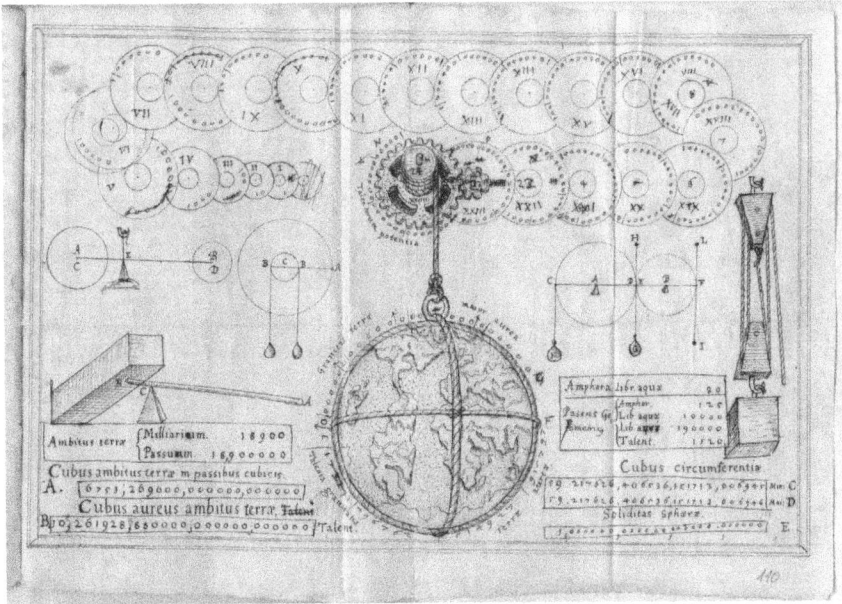

Figure 2.7 Christoph Grienberger, device for raising a golden earth by the force of one talent from Grienberger, *Problema, Terra Auream, Talenti potentia movere*. APUG Fondo Curia 2052

made entirely of gold, could be driven away from the centre [of the universe], by the force of only one Talent'.[76] The demonstration, later plagiarised by Kaspar Schott [**FIG 2.8**],[77] is preceded by a long passage extolling the virtues of mathematics that is anything but modest.

'The boldness of Mathematicians,' Grienberger begins, 'has always been great, as has their power, Most Religious Fathers and other most honourable members of the audience; and they possess so much spirit in a small number of people, that there is nothing in the whole universe either cloaked in darkness or buried in difficulties that has been able to escape their ingenuity and that has not been investigated with their machines.'[78] Although nobody could doubt that the motions of the heavens had been translated into the laws of mathematics [*leges Mathematicorum*], someone might still query the dominion of mathematicians over the elementary world. However, 'the elements themselves', the author continues, 'love to be governed by mathematics as much as they love their own dignities and powers, and prefer to be ornamented by the mathematicians than to be reduced to almost nothing by the natural philosophers'. The *Naturales* dress the elements poorly, in the different qualities of heat, cold, wetness and dryness, and imprison them in concentric spheres.

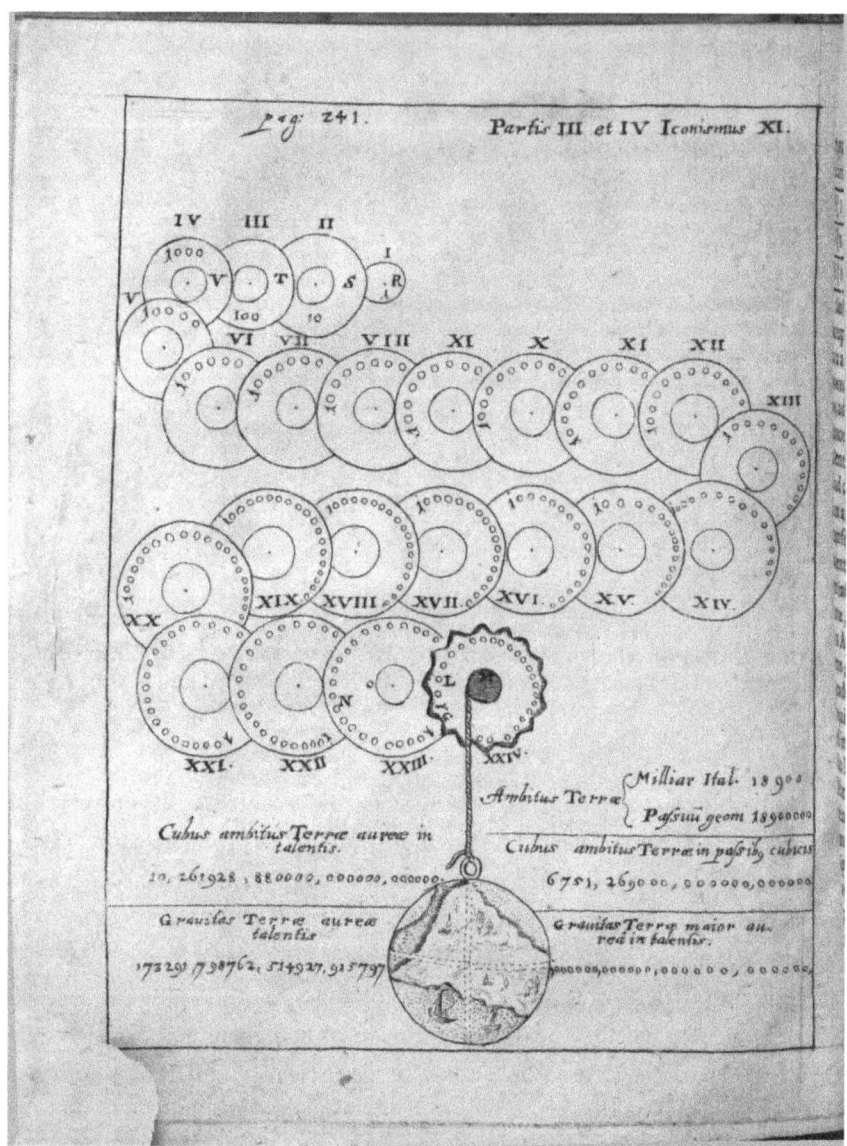

Figure 2.8 Earth-moving machine, from Kaspar Schott, *Magia universalis*, Bamberg: J. M. Schönwetter, 1677

Why should [the elements] not be miserable, then, being so poorly dressed, confined in prisons and constrained to serve people that treat them so badly. They dig into the earth with ploughs, and utterly disembowel it even to wrench out a handful of gold. They make water wash all the filthiest people; condemn air to the mills and grindstones, and fire to the furnaces. There is no service that is so vile that [the elements] are not subjected to it [...] It should not seem strange, then, if the elements would happily resort to the Mathematicians, who care for their dignity, and whose works often free them from prison, and bring them into the gardens and palaces of kings.[79]

The elements are happier under the dominion (*imperia*) of mathematicians than that of physicists, or natural philosophers. As the passage cited shows, in Grienberger's text, the social and cognitive status of the mathematical disciplines are inextricably entangled, and this is also true of the other mathematical problems. When Grienberger first arrived in the *Collegio Romano* in 1591, he delivered an oration on the mathematical disciplines, much of which was taken up with establishing the nobility of the family made up by the seven mathematical 'sisters': Arithmetic, Geometry, Music, Astronomy, Mechanics, Geodesy, Perspective and Practical Arithmetic (*Supputatrix*).[80] In the midst of a rather labyrinthine account of the resemblances and quasi-incestuous interrelations between the different 'sisters', he mentions an experiment to show that the study of perspective furnishes the causes of appearances that would otherwise remain a mystery, an experiment that is taken up and performed by the narrator of the following *Problema*.[81] The classical *topos* for discussions of mathematical power is the role of the war machines designed by Archimedes in the siege of Syracuse, and this event is cited repeatedly in many of Grienberger's *Problemata*. Mathematical wonders, however, need not be limited to the military domain, and Grienberger also describes a trick-picture, possibly an anamorphosis, which he had heard of, in which a forest landscape seen from one position is transformed into a picture of the emperor with his brother when one looks through a specially constructed hole.[82] As well as being an ancestral mathematical powermonger, Archimedes also provided a source for the credibility of the early modern mathematical practitioner, and Grienberger makes much of the story that Hieron, King of Syracuse, ordered that everything Archimedes said should be believed.[83]

Knowing the world through mathematics

He handed the world over to their disputation, but man could not grasp the work which God made from the beginning up to the end
   Ecclesiastes 3:11
      (Quoted in P. A. Foscarini, *Lettera sopra l'opinione de' Pittagorici e del Copernico*. Naples: Lazzaro Scoriggio; 1615)

Describing the audience of his 1595 oration to mark the beginning of studies to Clavius, Grienberger wrote that

> Our Reverend Father General was there, unexpectedly, along with several other unexpected people, and he seemed to apprehend the matter with some delight, as I understood afterwards from Father Pereira, who complained to me because I didn't invite him.[84]

Pereira was unhappy not to be invited to hear Grienberger's discourse (one of the very few public speeches that he seems to have given in person), and indeed the statements about mathematics made at the beginning of the 1595 oration were little short of anathema to Pereira's perception of the cognitive impotence of the mathematical disciplines, discussed above:

> You know that the whole of Philosophy is divided chiefly into three kinds of Sciences: Natural, Mathematical and that divine one that is called Metaphysical. The first one verifies for itself things immersed in matter, that is, abstracted neither from reality nor from reason. The last one assumes as its objects things that are utterly alien to matter. Even if the other two might seem to have all things distributed between them, the middle one, however (which, even by virtue of being the middle one can be said to be more excellent than the others), finds that in [treating] the same matters it ascribes them to itself in such a way that in its object it nevertheless in no way defrauds the other [sciences].[85]

Although mathematics considered quantity abstracted from any specific material incarnation, such abstraction rendered mathematical truths universal in their application, rather than inapplicable to the natural world as Pereira and others wished to suggest. The theme, later to be central to Giuseppe Biancani's *De mathematicarum natura dissertatio* (1615)[86] recurs frequently in the other *Problemata*, which Biancani may have had the opportunity to read during his time as one of the academicians in the *Collegio Romano*.[87] Mathematical conclusions made about quantity in general were applicable to any physical quantity, including motion; and Grienberger completes his oration with the suggestion that the possibility of incommensurable lengths implied the possibility of real incommensurable motions. The other *Problemata* put the application of mathematics to natural motions into action, and include one dealing with the motion of a weight attached to a rod, influenced by the medieval *calculatores* and Tartaglia, and another on the reality of the motions of the heavens described by astronomers.[88] In the latter, Grienberger, furthering an argument put forward in Clavius's *Commentary on the Sphere of Sacrobosco*, considers the motion of an ant on a moving table, to demonstrate, against the views of Pereira and the other 'homocentrists',[89] that a single body, could possess two real motions simultaneously without involving a contradiction. This allowed Grienberger to argue for the reality of the convoluted motions ascribed to the planets by astronomers, although he avoids confronting the vexed question of the Aristotelian distinction between natural and violent motion. As geometry pervades Grienberger's depiction of the natural world, so it inhabits the artificial domain of buildings and other institutions necessary to civic life. In one *Problema*, Grienberger writes that

> Without doubt that Bolognese structure [i.e. the Torre degli Asinelli in Bologna] had an outstanding mathematician as its architect [*delineator*] by whose vigilance Geometry has come to inhabit that tower.[90]

Another problem was prompted by the disagreement between a group of Spanish sailors and a group of Portuguese sailors who arrived simultaneously in Lisbon, having circumnavigated the world in opposite directions, and unable to decide which day was Sunday,[91] and indeed the Gregorian calendar, co-authored by Clavius, is an obvious example of an enormous mathematical artifice of a religious and civic nature.

Performing Physico-Mathematics – The Floating Bodies Debate

On February 5th 1612, Christoph Grienberger interrupted a letter that he was writing to Galileo to report the news of Clavius's death in "real time":

> While I pause from writing for a moment, behold here is someone who rushes to announce that our Clavius is about to be given his Travelling money, which he accepted this very evening at the first hour of the night. So do not be surprised that I break off this letter in a rather untimely fashion – such news does not allow me to linger any longer on these matters. You will learn more from the bearer of the letter, Father Odo van Maelcote, who, by returning to Flanders has shackled me once more to the mathematics classes.[92]

Grienberger's relationship with Galileo had been strengthening steadily since 1611, as the physical powers of his senior colleague Clavius decreased. The *Ad benevolum lectorem* introducing Grienberger's 1612 star-charts eulogises Galileo's telescopic observations in highly charged language.[93] The decline of Clavius brought the Austrian and the Tuscan ever closer; and after Galileo's triumphal visit to the *Collegio Romano* Grienberger spoke eagerly to Galileo of future reunions of the aging Clavian telescope of the *Collegio* with Galileo's instrument.[94] Galileo's anger at the criticism of his opinions on the heights of lunar mountains by a Jesuit in Mantua led him to write a long letter to Grienberger to defend his position in detail.[95] Replying on the "anniversary of the death of our most beloved Clavius", Grienberger displayed a prudence that brings into relief the boundary of the corporate culture within which he carried out his work:

> Do not be surprised that I am silent about your [letter]: I do not have the same liberty as you do.[96]

To have entered the dispute on Galileo's side would have constituted a breach of discipline for Grienberger, and would have been incompatible with his institutionalised *modus procedendi*. Instead, as ever, he breaks his silence through the words of others. A young former pupil of Galileo's studying in the *Collegio Romano*, Giovanni Bardi, wrote to him to describe a meeting with Grienberger:

> I visited Father Grienberger on behalf of Your Lordship and saluted him in your name. He returns your salutations doubled. I asked him for his opinion on that book [i.e. Galileo's *Sunspot letters*] which he had already seen and he said that he thought very well of it, and that on this subject, as on the other matter of things that float on water, he was of [the opinion] of Your Lordship.[97]

Galileo had spent much of 1612 embroiled in a bitter dispute with a group of Florentine Aristotelians about the cause of flotation of flattened bodies having a "weight" greater than that of water. As the early part of the debate has been discussed at length elsewhere[98] I shall limit my discussion to a brief summary. Vincenzo di Grazia's claim that ice was condensed water was attacked by Galileo, who pointed out that in this case ice would sink, as is patently contrary to experience. Di Grazia replied that ice floated because of its flat shape, and a dispute quickly flared up about the true cause of the flotation of bodies. Ludovico delle Colombe then joined the debate, and began performing experiments in public with chips of ebony to demonstrate that, in this case, shape, not heaviness, was the cause of flotation. Galileo's *Discorso intorno alle cose che stanno in sù l'acqua* was published in 1612,[99] and attempted to explain the flotation of the anomalous ebony chips in terms of a small dip in the surface of the water, leading the combined weight of ebony and air to be less than that of water.

Bardi complained to Galileo that, although Grienberger was very much in agreement with the Archimedean conclusions of the *Discorso*, students with only half a year of philosophy were pronouncing ridiculous judgements on the work, and the remaining professors were not yet discussing it.[100]

Grienberger's backstage participation in the polemic surrounding Galileo's *Discorso* illustrates both the possibilities and the limitations of the new public space for mathematical and experimental demonstrations provided by the *Problemata* in the Collegio Romano, a space that Grienberger had by now made his own. Previous *problemata*, particularly that concerning the nova of 1604[101] and the *Nuntius Sidereus Collegii Romani* recited by Odo van Maelcote in 1611[102] had demonstrated that the mathematicians of the Collegio Romano were prepared to risk conflict with the professors of Aristotelian natural philosophy by openly endorsing observations that challenged Aristotelian teachings concerning the incorruptibility of celestial matter. The dispute on *galleggianti* demonstrated that, under Grienberger's guidance and instruction, they were willing to extend the domain of conflict to the most Archimedean domain of hydrostatics.

Bardi served as Grienberger's public mouthpiece on this occasion. On 20th June 1614, Bardi wrote to Galileo sending him the text of the presentation in defence of Galileo's position that he was to make in the *Collegio Romano*.

> As one of these Problems had to be done, and it was allocated to me, Fr. Grienberger asked me what I would like to treat, proposing some other things to me. I told him that I would have liked to deal with some matter similar to this, so he took this, which I think will please you no small amount, because it conforms entirely with your opinion, or rather it is your opinion, with the addition of those two experiments that cannot but support your view. And Fr. Grienberger told me that if he hadn't had to have respect for Aristotle, whom they are not allowed to oppose in any way by order of the General, but must always save, *he would have spoken more clearly than he did*, because in this [matter] he is entirely on your side; and he told me that it is no wonder that Aristotle is in opposition, because he was most clearly mistaken in that which Your Lordship told me once about those two weights falling earlier or later[103]

The wording of Bardi's letter is slightly ambiguous with regard to the authorship of the text he was to recite three days later. Nonetheless, Bardi's claim that Grienberger "would have spoken more clearly than he did" clearly refers to the immediate context of the floating bodies debate, as shown by the remainder of the letter. Grienberger and Galileo had not met since the outbreak of the debate, and none of the previous surviving letters from Grienberger to Galileo mentions the dispute on floating bodies. I would like to suggest that Bardi is telling Galileo that Grienberger "would have spoken more clearly than he did" <u>in the enclosed *Problema*</u> due to be recited by Bardi, but in fact written by Grienberger on Bardi's suggestion. This interpretation is consonant with a significant amount of additional evidence. The only surviving manuscript of Bardi's presentation, entitled *De ijs quae vehuntur in aquis*, is preserved amongst Grienberger's papers, where it is bound between rough trigonometric tables and a draft of his *Problema* on the basic principles of algebra. The differences between this text and the printed version suggest its priority as the manuscript contains corrections which are incorporated in the printed text. The handwriting of this *Problema* is identical to Grienberger's other *Problemata*. Despite Bardi's professed studies with Galileo and Grienberger no other evidence of his mathematical ability exists besides this single *Problema* and before its appearance Galileo dismissed his hydrostatic concerns as puerile.[104] The most recent discussion of Bardi's text suggests that Bardi was subsequently unproductive, despite this prodigious beginning, due to eye-problems[105], but an attribution to Grienberger seems a more economic explanation. The sources cited in the *Problema*, including the works of Marino Ghetaldi and Juan Bautista Villalpando, are also cited elsewhere in Grienberger's writings. Stylistically, the *Problema De ijs quae vehuntur in aquis* is also entirely in accordance with Grienberger's other *Problemata*.[106] From what we have seen of Grienberger's behavioural patterns, it is unsurprising that Grienberger made no attempt to arrogate the work for himself, and indeed, as Bardi's letter indicates, the Jesuit system of censorship made it more convenient for such a work to appear under the name of a lay-person.

Grienberger's caution seems to have been grounded in fact. On 14 December 1613 the ageing Jesuit general Claudio Acquaviva had issued a lengthy *Ordinance for the solidity and uniformity of doctrine* to all of the Jesuit provinces. While strenuously criticising departures from Thomist theology, Acquaviva also condemned the introduction of new opinons in philosophy, and ordered the Provincials to ensure "that the opinions that are taught in philosophy are subservient to theology, and that our philosophers follow Aristotle alone, wherever his teachings are not at variance with catholic truth".[107] An attempt by the Jesuit mathematician Giuseppe Biancani to publish a similar work supporting Galileo's position in the floating-bodies debate fell afoul of the Jesuit censors shortly after Bardi's presentation because it was "an assault on, not an explanation of Aristotle". And the conclusion and arguments of the work are not those of the author, but of Galileo: it would have been enough to have read them in Galileo's work. To transcribe in the books of Ours [i.e. Jesuits] the discoveries of Galileo, especially those by which he attacks Aristotle, seems neither decent nor expedient[108]. These were accusations to which "Bardi"'s work would have clearly been equally prone, had they not avoided Jesuit censorship altogether.

Mathematics and Modesty

Bardi, I am suggesting, did little more than provide the occasion for Grienberger to give public legitimation to Galileo's explicitly anti-Aristotelian conclusions in the Collegio Romano in the courtly and collegiate context of a *Problema*. Without delaying more on the question of attribution, I would like to move on to the content of Bardi's theatrical hydrostatic performance. Bardi described the planned event in some detail in his letter to Galileo

> There will be, in addition to the paintings [*dipinte*] and printed sheets [*stampate*], all of these experiments on a table, so that they can be seen by everybody, in such a way that they cannot deny what they see with their eyes[109]

The reference to *dipinte* suggests either a large panel bearing the picture [**fig 2.9**] appended to the end of the manuscript, or possibly small copies distributed to each member of the audience which could be taken away as souvenirs. Bardi's reference to *stampate* is not so clear, but may refer to a printed list of his Archimedean conclusions, handed out to the members of the audience, possibly accompanied by schematic diagrams bearing the letters that he cites in his talk. Shortly after the event Stelluti wrote to Galileo giving a full report of Bardi's performance. He described his delight in seeing Galileo's opinion defended to rapturous applause, and admired the "experiments made in the presence of everybody by Father Christoph Grienberger, after he had brought all of the instruments which you can see in the enclosed picture into the room where the Problem was recited. Stelluti observed that "although there was the odd Peripatetic who shook his head ... everything was made quite clear by the end". Stelluti also provides crucial information on the audience, which included, as well as Stelluti himself and Federico Cesi, the brother of the latter, Bartolommeo Cesi, the mathematicians Luca Valerio[110] and Johannes Faber and other Prelates and "signori letterati". All of these spectators were, Stelluti continues, "extremely satisfied to see such a "good Jesuitical demonstration" towards Galileo, to the annoyance of his imitators.[111]

The opening of Bardi's talk recounts Galileo's recent metamorphosis from sidereal messenger into Neptune. Announcing his aim to "uncover the cause by which things that should sink in water [...] are discovered to float in water in accordance with Galileo's explanation, Bardi frames the polemic in distinctly violent terms:

> [F]rom this dissertation of ours it is to be hoped that every victory and every trophy of truth will be in your possession. The material will be abundantly supplied by Experience which, as it fights for the cause of this serious dispute, as

in a battle, gathers soldiers, provides them with weapons, and urges them to war, and, like anyone who wishes to encourage people to fight vigorously, must take up a position on the front line itself.[112]

The wonderful drawing appended to Bardi's presentation (see **fig 2.9**) illustrates the extent to which Grienberger's mathematical *problemata* drew on Jesuit traditions in emblematics. The use of *putti* to perform experiments, later to become widespread, was a convention first adopted a year previously by Rubens in his illustrations for the Jesuit Francois Aguilon's collosal work on optics, the *Opticorum libri sex*.[113] Rubens appears to have drawn some of his inspiration from the iconographic conventions of the profusion of Jesuit emblem books representing divine and profane love that flourished in the late sixteenth century.[114] In our case, as we are informed that the experiments described by Bardi were actually carried out by Grienberger, the performing putti might be said to represent his ultimate act of iconographic self-effacement–Grienberger's only surviving self-portrait, one might say.

Pictures, as well as words, formed essential tools for persuasion in the Jesuit rhetorical tradition.[115] Works such as Jeronimo Nadal's *Evangelicae Historiae Imagines* used carefully crafted engravings to reinforce the gospel message, and played a crucial role in Jesuit missionary work.[116] Grienberger's lavish illustration, far more elaborate than the few schematic diagrams present in Galileo's own *Discorso*, also formed an integral part of the persuasive artillery of Bardi. Central

Figure 2.9 Christoph Grienberger, Hydrostatic experiments, from *De ijs quae vehuntur in aquis* recited by Giovanni Bardi, 23 June 1614, APUG Fondo Curia 2052.

to the illustration is the phenomenon that formed the focus of Galileo's dispute with the Florentine Aristotelians: a flat metal disc floating in a circular dish full of water, accompanied by two *putti* locked in discussion. The Galileian conclusion " namely that the plate floats because of a small dip in the water's surface "a well" (*puteus*) in the language of Bardi/Grienberger, is assumed in the diagram, which gives the water's surface a conspicuous hollow. Such a minuscule experimental phenomenon is unlikely to have been easily visible to the grouped aristocratic and ecclesiastic spectators, unless their gaze was carefully shepherded. While Galileo's *Discorso* legitimated its own existence by arguing that writing was more efficacious than speech as a means of distinguishing truth from falsehood,[117] Bardi's oration endorsed direct, shared ocular experience, assisted by instruments.

> Some things are heavier than others, others are lighter, and some are of the same weight [as each other]. It is by means of balances [*bilances*], scales [*trutinae*] and steelyards [*staterae*] that weights are conferred on heavy bodies. Although these are common [devices] neither Philosophy nor Mathematics abhorrs them.[118]

Grienberger's rough manuscript notes and calculations reveal that he was no stranger to the balance, and conducted a series of measurements of the specific gravities of different metals in the wake of Marino Ghetaldi's *Archimedes Promotus*. A representative excerpt, in coarser style than the public *Problemata*, reads as follows:

> The weight of the cylinder in air according to one of my observations is 1 pound and 11.1/4.1/8.1/128 ounces. In water, however, it is 1 pound and 8 1/8.1/16.1/64 ounces, or, in 128th parts of an ounce, 2993 parts in air and 2586 parts in water, and since the difference is 407 parts a cylinder of water equal to the cylinder of tin will thus be of 407 parts, and the proportion of the weight of tin to the weight of water will thus be 2993 to 407.[119]

Returning to the public stage of Bardi's *Problema*, it might be opportune to consider the remaining experiments illustrated in his diagram. On the left we have a floating brass "boat" (*scapha*), which the weight of the *putto* is insufficient to submerge. On the right is a further "miracle of nature": a cylindrical tube, beneath which a lead plate remains suspended when plunged into water. Grienberger suggests that this additional experiment, taken from Simon Stevin, provides additional evidence that a cylindrical "well" of air can lead bodies heavier than water not to sink. The device in the upper-middle is another experiment derived from Stevin "a scales in which 10 pounds of lead are balanced by only 1 pound of water, through the insertion of a cylinder fixed to the wall which occupies the space of 9 pounds of water. Grienberger/Bardi suggests that since the shape of the immersed body is arbitrary, this provides further evidence in favour of Galileo, and in direct opposition to the Florentine Aristotelians, that "in similar experiments no account whatsoever should be taken of shapes or resistances of the medium."[120]

Bardi hoped that the publication of "his" text would provide a non-Italian public with a Latin compendium of the central teachings of Galileo's *Discorso*.[121] Federico Cesi, to whom Bardi wished to dedicate the work was unhappy with the dedicatory letter, complaining obliquely to Galileo that "when one is dealing with men who are truly great, I would like them to be treated in a fitting manner".[122] Stelluti elaborated to Galileo that Cesi objected to the dedication "both because he did not state that it was recited in the said College [i.e. the Collegio Romano] and because he does not give your Lordship the mention that your valour deserves, passing over it with most languid expressions".[123] Again, Grienberger speaks through Bardi to Galileo to emphasize his subjection to Jesuit discipline:

> And [Grienberger] asked me to send you his greetings, and to say to you that if he could have spoken in his own way, he would have said even more, but that he could not do otherwise, and perhaps he may have done more than he should have. For this reason he avoided becoming at all involved in the printing process, and I was obliged to show myself to be resolved to print it, because otherwise it might easily not have come about, because there were some who inclined more towards "no" than a "yes".[124]

Deprived of the essential diagram and, perhaps even more importantly, the endorsement of the Collegio Romano of the Society of Jesus, Bardi's *Problema* made far less stir as a publication than as a public performance, and Grienberger retreated once more from the public sphere.[125]

Conclusion: Anti-Galileo

This chapter has aimed to demonstrate Grienberger's deployment of pedagogy, bureaucracy and anonymity to effect significant changes in the way the natural world was understood by his peers. In spite of the increasing dangers involved in public opposition to Aristotle in the Jesuit context, Grienberger was prepared to use the public forum offered by the mathematical *problemata* to attack a number of key tenets of Aristotelian natural philosophy, and to defend the legitimacy of mathematics as a key instrument for the investigation of the natural world. Institutionalised Jesuit modesty provided Grienberger with a powerful resource which he used to great effect to discredit Aristotelian natural philosophy from behind the scenes, to strengthen the arguments of the books of Jesuit mathematicians and to perfect the observational and mathematical instruments of his colleagues from his *cubiculum* in the *Collegio Romano*.

George Fortescue, an alumnus of the English college in Rome, provides a rare example of a literary portrait of Grienberger which may serve us as a fitting closing image. Fortescue, who resided in Rome between 1609 and 1614, stages an imaginary meeting between Galileo, Grienberger and Clavius in the Roman residence of the Mantuan Cardinal Ferdinando Gonzaga. After Galileo had finished displaying the two magical wonders he brought with him to Rome, the phosphorescent Bologna stone and his "new opticon"–the telescope – it was Grienberger's turn to show his wares:

> As I see that this is an arena of subtleties I bring to you an experiment on the most clear, distinct and eloquent Voice that I recently imparted to a statue, while removed from more serious studies. The statue was made not of brass or of solid marble, but of plaster, in which the coiled receptacles of the voice, as if contained in a cavity, received the percussions of sounds and rendered them more felicitous. When words were introduced into this voice-duct, punctuated by breathing, and I transmitted others to the machine successively after similar breathing intervals and closed the point of entrance of the voice tightly, then at length, after the various by-roads, turnings and hinderances that really gave assistance, the oration arrived at the head [of the duct] and the boundary of the statue. The acute power of the words and the succession of breaths moved the gullet and mobile tongue most easily to produce the diversity of syllables. But how much I see myself to disappoint my companions, equal to the stars, who perhaps smile at their Grienberger, studiously forming voices from plastered blocks[126]

Surprisingly, Fortescue's attribution of the invention of a speaking-statue to Grienberger is somewhat plausible. Much later Athanasius Kircher had such a device in his bedroom in the Collegio Romano, which he later transferred to his museum, where it was baptised the Delphic oracle. In response to English claims to priority in the invention of the speaking-tube, Kircher never claimed its invention but merely stated that it had been invented years before "in the Collegio Romano".[127] Whether or not it was really built by Grienberger, however, is of little importance – a mechanical orator, a machine which moved its tongue and appeared to speak, amplifying the voice of the hidden operator, could hardly have been more appropriate to his favoured mode of self-expression.

Notes

1. A version of this chapter appeared as M.J. Gorman, 'Mathematics and Modesty in the Society of Jesus: The Problems of Christoph Grienberger', in *The New Science and Jesuit Science: Seventeenth Century Perspectives*, edited by Mordechai Feingold, Dordrecht: Kluwer, 2003, 1–120. I am grateful to Mordechai Feingold for permission to reproduce here.
2. The Vulgate translates κενοδοξία as *'inanis gloriae cupido'* (Gal. 5, 26, Phil. 2,3).
3. Jakob Bidermann, *Cenodoxus*, in Bidermann, *Ludi theatrales sacri, sive Opera comica posthuma*, Munich: J. W. Schell; 1666, 2 vols. Reprinted (Herausgegeben von Rolf Tarot) Tübingen: Max Niemeyer Verlag; 1967, Band 1, pp. 78–159. On *Cenodoxus*, see also Roland Mayer, *Personata Stoa: Neostoicism and Senecan Tragedy*. Journal of the Warburg and Courtauld Institutes, 1994; 57: 151–174, on p. 166.
4. Bidermann, *Ludi theatrales*, cit., I, sig. [(†) 8] v – sig. (††)1 r.
5. For pointers to the more important Ignatian sources, see the excellent introduction to Rolf Tarot's critical edition of *Cenodoxus*, cit., particularly pp. XXI–XXIII.
6. See Dionysius Fernández Zapico (ed.), *Regulae Societatis Iesu (1540-1556)*. Monumenta Historica Societatis Iesu. Rome; 1948. pp. 514–527.
7. In a critique of court-based accounts of the development of European civility, Dilwyn Knox situates the Jesuit rules of modesty in the context of a medieval monastic tradition of *disciplina*. See Dilwyn Knox, *Disciplina: The Monastic and clerical origins of European Civility* in John Monfasani and Ronald G. Musto, eds. *Renaissance society and culture: Essays in honour of Eugene F. Rice, Jr.* New York: Italica Press; 1991: 107–135, especially on pp. 126–8.
8. Unpublished versions extended this rule to the written word, see Zapico, cit., p.526 (*Regularum modestiae complementum* (1555): 'In loquendo vel scribendo nulla detur significatio arrogantiae', but the published versions of the *Regulae* restricted their attention to the body.
9. Pedro Ribadeneira, *Vita Beati Patris Ignatii Loyolae*, Antwerp, 1610. In so far as Ignatius's life story and spiritual disciplining after his injury at the battle of Pamplona came to serve as a model for those entering the order, it is perhaps worth mentioning in the context of self-effacement that Ignatius never allowed his portrait to be painted while General of the Society – future portraits had to rely heavily on sketches made at his deathbed and several death masks. See Thomas M. Lucas (ed.), *Saint, Site and Sacred Strategy: Ignatius, Rome and Jesuit Urbanism*, Vatican City: Biblioteca Apostolica Vaticana; 1990, p. 63 and *Fontes Narrativi de S. Ignatio de Loyola et de Societate Iesu* (Monumenta Historica Societatis Iesu), 4 vols., Rome: Institutum Historicum Societatis Iesu; 1943–1965, Vol. III, pp. 240–1.
10. Steven Shapin, *A Social History of Truth: Civility and Science in Seventeenth-Century England*, Chicago and London: University of Chicago Press; 1994, especially

Chapter 4, and Mario Biagioli, *Galileo Courtier: The practice of science in the culture of absolutism*, Chicago: University of Chicago Press; 1993. Through a peculiarly reflexive twist, both Biagioli and Shapin have been castigated for intellectual deportment inappropriate to the exalted station of the historian. On Biagioli, see Michael Shank, *Galileo's Day in Court*, Journal for the History of Astronomy, 25 (1994), 236–243, answered in Mario Biagioli, *Playing with the Evidence*, Early Science and Medicine, 1 (1996), 70–105, followed by Shank's lengthy rejoinder, *How shall we practice history? The case of Mario Biagioli's* Galileo Courtier, Early Science and Medicine, 1:1 (1996), 106–150. On Shapin, see Mordechai Feingold, *When Facts Matter*, Isis, 1996, 87: 131–139, Peter Dear's response (Isis, 1996, 87: 505–6), Shapin's response (Isis, 1996, 87: 681–4) and Feingold's rejoinder (Isis, 1996, 87: 684–7).

11 The work of Roger Chartier, in particular, has provided a renewed critique of the problems related to early modern authorship raised by Foucault's celebrated 1969 essay *Qu'est-ce qu'un auteur?* See especially Roger Chartier, *Figures of the Author*, in Chartier, *The Order of Books: Readers, Authors, and Libraries in Europe between the Fourteenth and Eighteenth Centuries*, Stanford, CA: Stanford University Press; 1994.

12 See Jacob Bidermann, *Cenodoxus*, ed. and transl. D.G. Dyer, Edinburgh: Edinburgh University Press, 1975, p. 8.

13 On Scheiner's troublesome courtly deportment, see Steve Harris, *Les chaires de mathématiques*. in Luce Giard, *Les jésuites à la Renaissance. Système educatif et production du savoir*. Paris: Presses Universitaires de France; 1995: pp. 239–261, and M.J. Gorman, *A Matter of Faith? Christoph Scheiner, Jesuit censorship and the Trial of Galileo*. Perspectives on Science. 1996; 4(3): 283–320.

14 Studies of Ignatian spirituality include David Lonsdale, *Eyes to See, Ears to Hear: An Introduction to Ignatian Spirituality*, London: Darton, Longman & Todd; 1990. For the connections between the development of the Jesuit spiritual programme with the organizational structure of the order see especially John W. O'Malley, *The First Jesuits*, Cambridge MA, Harvard University Press, 1993.

15 Of course, self-abnegation, or 'self-cancellation', in Greenblatt's terminology, is really just a form of self-fashioning. See Stephen Greenblatt, *Renaissance self-fashioning from More to Shakespeare*, Chicago and London: The University of Chicago Press; 1980.

16 See Gorman, *Mathematics and Modesty*, pp.32–120.

17 A cursory examination of the *Edizione Nazionale* of Galileo's *Works* and the Clavius correspondence revealed no less than nineteen current variants. See OG, *s.v.* 'Grienberger' and CC, *passim*.

18 *Catalogus veteres affixarum Longitudines ac Latitudines conferens,* Romae: Apud Bartholomaeum Zannetum; 1612, and *Euclidis sex primi Elementorum Geometricorum libri,* Romae, apud Haeredes Bartholomaei Zanetti, 1629 respectively.

A manuscript version of the latter work, is in the Biblioteca Nazionale Centrale in Rome (BNR, Fondo Gesuitico 594 (2723) A manuscript of the second part of the latter work, *Elementa trigonometrica, id est sinus Tangentes Secantes in Partibus Sinus totius 100000. Opusculum Secundum*. Romae, per Haered. Barthol. Zannetti, 1630, containing Muzio Vitelleschi's original letter of approval, is in the Biblioteca Medicea-Laurenziana in Florence (Ms. Ashburnam 1650). For information on later editions of both works see Sommervogel, Vol. III, coll. 1810–1811.

19 A useful summary biography has already appeared in CC I.2, pp. 55–7. Other information concerning Grienberger may be found in Ugo Baldini, *Astronomia e meccanica: La corrispondenza Grienberger-Burgo sull'idrostatica galileiana*, in Baldini, *Legem*, cit., pp. 183–216 and Baldini, *Dal geocentrismo alfonsino al modello di Brahe. La discussione Grienberger-Biancani*, in Baldini, *Legem*, cit., pp. 217–250.

20 Apart from the letters published in OG and CC, the codex APUG 534 contains numerous unpublished letters sent to and from Grienberger, a list of which is published in Ugo Baldini, *Legem impone subactis. Studi su filosofia e scienza dei gesuiti in Italia, 1540–1632*. Rome: Bulzoni; 1992, p. 200. For further details see CC I.2, p. 57.

21 See Paul P. Bockstaele, *Four Letters from Gregorius a S. Vincentio to Christopher Grienberger*. Janus. 1969; 56: pp. 191–202 and William B. Ashworth Jr, *The Habsburg Circle*. in Bruce T. Moran, ed., *Patronage and Institutions. Science, technology and medicine at the European Court 1500–1750*. Rochester, New York: The Boydell Press; 1991: 137–167.

22 Mario Bettini, *Apiaria Universae Philosophiae Mathematicae* Bononiae: Io. Baptistae Ferronij; 1645.

23 See Stefano Ghisoni to Giannantonio Rocca, Bologna; 23 November 1636, 'Ed avendolo io sentito molto lamentarsi, che sia in così poca stima la Matematica, dal modo di esagerar questo punto, ho congetturato, che non abbia trovato persona, che faccia la spesa per il suo libro', published in *Lettere d'uomini illustri del secolo XVII a Giannantonio Rocca*. Modena: Società Tipografica; 1785, pp. 62–4.

24 Bettini, *Apiaria Universae Philosophiae Mathematicae* Bononiae: Io. Baptistae Ferronij; 1645, Sig. C2 *recto*.

25 Bettini, cit, Apiarium V, Caput VI, pp. 44–6.

26 Bettini, *Aerarium Philosophiae Mathematicae,*. Bononiae: Io. Baptistae Ferronij; 1648, p. 75. Bettini provides much information on Grienberger's manuscript heritage here that Sommervogel incorrectly ascribes to Kaspar Schott.

27 'Singulare in eo id fuit, quod Archimedis exemplo acutissimas theorias mirificis praxibus iungebat', Bettini, *Aerarium*, loc. cit.

28 E.g. Juan Bautista Villalpando, 'Is vero in examinandis mechanicis instrumentis tanta est sollertia praeditus [Griembergerus], ut nemini debeat haberi secundus' (J. B. Villalpando, *Apparatus Urbis ac Templi Hierosolymotani*. Rome; 1604, p. 436)

and Kaspar Schott, describing an oil-spouting lantern designed by Grienberger (G. Schott, *Mechanica Hydraulico-Pneumatica*. Würzburg: Henricus Pigrin; 1657, p. 290).

29 See Christoph Scheiner, *Pantographice seu Ars delineandi res quaslibet per parallelogrammum lineare seu cauum, mechanicum mobile*, Romae; 1631.

30 Bettini, *Apiaria*, cit., Apiarium V, p. 44.

31 It seems highly plausible that the engravings were made by Grienberger himself, who's published *Catalogus* shows him to have been an extremely accomplished draughtsman, as do the drawings and engravings that accompany several of the *Problemata* discussed later in the present chapter.

32 Bettini, *Apiaria*, cit., Apiarium V, Caput VI, pp. 44–6.

33 Christoph Scheiner, *Rosa ursina, sive, Sol, ex admirando facularum et macularum suarum phaenomeno varius*. Bracciani: Apud Andream Phaeum Typographum Ducalem; 1630.

34 The relationship between Scheiner and the Orsini household was not so harmonious as this vignette might seem to suggest, as revealed by letters sent by Scheiner to Archduke Leopold of Austria during this time. See Franz Daxecker, *Briefe des Naturwissenschafftlers Christoph Scheiner SJ an Erzherzog Leopold V von österreich Tirol 1620–1632*, Innsbruck: Publikationsstelle der Universität Innsbruck, 1995, pp. 135, 152, 156, 159.

35 On Scheiner's 'collectivist' approach to natural investigation (apparently cramped somewhat by his unpopularity amongst other Jesuits) see Rivka Feldhay, *Galileo and the Church: Political Inquisition or Critical Dialogue?* Cambridge: Cambridge University Press; 1995, p. 288.

36 Christoph Scheiner, *Rosa ursina, sive, Sol, ex admirando facularum et macularum suarum phaenomeno varius*. Bracciani: Apud Andream Phaeum Typographum Ducalem; 1630, p. 348

37 See C. Grienberger to Ricardo de Burgo, [Rome], [June–July 1619], 'Rebus denique Galilaei vellem me non immiscere si possem postquam tam male de Mathematica Collegii Romani est meritus a qua non semel et quidem in praesentia tam bene quam sincere est habitus', published in Baldini, *Legem impone subactis*, cit., pp. 194–5, on p. 195. Surprisingly, before the election of Urban VIII to the Holy See, Federico Cesi made the rather strange suggestion that Galileo's *Il Saggiatore*, a work that was far more critical of the mathematicians of the Collegio Romano, should be dedicated to none other than Grienberger. See Giovanni Ciampoli to Galileo, Rome; 1620 Jul 17 inOG XIII, p. 44 'Il Sig.r Pinc.e Cesi mi ha mandato aperta l'inclusa: vi era una poliza, nella quale adduceva alcune ragione per le quali giudicava bene il dedicar l'opera al P. Bamberger [i.e. Grienberger], e rimette a noi il mandarla, i quali, essendo qua in paese, assolutamente non giudichiamo bene il farlo per non mettere in fastidi quel povero Padre, come certamente sappiamo *ab exemplo* che seguirebbe.'

38 After Galileo's trial, Grienberger is reported to have said that 'If Galileo had maintained the affection of the Fathers of this College, he would live gloriously to the world and none of his disgraces would have come about. He would have been able to write freely on any subject, even on the motion of the earth etc.'. Galileo interpreted this statement as a confirmation of his suspicions that the Jesuits had engineered his downfall, but the material presented here might suggest an alternative interpretation – Grienberger had hoped that together Galileo as 'author' and the Jesuit astronomers as 'expert' corroborators of his observations might produce a reformed cosmology sanctioned by the catholic church. See Galileo to Elio Diodati, Florence; 25 July 1634, in OG XVI, p.117. Grienberger's remarks in his notes for the censorship of Giuseppe Biancani's cosmography (*Sphaera mundi*, Bononiae: Typis Sebastiani Bonomij; 1620) also suggest that Grienberger was in favour of a new cosmography being written by someone outside the order, to replace the outdated commentaries on the Sphere of Sacrobosco: 'Laudabile est in primis studium Auctoris, quod potissimum tyronibus in rebus mathematicis prodesse conatur, quibus Cosmographia nova necessaria videtur, eo quod vetus plurimum hoc tempore immutata sit eique non pauca accesserint ornamenta. Sed dubium est expediat necne, *ut id per Nostros fiat*', ARSI FG 655, f. 118r, published in Baldini, *Legem impone subactis*, cit., p. 235 (emphasis added).

39 See M.J. Gorman, *A Matter of Faith? Christoph Scheiner, Jesuit Censorship and the Trial of Galileo.* Perspectives on Science. 1996; 4(3): 283–320 on p. 312.

40 See CC I.2, pp. 55-7.

41 Grienberger to Clavius, Vienna, 15 September 1590, in CC II.1, pp. 158-162, on p. 158.

42 Christoph Clavius, *Euclidis Elementorum libri XV. Accessit XVI. de Solidorum Regularium comparatione. Omnes perspicuis demonstrationibus, accuratisque scholiis illustrati*, Rome: Apud Vincentium Accoltum; 1574, Christoph Clavius, *In Sphaeram Ioannis de Sacrobosco Commentarius*, Rome: Apud Victorium Helianum; 1570 (first edn.). See CC I.3 pp. 5–11 for a full bibliography of Clavius's works.

43 Grienberger to Clavius, Wien, 15 September 1590, cit., on p. 158.

44 Mario Biagioli has pointed to the importance of the homosocial bond in the activities of the *Accademia de' Lincei*, but the connection between homosociality and epistolary links, clearly evinced in the Jesuit order well before the creation of the Lincei and stemming from the Ignatian conception of the importance of epistolary relationships to the 'union of hearts' in the Society, has yet to receive adequate attention. See M. Biagioli, *Knowledge, Freedom, and Brotherly Love: Homosociality and the Accademia dei Lincei.* Configurations. 1995; 2: 139–166.

45 Biancani sent Clavius a proposed solution to the problem of measuring longitude at sea. He suggested that a large number of accurate clepsydras could be used to constitute a shipboard clock which, in conjunction with accurate measurements of a particular fixed star, ideally lying close to the equator, could be used to

calculate longitudes accurately. The Rector of the Jesuit College of Padua sent an accompanying letter pleading with Clavius to help Biancani, to allow the College to have a mathematics teacher, which it had lacked since Marc' Antonio de Dominis, Biancani's teacher, had become Bishop of Segna. See Biancani to Clavius, Padova, 28 February 1598 in CC IV.1, pp. 34–37. Maelcote sent an astrolabe of his own design, prompting Clavius to suggest his transfer to Rome. See Clavius to Maelcote, Roma, 16 February 1601, in CC IV.1 pp. 124–5 on p. 124 'Si tui Superiores cum P. N. Generali agerent, ut in urbem vocaveris, donec vivo, res mihi esset gratissima'.

46 Grienberger to Clavius (in Naples), Rome, 6 October 1595, in CC III.1, pp. 137–8, on p. 137.

47 ARSI Rom. 150, I. 36r, cit. in CC III.2, pp. 54–5, note 2.

48 Such 'mathematical museums' were also to be found in other Jesuit colleges, including the college of Prague, where Jakob Johann Wenceslaus Dobrzensky de Nigro Ponte saw a a hydro-magnetic clock-fountain made by Kircher's disciple Valentin Stansel. See J.J.W. Dobrzensky de Nigro Ponte, *Nova, et amaenior de admirando fontium genio (ex abditis naturae claustris, in orbis lucem emanante) philosophia*. Ferrara: Alphonsum, & Io. Baptistam de Marestis; 1657, p. 46.

49 'Exspecto proximam studiorum interruptionem, ut diligentius perquiram et conscribam fabricam cuiusdam Machinae quam iam ab aliquot mensibus mecum habeo in cubiculo'. Grienberger to Clavius (in Rome), Lisbon, 24 March 1601, CC IV.1, pp. 136–9 on p. 137.

50 On the different levels of mathematical tuition in the *Collegio Romano*, ranging from the private mathematical academy for the training of future teachers to the normal public classes, see CC I.1 59–89.

51 See CC III.1, p.138.

52 This practice was formalised in 1636, when Muzio Vitelleschi wrote to Francesco Piccolomini, the Roman Provincial, that the papers of a Jesuit should be examined by the Superiors after his death and only papers of special interest should be preserved. According to Baldini and Napolitani, this was more or less what went on beforehand. See Muzio Vitelleschi to F. Piccolomini, Rome; 9 August 1636, cited in CC I.1, p. 30.

53 Athanasius Kircher, *Magnes, sive de arte magnetica opus tripartitum*, Romae: Ex Typographia Ludovici Grignani, 1641, Lib. II, Cap. II, p. 431, '[P]artim è literis ab ijs, qui iter in Indias susceperant, vel oretenus ab ijs, qui inde peregrini Romam advenerant; partim ex literarum Mathematicarum è diversis orbis terrae partibus ad *Clavium, Grimbergerum, aliosque Romanos Societatis IESU Mathematicos praedecessores meos datarum, quod penes me est, Archivio*; multas sanè, circa declinationes Magneticas haud spernendas observationes collegi', Kaspar Schott, *Mechanica Hydraulica-Pneumatica*, Würzburg: Pigrin, 1657, pp. 300 and, esp., 339: 'In Manuscriptis doctissimi viri P. Christophori Grünbergeri, olim in Romano Collegio

Mathematicae Professoris, quae in *Archivio Clavij & Grünbergeri* reperi, haec habentur verba circa praesentem Bettini Machinam, & de motu perpetuo opinionem...' [emphasis added].

54 Grienberger dismisses the compass, celebrated by Giordano Bruno in two poems, as a pretty plaything: 'Nempe in rebus exactioribus esse inutile instrumentum at in apparentia et operatione vulgari pulchrum simul et iucundum immo quod multum faciat mirari spectantes', Grienberger to Clavius (in Naples), Rome, 23 February 1596, CC III.1, pp. 161-9, on p.164.

55 E.g. 'O felices barattolas, o dulces mustacciulos. eoque feliciores quo minus deficient', Grienberger to Clavius (in Naples), Rome, 27 October 1595, in CC III.1, pp. 139-142 on p. 139, 'Saltem optarem, ut ne Roma una cum Clavio abisse videatur Mathematica, et Neapolim veluti in novam Coloniam transmigrasse. Quam spero redituram propediem ubi nimia quae illic est dulcedo nauseam attulerit. Quod si ita dulcedo delectat, ut eius satietas sit nulla, saltem meminerit extra patriam se vicere...' Clavius's gluttony was later lampooned by Joseph Scaliger: 'Clavius qu'on m'avoit dit estre un grand personnage & que i'ay trop loué, est une beste. Monsieur Dabin m'a dit qu'il luy faut tous les matins un morceau de jambon & un verre de vin Grec. C'est un gros ventre d'Aleman...', Joseph Scaliger, *Scaligerana. Editio altera. ad verum exemplar restituta*, Coloniae Agrippinae: Apud Gerbrandum Scagen; 1667 p. 51 (*s.v.* 'Clavius').

56 Grienberger to Clavius (in Naples), Rome, 12 January 1596, CC III.1, pp. 146-50.

57 Grienberger to Clavius (in Naples), Rome, 23 March 1596, in CC III.1, pp.170-3, on p. 170. It is worth mentioning that it is just possible that 'Horatius' may have been the young Orazio Grassi (though he came from Savona rather than Perugia), who joined the Jesuit Novitiate at S. Andrea al Quirinale in 1600 at the age of eighteen.

58 'Sed exclusus est forte [Clavius] cubiculo? immo vero ita aptum est ut vel nolentem invitet. eodem libenter. facile enim ut puto aliud inveniam quod aeque sit frigidum, nisi forte omnia sint frigida in quibus ego frigidissimus inhabito'. Grienberger to Clavius (in Naples, Rome, 24 November 1595, in CC III.1, pp. 142-145 on p. 143

59 Grienberger to Clavius (in Naples), Rome, 26 January 1596, CC III.1, pp. 151-2: 'Nunc parum quiescere cogor, quod mihi in cubiculum allatus sit lapis ut in eo describam horologia pro Cardinali Lanceloto'.

60 Grienberger to Clavius (in Naples), Rome, 23 March 1596, in CC III.1, pp. 170-3, on pp. 170-1.

61 Grienberger to Clavius (in Naples), Rome, 23 March 1596, in CC III.1, pp. 170-3, on p. 172

62 ibid., loc. cit.: 'Scilicet Clavius Gustum servat sibi nobisque mittit odorem, ut hinc conclusimus quam omnia plana sint in cubiculo Clavii cum etiam chartae dulcem illum odorem hauriant et Romam usque deferant'.

63 Grienberger to Christoph Clavius (in Naples). Rome; 24 November 1595, CC. III.1, pp. 142-46, on p. 143.

64 Giuseppe Biancani, *De mathematicarum natura dissertatio una cum clarorum mathematicorum chronologia*. Bononiae; 1615 (bound with idem., *Aristotelis loca mathematica*, Bononiae, 1615).
65 Bettini, *Aerarium*, cit., Def. 10, §3, p. 75
66 MP V, p. 110.
67 MP V, p. 284. This suggestion was previously made in the unpublished 1586 version of the Ratio Studiorum, MP, V, p. 177.
68 Gorman, *Mathematics and Modesty*, pp. 32–100.
69 Louise Rice, *College Art: Prints, Poetry and Music for the Academic Defense at the Collegio Romano*, paper given at the conference *The Jesuits: Culture, Learning, and the Arts, 1540–1773*, May 28– June 1 1997, Boston College, Chestnut Hill, MA.
70 Zannetti's close links with the Collegio dated from the extraordinary success of his edition of Bellarmine's 1578 Hebrew grammar. In 1598, the publishing house moved to new premises located adjacent to the *Collegio* to facilitate the collaboration, and when Bartolomeo Zannetti died in 1621, he left all his printing equipment to the Jesuits. Mascardi moved to occupy his premises shortly after this time, taking over a lucrative and spiritually edifying business relationship between College and printing-house. See Saverio Franchi, *Le Impressioni Sceniche. Dizionario bio-bibliografico degli editori e stampatoi Romani e Laziali di testi drammatici e libretti per musica dal 1579 al 1800*. Roma: Edizioni di Storia e Letteratura; 1994, pp. 780–805.
71 Carlo Ginzburg, *The High and the Low: The theme of forbidden knowledge in the sixteenth and seventeenth centuries* in Ginzburg, *Clues, Myths, and the Historical Method*, Baltimore and London: The Johns Hopkins University Press; 1989 p. 60–76.
72 In addition to the confessions of bodily inability found in Grienberger's letters to Clavius cited above, see Christoph Grienberger, *Catalogus veteres affixarum Longitudines ac Latitudines,* Romae: B. Zannetti, 1612, *Ad Benevolum Lectorem*: 'Quam cum etiam ipse probe perspectam heberem, videremque desiderium meum plurimorum esse; neque spes ulla affulgeret inueniendi ea apud alios quibus levari diuturna nostra sitis posset, *ipse meos imbecilles humeros,* tandem huic oneri utilitate gravissimo submittendos duxi, & aquam quae meae aliorumque siti extinguendae sufficeret, primo DEO dante & expensas faciente, domum detuli, tum foras ductis rivulis ob commune studium, etiam ad irrigandos aliorum hortulos eduxi.' [emphasis added]
73 *Imago Primi Saeculi Societatis Iesu A Provincia Flandro-Belgica eiusdem Societatis Repraesentata.* Antwerp: Balthasar Moretus; 1640.
74 Marc Fumaroli, *Baroque et Classicisme: L'Imago Primi Saeculi Societatis Jesu (1640) et ses adversaires.* in Fumaroli, *L'Ecole du Silence: Le sentiment des images au XVIIe siècle*. Paris: Flammarion; 1994: 343–365.
75 Paul Guldin, *De Centro Gravitatis Trium specierum Quantitatis continuae*, Viennae: Gregorii Gelbhar, pp. 137–148, *Dissertatio Phisico-Mathematica de Motu Terrae,* on

p. 137, Guldin originally delivered this *dissertatio*, which discusses the trepidation of the earth through small shifts in the position of its centre of gravity, while in the Collegio Romano. The text bears similarities to Grienberger's *Problemata* in style and content, as does the accompanying diagram.

76 'Dico igitur rotis non amplius 24 et solidem axibus dentatis Globum terrestrem quamuis aureus foret totus, extra centrum propelli posse, uel ab ea potentia, quae Talentum', Problema: Terram auream, Talenti potentia movere, 5 November 1603, APUG Fondo Curia 2052, published in Gorman, *Mathematics and Modesty*, pp. 77-87.

77 See Schott, *Magia universalis,* Bamberg: J. M. Schönwetter, 1677^2 (4 vols.), pars III, pp. 219-228, 'Machina II: Glossocomum nostrum, quo talenti potentia movetur Terraqua, si aurea foret'. Schott makes no mention of Grienberger's authorship of the *problema*, but makes only minor changes to the original. Schott's diagram, almost identical to Grienberger's, is reproduced in William B. Ashworth, Jr, Iconography of a new physics, History and Technology, 4, 1987: 267-297

78 Ibid.

79 Ibid.

80 See *Praefatio*, [November 1591], APUG Fondo Curia 2052, in Gorman, *Mathematics and Modesty*, pp. 33-40.

81 See *Fieri posse [. . .] in aliqua mensa lumine*, APUG Fondo Curia 2052, in Gorman, *Mathematics and Modesty*, pp. 41-48.

82 *Praefatio*, [November 1591], APUG Fondo Curia 2052, published in Gorman, *Mathematics and Modesty*, pp. 33-40.

83 'In omnibus hisce vel maxime excelluit Archimedes. [. . .] Per mirabilia vero opera quae efficit illud tandem ab Hierone Rege privilegium est consequutus, ut quidquid tandem affirmaret omnino sibi fides haberetur', ibid.

84 Grienberger to Clavius (in Naples), Rome; 24 November 1595, in CC III.1, pp. 142-146.

85 [*Problema. De Dimensione Circuli*], [27 October-24 November 1595?], APUG Fondo Curia 2052, in Gorman, *Mathematics and Modesty*, pp. 49-55.

86 Giuseppe Biancani, *De mathematicarum natura dissertatio una cum clarorum mathematicorum chronologia*. Bononiae; 1615 (bound with idem., *Aristotelis loca mathematica*, 1615), English translation in Paolo Mancosu, *Philosophy of Mathematics and Mathematical Practice in the Seventeenth Century*, New York, Oxford, Oxford University Press, 1996, pp. 178-212

87 Biancani is recorded as an academician in 1599-1600 (ARSI, Rom. 54 ff. 2v, 12v, 77r, cit. in CC I.2 pp. 18-19).

88 Grienberger, *Problema Mechanicum Circa motus ponderum*, [January 1596?], APUG Fondo Curia 2052 in Gorman, *Mathematics and Modesty*, pp. 66-69, Grienberger, *Problema Circa motus caelorum*, APUG Fondo Curia 2052 in Gorman, *Mathematics and Modesty*, pp. 55-66.

89 On Clavius's response to homocentric astronomy see James M. Lattis, *Between Copernicus and Galileo: Christoph Clavius and the collapse of Ptolemaic cosmology.* Chicago and London: University of Chicago Press; 1994, pp. 87-94.

90 Grienberger, *Problema Datis excessibus quibus diameter Quadrati aut figurae*, APUG Fondo Curia 2052, in Gorman, *Mathematics and Modesty*, pp. 70-74.

91 *De errore qui in denominandis numerandisque diebus in Indicae navigatione commititur*, APUG Fondo Curia 2052, in Gorman, *Mathematics and Modesty*, pp. 75-76.

92 Grienberger to Galileo, Rome, 5 February 1612, in OG XI, 272-4, on p. 273

93 Grienberger, *Catalogus veteres affixarum Longitudines ac Latitudines, conferens cum novis. Imaginum Coelestium Prospectiva duplex.* Romae: Apud Bartholomaeum Zannetum; 1612, Sig. A2 verso

94 Grienberger to Galileo, Rome; 24 June 1611, in OG XI, pp. 130-1.

95 Galileo to Grienberger, Florence; 1 September 1611, in OG XI, pp. 178-203.

96 Grienberger to Galileo, Rome; 5 February 1613 in OG XI, pp. 479-80.

97 Giovanni Bardi to Galileo, Rome; 24 May 1613, in OG XI, pp. 512-513.

98 On the debate about the *Gallegianti*, see especially Stillman Drake, "The Dispute over Bodies in water", *Galileo Studies*, Ann Arbor: University of Michigan Press; 1970, pp.159-176, Mario Biagioli *Galileo Courtier: The practice of science in the culture of absolutism.* Chicago: University of Chicago Press; 1993, Ch.3, pp.159-209 and Francesco de Ceglia, *Reazioni romane: L'idraulica galileiana negli scritti di Giovanni Bardi e Giuseppe Biancani*, Bari: Laterza, 1997.

99 Despite the Archimedean conclusions of this work, Galileo's attempts to produce an Archimedean demonstration foundered, due to his assumption that the volume of the liquid displaced is equal to the submerged bulk of the body. To be consistent with Archimedes, he should have assumed that the volume of liquid displaced is equal to the bulk of the body below the <u>original</u> level of the liquid. This slip meant that Galileo ended up using the method of the Pseudo-Aristotelian mechanical problems. See William R.Shea, *Galileo's discourse on floating bodies: Archimedean and Aristotelian elements*, Actes du XIIe Congrès International d'Histoire des Sciences, 1968, Paris: Blanchard;1971, IV, pp. 149-153.

100 Bardi to Galileo, Rome 24th May 1613, cit.

101 BNR, Fondo Gesuitico 1186, ff. 108r-114v. See Baldini, *Legem impone subactis*, cit. pp. 155-182. Although Baldini suggests that this *Problema* is probably due to Odo van Maelcote there is substantial evidence that it was composed by Grienberger. See also Gorman, *Mathematics and Modesty*, pp. 87-91.

102 OG III, 1, 291-298. While there is evidence that Maelcote recited this oration, there is no independent evidence that he was responsible for its composition. The copy preserved in the Vatican library (BAV Barb. Lat. 231, ff. 177r-182r) is not in Grienberger's hand.

103 Bardi to Galileo, Rome, 20 June 1614, OG XII p. 76.
104 See OG IV, p. 195
105 De Ceglia, op. cit., p. 44
106 See also Gorman, *Mathematics and Modesty*, cit., pp. 96–97.
107 Claudio Acquaviva S.J., *Ordinatio pro soliditate et uniformitate doctrinae, ad omnes praepositos provinciales S.I.*, Rome, 14 December 1613, in MP VII, pp. 660–664, on p. 663. On the issue of uniformity and solidity of doctrine, see Baldini, *Legem impone subactis*, ch. 1 and 2.
108 Giovanni Camerota, censure of G. Biancani, *Brevis tractatio de iis quae moventur in aqua*, 16 February 1615, ARSI FG 662 f. 166r, published in Baldini, *Legem impone subactis*, cit., p. 232.
109 Bardi to Galileo, Rome 20th June 1614, OG XII, p.76.
110 On Valerio, who left the Society of Jesus in 1580, but maintained close contact with Clavius see Ugo Baldini and Pier Daniele Napolitani, *Per una biografia di Luca Valerio*. Bollettino di Storia delle Scienze Matematiche. 1991; Anno XI(1): pp. 3–157.
111 Francesco Stelluti to Galileo, Rome, 28 June 1614, in OG XII, p. 78.
112 *De ijs quae vehuntur in aquis*, APUG Fondo Curia 2052 VIII 47r–54r, on f. 47r.
113 *Francisci Aguilonii e societate Iesu Opticorum libri sex. Philosophis iuxta ac Mathematicis utiles*. Antverpiae, ex officina Plantiniana: Apud Viduam et Filios Io. Moreti; 1613.
114 On Rubens' illustrations see August Ziggelaar, *François de Auguilon S.J. (1567–1617). Scientist and Architect*. Bibliotheca Instituti Historici S.I., Vol. 44. Rome: Institutum Historicum S.I.; 1983. On his relationship to the Jesuit emblem-book tradition through his master Otto Vaenius, see Charles Parkhurst, *Aguilonius' Optics and eoRubens' Color*. Nederlands Kunsthistorisch Jaarboek. 1961; 12: 35–49. Anatomical texts using *putti* may also have provided Rubens with a model. On the Jesuit emblem book tradition in general see Karl Josef Höltgen, *Emblem and meditation: Some English emblem books and their Jesuit models*. Explorations in Renaissance Culture. 1992; 18: 55–91, Loretta Innocenti, *Vis eloquentiae: Emblematica e persuasione*. Palermo: Sellerio; 1983 and Mario Praz, *Studi sul Concettismo*. Milano: Soc. Editrice "La Cultura"; 1934, especially ch. 4, pp. 134–164 (English translation: *Studies in seventeenth century imagery*. Rome: Edizioni di Storia e Letteratura; 1964). The author of this study points out that the Jesuits transformed *eros* into *amor divinus*, turning profane love emblems into instruments of religious propaganda.
115 For a contextualised appraisal of the Jesuit rhetorical tradition, the most comprehensive study remains Marc Fumaroli, *L'Age de l'éloquence: Rhétorique et res literaria de la Renaissance au seuil de l'Époque classique*. Geneva: Librairie Droz; 1980.

116 Jeronimo Nadal, *Evangelicae historiae imagines*, Antwerp: Martin Nutius, 1593. On this work see Pierre-Antoine Fabre, *Ignace de Loyola: Le lieu de l'image*, Paris: Vrin, 1992, and the Spanish critical edition, *Jeronimo Nadal, Imagenes de la Historia Evangelica, con un estudio introductorio por Alfonso Rodriguez G. de Ceballos*, Barcelona: Ediciones El Albir, 1975.
117 Galileo Galilei, *Discorso intorno alle cose, che Stanno in sù l'acqua, e che in quella si muovono*, Firenze: Cosimo Giunti, 1612, in OG IV pp. 57–141, on p. 65.
118 Anon. [Christoph Grienberger], *De ijs quae vehuntur in aquis*, loc. cit.
119 APUG FC 2052 VIII f.130r. A series of experimental notes in Grienberger's hand can be found in APUG FC 2052 VIII f.130r–132v. On Ghetaldi and Villalpando's investigations of specific gravities, see P.D. Napolitani, *La Geometrizzazione della realtà fisica: il peso specifico in Ghetaldi e in Galileo*. Bolletino di Storia delle Scienze Mathematiche. 1988; VIII: pp.139–237.
120 *De ijs quae vehuntur in aquis*, cit., f. 51r.
121 Bardi to Galileo, Rome, 2 July 1614, OG XII, pp. 79–80.
122 Cesi to Galileo, Rome, 16 August 1614, OG XII, pp. 95–96.
123 Stelluti to Galileo, Rome, 2 August 1614, OG XII, pp. 90–91.
124 Bardi to Galileo, Rome, 2 July 1614, OG XII, pp. 79–80.
125 G. Bardi, *Eorum quae vehuntur in aquis experimenta a Ioanne Bardio Florentino ad Archimedis trutinam examinata*, IX. Kalend. Iul. Anno Domini M.DC. XIV. Romae, ex Typographia Bartholomaei Zannetti. M.DC. XIV. Even the title of the book, while giving the date of the original recital, avoided mentioning that it took place in the Collegio Romano.
126 George Fortescue, *Feriae Academicae. Auctore Georgio de Forti Scuti Nobili Anglo*. Duaci: Officina Marci Wyon sub signo Phoenicis; 1630, pp. 140–145.
127 See G. de Sepi, *Romani Collegii Societatis Iesu Musaeum Celeberrimum*, Amsterdam: Jannson-van Waesberghe; 1678 p. 60 and Kircher, *Phonurgia nova*, Campidoniae: Dreherr, 1673, pp. 112–113.

3

Discipline, Authority and Jesuit Censorship: From the Galileo Trial to the *Ordinatio Pro Studiis Superioribus*[1]

Clavius's vision of a distributed community of trustworthy Jesuit mathematical practitioners, entrusted with the education of the sons of princes, respected by natural philosophers and theologians, and maintained in a fraternal bond through the frequent exchange of letters, was clearly well-suited to the structure of the Society of Jesus as a whole. Nonetheless this chapter will argue that the relationship between being a Jesuit and being a mathematical practitioner was made far more complex by the enormous differences that existed between the various contexts in which Jesuit mathematicians carried out their work. Although Christoph Grienberger, cocooned within the protected space of the Collegio Romano, might manage to reconcile exemplary Jesuit behaviour with mathematical work, for members of the Order who were obliged to live their lives in less shielded circumstances strong tensions could emerge between the types of behaviour expected of them as Jesuits and the type of behaviour required of courtly mathematical practitioners. In 1631 Muzio Vitelleschi could chasten Paul Guldin and Henricus Philippus for possessing mechanical clocks [*horologij rotati*], seen by Vitelleschi as 'utterly incompatible with the religious poverty of the Society of Jesus', despite the attempt by the mathematicians to advise the General that such clocks were a necessary item of astronomical equipment, rather than a luxury item.[2] The tension could run deeper, however, as demonstrated by the participation of two Jesuits, Christoph Scheiner and Melchior Inchofer, in the events surrounding Galileo's trial of 1632–33. Inchofer's hopes to standardise the philosophical and theological doctrines taught in Jesuit colleges and printed works by members of the order, as part of a general attempt to reform the corruptions of the Society, culminated in a lengthy process of disciplinary enforcement in the 1640s, leading to the *Ordinatio pro Studiis Superioribus* of 1651.

Christoph Scheiner, the Galileo Trial and the credibility of the Jesuit mathematical practitioner

On 17 August 1633, less than two months after Galileo had publicly abjured the central tenets of the Copernican theory in Rome, Nicholas Fabri de Peiresc wrote from Aix to Athanasius Kircher in Avignon:

> I am more than a little sorry for Father Scheiner's poor adversary, and do not know how this can have come to pass as he only treated the question problematically, without declaring himself in favour of one opinion or the other. That is the way of the world. Certainly, he cannot excuse himself for treating Father Scheiner as badly as he did, but this would have been sufficiently repudiated in the Volume *Sol Mobilis*, which would even have reduced it *ad metam non loqui* as we say. For he had only to contradict him, which would have satisfied anyone else and would have allowed him to live in peace for the few days which remain to him. But God has reserved that mortification for the Glory which he had in discovering so many marvels in the heavens which were unknown before him. All because he envied Father Scheiner the discovery of sunspots and their motion, which then allowed him to induce the movement of the body of the Sun about its centre, which I regard as extremely important for the knowledge of Nature.³

This letter suggests that for Peiresc there was little doubt that the fate of Galileo was directly linked to his long-standing and bitter controversy with the Jesuit Christoph Scheiner over the discovery of sunspots.⁴ As Peiresc had been keeping a close eye on the activities of Scheiner's *pauvre adversaire* ever since his meetings with Galileo in Padua in 1601, his opinion on the matter carries some degree of weight.⁵ His principal sources of information on this point seem to have been Gabriel Naudé⁶ and Kircher himself. Naudé, then the librarian of Cardinal Giovanfrancesco dei Conti Guidi di Bagno, was very quick to cast suspicion on the involvement of the Jesuits in the trial of Galileo. He wrote to Gassendi from Rome in April 1633 to say that 'the mob is bankrupt of [the *Dialogo*] in this country, because of the curse pronounced on [Galileo] by the Court of Rome, where Galileo has been summoned through the machinations of Father Scheiner and other Jesuits, who want to ruin him, and would assuredly have done so if he wasn't powerfully protected by the Duke of Florence'.⁷

Kircher, on 9 August 1633, amplified the rumours that were already beginning to circulate in France by sending Peiresc a copy of a letter sent to him by Scheiner three weeks previously. In this letter Scheiner had described Galileo's act of obeisance to the Holy Office:

A few days ago Galileo abjured and damned his opinion on the immobility of the sun and motion of the earth, before the Inquisitor and in the presence of 20 witnesses, *de vehementi*, that is to say vehemently suspected of heresy. His book will be banned.[8]

The tradition of ascribing the downfall of Galileo to the unseen hand of his Jesuit adversaries is one which has persisted in the historiography of the trial from the outset. Descartes was expressing an opinion that was already widely accepted when, in February 1634, he confessed to Mersenne that 'I have allowed myself to be persuaded that the Jesuits aided in the condemnation of Galileo'.[9] Such a hypothesis finds a certain amount of circumstantial support from some of the surviving documents of the time, such as the reported claim made by Christoph Grienberger that 'If Galileo had remained a friend of the Fathers of this College, he would be honoured in the world and would not have incurred any of his misfortunes. He could have gone on writing freely on any subject, including the motion of the earth'.[10]

Amongst the Jesuits suspected of having formulated a denunciation against Galileo's *Dialogo*, initiating the process that led to the events of 22 June 1633, none has seemed a more likely candidate for the deed than Scheiner.[11]

His acrimonious rivalry with Galileo is legendary, and his furious white-faced reaction to the publication of the *Dialogo* has been frequently cited.[12] His proclamation in a letter to Gassendi in February of the year of Galileo's trial that 'I am preparing to defend myself and the truth'[13] has been read as pregnant with grim significance. The mention of Scheiner's name in the report to the Holy Office on the *Dialogo*, as a part of the Inquisitional proceedings against Galileo, made by another Jesuit, Melchior Inchofer, has also been remarked upon,[14] as has Scheiner's departure from Rome for Vienna, allegedly shortly after the conclusion of the trial.[15]

Such a variety of suggestive 'clues' has been enough to fuel a long-standing suspicion of Scheiner as the engineer of the downfall of Galileo and the ensuing crisis of conscience for Roman Catholic practitioners of astronomical investigation.[16] The lack of documentary evidence for Scheiner's role can also be conveniently explained away in terms of a Jesuit conspiracy of silence, whereby any incriminating documents would not be permitted to surface, lest they damage the reputation of the Order.

However, the fragile documentary basis for such a reading of the 1632-33 trial is easily seen from a close examination of the origins of any single piece of evidence. For example, a document in the Peiresc correspondence in Paris demonstrates that Scheiner's departure from Rome for Vienna, at the invitation of Ferdinand II, must have taken place after 1 December 1633, and thus cannot be plausibly interpreted as a convenient escape from the scene of the 'crime'.[17] The document is a letter which Kircher sent to Peiresc on his arrival in Rome, where he was to become the most famous professor and, later, *scriptor* of the Collegio Romano,[18] unlike Scheiner who curiously never held a teaching post during his nine years in Rome.[19]

The letter bears a friendly greeting to Peiresc in the hand of Scheiner:

> I too offer most humble greetings and dutiful servitude to your most Illustrious Lordship. Christoph Scheiner.[20]

Furthermore, Scheiner's letters from Rome to Archduke Leopold of Austria demonstrate that, although he was delayed by the slow publication of his *Rosa Ursina*, he had hoped to leave Rome as early as 1627,[21] and had packed almost all of his belongings into boxes by 2 June 1628.[22]

However, evidence for Scheiner's 'innocence' also seemed to be in short supply, except for an intriguing claim made by Fr Adolf Müller S.J. that Scheiner's lack of involvement in Galileo's trial was demonstrated by the 'unpublished and confidential correspondence of Scheiner with the General of his Order'.[23] The Galileo scholar Antonio Favaro was sceptical of Müller's claim in his essay on Scheiner, stating that '*questa pretesa prova non persuaderà alcuno*'. To these strong words Favaro adds that Scheiner and General Muzio Vitelleschi were both in Rome during the period of the trial[24] so there was no clear reason why they should correspond, but such internal correspondence, especially on official matters, is far from being as rare as Favaro suggests.

Unfortunately, no further evidence of the existence of these letters has appeared since Müller's book, despite Favaro's demand for hard evidence, leading others to reiterate the importance of these documents for a clear understanding of at least one aspect of the 1633 trial.[25] A search of the more likely parts of the Roman Archives of the Jesuits did not reveal any letters between Scheiner and General Vitelleschi for this period.[26]

However, a document which, despite not being, except in the loosest sense of the word, a letter from Scheiner to the General,[27] seems to makes it difficult to sustain any suspicion of Scheiner's involvement in the proceedings of the Holy Office against Galileo.

This document is Scheiner's *censura* of a book entitled *Tractatus Syllepticus*[28] written by Melchior Inchofer,[29] the very Jesuit who had been called to submit his opinion on Galileo's *Dialogo* to the Holy Office.[30] Inchofer's book, which appears to have been composed around the same time as his report on Galileo's *Dialogo*[31] is an extended series of arguments for the immobility of the earth and the motion of the sun based on Scripture and the Church Fathers.[32] Inchofer emphasises that a large number of biblical passages, interpreted literally, suggest that the earth is stationary and that the sun moves, whereas the few passages which seem to imply the possibility of terrestrial motion are really cases of the word 'earth' being written to signify the inhabitants of the world.[33] Eighteen years previously, on 12 April 1615, Cardinal Robert Bellarmine had written to the Carmelite friar Paolo Antonio Foscarini on precisely this issue, arguing that the Council of Trent had prohibited scriptural exegesis that did not conform with the common opinion of the Church Fathers, and that the latter were in agreement that the motion of the sun and immobility of the earth were to be interpreted literally. For Bellarmine, perhaps the principal authority on such matters for later Jesuits such as Inchofer,[34] the fact that these propositions did not concern the central mysteries of the faith, and were not *de fide ex parte obiecti* did not permit their denial, as they were *de fide ex parte dicentis*; that is, they were to be considered as matters of faith because of the sanctity of their author, who was none other than the Holy Spirit, speaking through the mouths of the Prophets and Apostles.[35] Thus, a person who denied that Abraham had two sons and Jacob had twelve was no less a heretic than a person who denied that Christ was born of a virgin.[36]

A series of *censurae* of Inchofer's *Tractatus* is preserved in the Jesuit Archives in Rome.[37] Before examining them, a brief review of the system of Jesuit censorship might be in order.[38] Despite the emphasis on doctrinal uniformity in the Jesuit order even since the 1550 Ignatian *Constitutiones*, it was only in 1597, during the Generalate of Claudio Aquaviva, that the policing of such uniformity, previously carried out at provincial level, was centralised in Rome and institutionalised in the form of a College of *Revisores*, based in the Collegio Romano.

The role of the *Revisores* was clarified in a set of rules[39] dating from 1601, and consisted in judging whether books written by members of the Society were fit for publication, based on a number of criteria including consonance with the Jesuit *Constitutiones*, Christian piety and the absence of anything that might bring the Society into disrepute. Books that did not concern serious issues need only be read in their entirety by two of the Revisores, and could be read in part by the remaining three, and the *censurae* could be produced as a result of a discussion. Books that impinged upon matters of theological and philosophical import were to be read completely by at least three Revisores and their *censurae* were to be sent to the General, who would decide on whether the book was to be published as it stood, amended prior to publication or left unpublished.

Muzio Vitelleschi,[40] General of the Society at the time of Galileo's trial, was responsible for a number of reforms to the system, including requiring the censorship of dedications and translations of previously approved works and, in 1621, prohibiting the publication of anonymous and pseudonymous publications.[41] He also ordered, in 1623, that all printed books should be sent to the library of the Collegio Romano, though this was not always observed.[42] These reforms took place as a reaction to the increasing problem of maintaining the doctrinal uniformity desired by Loyola throughout the global network of Jesuit colleges, a problem accentuated by the difficulty of communicating with colleges outside the Italian Province as fresh confrontations with *novatores* in philosophical and theological matters arose. In 1632 Vitelleschi wrote a letter to the Roman Provincial, in which he insisted on the importance of censorship in ensuring the 'common good and peace of the Society'.[43] Vitelleschi's interest in reinforcing the system is also displayed in his insistence on the production of lists of approved and prohibited philosophical propositions, a practice which eventually culminated in the 1651 *Ordinatio pro studiis superioribus*[44] prepared under the Generalate of Francesco Piccolomini, which prohibited a large number of propositions, including many related to corpuscularianism and astronomy, from being taught in Jesuit colleges.

From this brief sketch, it can be seen that Vitelleschi was extremely concerned with the practice of censorship. His opposition to deviations from Aristotelian cosmology is attested by the position which he adopted in the debate on celestial fluidity, which began within the Society during his early years as General.[45] Despite the general acceptance of celestial fluidity and the Tychonic world-system amongst the astronomers of the Jesuit order after the publication of Giuseppe Biancani's *Sphaera* in 1620 Vitelleschi maintained his opposition at least until the 1630s. A letter that he sent to the Rector of the Jesuit College of Avignon, Father Claude Bonyol, in 1631 stated in no uncertain terms that:

> I received with your letter a certain writing on the fluidity of the heavens, and, as you are in a position of authority over higher studies, it will be up to you to make sure that this opinion is not proposed or defended in any way in the theses of our pupils. If at some time that opinion, put forth by others, becomes common and receives the approval of the majority, then we will easily allow our members to follow what is seen to be more probable in that matter. At the moment, however, in accordance with the Constitutions, the teaching of Aristotle must be followed.[46]

Given his attachment to Aristotelian cosmology, it seems likely that Vitelleschi would have paid close attention to the *censurae* of Inchofer's book, especially in light of the events which had taken place less than two months previously.

Two of the three surviving *censurae* approve the *Tractatus Syllepticus* without reservation, and without significant deviations from the standard codes of approval.[47] Scheiner's *censura*, however, dated 9 August 1633, is of particular interest as after stating his approval of the book for publication and signing his name, he goes on to say:

> It appears that the author asserts too absolutely at the beginning of Page 34 that the motion of the Sun and the immobility of the Earth are matters of Faith, which should be modified, as they are in Question, and are not thought to be a true matter of Faith. Moreover he should also indicate briefly for what reason this might be a matter of Faith. I believe that similar considerations ought to be taken into account with respect to the circular motion of the Sun and the Centre of the Earth being in the middle of the Universe.[48]

This statement, made such a short time after Galileo's trial, seems to disqualify Scheiner from having denounced the *Dialogo* for doctrinal heterodoxy.

It simply wouldn't make sense for someone who had recently formulated a denunciation of Galileo's Copernicanism to criticise Inchofer for asserting absolutely that the motion of the sun and stability of the earth are matters of Faith, and to claim instead that they are 'in Question', rather than truly *de fide*. If Scheiner regarded the heliocentric view as formally heretical and incompatible with a matter of faith he would, like the other Revisores, have given unconditional approval to Inchofer's book.

The tentative tone of his criticisms suggests that the issue was not entirely clear cut to his peers, but his position echoes that of other Jesuit mathematicians of the time, notably Orazio Grassi, another traditional scape-goat for the events of 1632–1633, who had argued in his response to Galileo's *Il Saggiatore*, the *Ratio ponderum Librae et Simbellae*, that

> to tell the truth, what has not been granted for the opinion on the earth's motion, *although its immobility is not considered among the fundamental points of our Faith*, will be even less permissible, if I am not mistaken, for that which constitutes the essential point of faith or contains all other essential points.[49]

It seems possible that Scheiner's *censura* might have sparked off a correspondence with Muzio Vitelleschi on the very issue of the status of the Copernican theory, which might be the correspondence cited by Müller. Vitelleschi's *imprimatur* for the *Tractatus* is dated 18 August, just nine days after Scheiner's *censura*, and states that 'Three theologians of the Society [of Jesus], [...] reviewed it and approved for publication,'[50] thus making no mention of Scheiner's misgivings. The question of the jurisdiction of the Holy Office in deciding on matters of faith was still very much alive even in 1642, as attested by a letter from the Milanese Capuchin Valeriano Magni to Giovanni Barsotti of 3 August 1647:

> Of all this I gave the example that the case of Galileo, that is, the question whether the motion of the Sun and the immobility of the Earth asserted in Holy Writ is meant as real or apparent, has never been proposed, let alone defined, in the Church of God. When rumour had it that the Holy Office had declared Galileo's opinion to be heretical, I was wounded to the heart, because I feared that the heretics, not distinguishing the absolute authority of the Pope from the authority of the Holy Office, would presume to be able to oppose some geometrical demonstration to Papal authority, thought by the Catholics to be infallible. For this reason, in the month of August of 1642 I went to see the Master of the Holy Palace [...] who told me the qualification with which the proposition of the Earth's motion and the Sun's immobility was designated by the Holy Office. He showed me a printed book, which qualified that proposition as temerarious, which gave me great consolation[51]

Magni had little in common with his Jesuit contemporaries on other matters of natural philosophy, as discussed in the following chapter,[52] but on this issue it seems there was a degree of consonance between their positions.[53] The standpoint of General Vitelleschi on this issue is less clear, but it is worth remarking that Scheiner claimed in his letter to Kircher of 16 July that Vitelleschi was among those who had urged him to write a response to Galileo's *Dialogo*, his *Prodromus pro Sole Mobili*:

> After my *Prodromus* [Forerunner] against Galileo, entitled «*Christophori Scheiner e Soc. Iesu, Pro sole mobili, terra stabili, Prodromus, oppositus suo censori, terrae motori, solis statori*» which is now in the hands of the most Reverend Master of the Holy Palace, and once I have the latter's approval, I will shortly leave for Germany. I have been called there to his Holy Imperial Majesty (one could wonder what I earn for the Emperor with mathematics). When the *Prodromus* has been completed, God willing, I say that I will go to every trouble to defend common astronomy against Galileo, as I am urged by the Pope, our General, and his Assistants, all of whom have followed the better [path].'[54]

The idea that Galileo's *Dialogo* was nothing more than a personal attack on Scheiner, a response to Scheiner's anti-Copernican work on sunspots, *Rosa Ursina*,[55] may now seem absurd, but undoubtedly had a certain currency in the 1630s,[56] as we have seen above from Melchior Inchofer's report to the Holy Office. Indeed, Galileo did not mince his words in describing the 'vain and foolish ideas' of Scheiner in the *Dialogo*,[57] and Scheiner suggests in his letter to Kircher that he was perceived by many members of the Roman hierarchy, including Urban VIII himself, as the champion of 'common astronomy' against the Galileian threat.[58] By precipitating an external resolution to the debate, a denunciation of the *Dialogo* by Scheiner would have denied him the pleasure (for which he was willing to part with ten gold *scudi*) of refuting Galileo's book immediately with astronomical and physical arguments.[59] To silence his opponent in such a way would have rendered such an answer superfluous.[60]

Melchior Inchofer, on the other hand, was so fervent in his anti-Copernican zeal that he wished to follow up his *Tractatus Syllepticus* with another work, vindicating the Holy Office and Congregation of the Index against the *Terrae motores*, but his fifteen minutes of fame had passed by this time, and he only succeeded in provoking the anger of the Jesuit *Revisores* against a book with the portentous title of *Vindiciae Sedis Apostolicae, SS. Tribunalium et Auctoriatum adversus Neopythagoreos Terrae Motores et Solis statores*. This work was not considered fit for publication, as firstly its title was felt to suggest to the unwary reader that, rather than being the work of an individual it was written on behalf of the institutions which it defended, and secondly because the (mathematical) arguments of Inchofer were not felt to be worthy of this weighty and dangerous subject.[61] Inchofer tried to repeat the technique by which he had got his *Epistola B. Mariae Virg. ad Messanenses Veritas Vindicata* past the Index, by making a small change to the title, but the Jesuit *Revisores* were not impressed.[62]

The second remark made by Scheiner on Inchofer's book is also extremely relevant, and argues further for the view that, despite Inchofer's apparent reliance on Scheiner in his preparation of his report on the *Dialogo* the latter was more than a little uneasy about Inchofer's insistence on the authority of Scripture, interpreted literally, over any human enquiry. Scheiner suggests that Inchofer should temper, or qualify 'the passage in which he says that the Authority of the Book is greater than the capacity of any human mind'.[63] Despite the tentative nature of these criticisms, they strike at the very heart of Inchofer's project, and show that Scheiner had fundamental objections to this project, and is thus arguably eliminated from having formulated a denunciation of the *Dialogo* which would necessarily have invoked the kind of arguments from Scripture which are so numerous in Inchofer's book.

There are, however, many aspects of this episode which remain to be clarified. One aspect is the curious fate of Scheiner's *Prodromus*.[64] Although a manuscript version of the work was completed and in the hands of Father Niccolò Riccardi, Master of the Sacred Palace, by 16 July 1633,[65] the book had to wait until 1651, after Scheiner's death, for publication. This is even more curious in light of the fact that a printing of the book seems to have been carried out in 1642.[66]

Unfortunately no trace of a debate concerning the book's publication seems to remain in the *Censurae librorum* of the Jesuit archives.

In his 1633 *Apes Urbanae*, a 'Who's Who' of Roman *literati* under the Barberini pontificate,[67] Leone Allacci mentions the manuscript of the work, which he describes as 'A Forerunner for the stability of the earth against the same writer of Dialogues, in which Galileo's errors in logic, physics, mathematics, ethics, theology and in sacred matters are advantageously brought together so that everybody can see the mask pulled away from all of these to reveal doctrine constructed from ignorance'.[68] The final published version, despite containing personal slights on Galileo, generally avoids theological arguments in favour of astronomical ones. However, this detail, despite its convenience for a hypothesis of a removal of the theological parts of Scheiner's work prior to publication, should not be overestimated, as, firstly, in all likelihood, Allacci had not read the manuscript and, secondly, the unimpeded publication of Inchofer's *Tractatus Syllepticus* and Scheiner's criticisms of this work, argue against the suppression of Scheiner's work for reasons of theological content.[69]

A significant factor in the delayed appearance of the *Prodromus* may have been the changing attitudes of the successive Emperors Ferdinand II and Ferdinand III to Scheiner. According to Francesco Piccolomini, in a letter written from Presburg to Galileo on 5 February 1638, the new Emperor Ferdinand III had exclaimed to him two weeks previously that 'Father Scheiner has neither the knowledge nor the ability to write books against Galileo'. Moreover, according to Piccolomini, 'It seems to his Imperial Majesty that Scheiner's book is wasted paper and otiose scribblings without conclusion'.[70] The popularity which Scheiner enjoyed under Ferdinand II and his brothers Archdukes Karl and Leopold, displayed by handsome donations to aid him in the foundation of the college of Neisse, seems to have vanished with the accession of Ferdinand III to the Imperial throne in 1637. Shortly before the death of Ferdinand II, Scheiner left Vienna for Neisse, never to return to the Imperial court. The expensive mathematical instruments with which he had equipped the college of Neisse, mostly the gifts of Habsburg patrons,[71] were almost all destroyed or lost during his absence in Rome, and his lack of astronomical production at the end of his life suggests that they were not replaced.

The general desire among Jesuit mathematicians, including Christoph Scheiner, seems to have been one of trying to avoid collapsing the boundary which allowed the peaceful coexistence of revealed truth and astronomical research, under threat as much from the Galileian programme of scriptural exegesis informed by natural knowledge[72] as from the 1616 decree against

Copernicus.⁷³ This desire to maintain a space for the pious investigation of the natural world is manifest in Scheiner's criticism of Inchofer's assertion that solar motion and terrestrial stability are *de fide*. Such an assertion, coupled with Inchofer's style of argument from Scripture and the Church Fathers rather than from astronomical observations and syllogistic demonstration threatened to paralyse natural knowledge, and to place the mathematicians of the Jesuit order outside a global republic of astronomical practitioners that they had helped to create.

Throughout the early seventeenth century Jesuit mathematical practitioners and theologians negotiated together to construct a stance of epistemological modesty, or 'mathematical phenomenism',⁷⁴ which permitted the adoption of any hypothesis, provided that it was taken as a hypothesis, and that nothing more than probability was claimed for it. This is the conception of science which is advocated in the more famous part of Bellarmine's 1615 letter to Foscarini,⁷⁵ and is also to be found in a later remark of another Jesuit-turned-Cardinal, Marchese Sforza Pallavicino, who argues as follows in a letter to Monsignor Carlo Roberti written around 1665:

> As regards the system of the world, St. Thomas spoke better than anyone when he told us that Astronomers do not intend to prove that one or other astronomical system is true, but that the system does not conflict with the appearances that we see, as innumerable other systems could be found which would also not conflict with appearances. And which of these happens to be true is only known by those who are in heaven, and not by us, mere little worms, distant from heaven by many thousands of miles, and who change the system every day to agree with the new appearances that arise. Who, living in Genoa and knowing nothing of Corsica except whatever he saw from there with a telescope, would boast of being able to describe [Corsica] in detail? And yet such a boast would be far more modest, as the thing in question is so much nearer and much smaller.⁷⁶

The definitive list of corrections to Copernicus's *De Revolutionibus*, made by Francesco Ingoli, was presented to the Cardinals of the Congregation of the Index on 2 April 1618. The document, submitted for the approval of the mathematicians of the Collegio Romano including Christoph Grienberger and Orazio Grassi and used as the basis for the published decree of 1620,⁷⁷ proposed a *via media* by which astronomical systems and theological doctrine could peacefully coexist and by which apparent contradictions between them could be resolved. This document characterised the proper method of astronomy as the use of 'false and imaginary principles in order to save celestial appearances and phenomena', and added that 'it is customary for the science of Astronomy in particular to make use of false suppositions'.⁷⁸ This attitude allowed the more unpalatable passages of Copernicus to be construed simply as breaches of astronomical etiquette.

However after the 1633 trial, as a consequence of conflicting visions of the relationship between astronomical investigation and cosmological truth and Galileo's 'fall from grace', the *via media* proposed by Ingoli had been closed off, and the Jesuit vanguard of Catholic astronomy was prevented from invoking the key tenets of the Copernican theory even as hypotheses in published works.⁷⁹

Economies of truth

Christoph Scheiner cultivated close relationships with the most powerful members of the Habsburg dynasty including Archduke Karl, Archduke Leopold and Emperor Ferdinand II. He used his astronomical discoveries and instruments such as the pantograph,⁸⁰ which allowed an unskilled person to make accurate drawings from life, to secure his favour with Archduke Maximilian and Archduke Karl (who invited Scheiner to be his personal confessor), and to acquire lavish Habsburg funding from Karl and Ferdinand II for Jesuit projects such as the church in Innsbruck (designed by Scheiner) and the new College of Neisse.⁸¹

In 1624, just after Scheiner's arrival in Rome, Karl died in Madrid.⁸² Scheiner, who had already done some minor intelligence work for Leopold,⁸³ was thus led to rely ever more on the Archduke's favour in his attempts to secure his position on his return to Austria and to ensure the continuation of the architectural projects begun under his supervision in Neisse and Innsbruck. From Rome, he reported on the inefficiency of Leopold's recently deceased intelligence agent at the papal court, Michael Will.⁸⁴ Scheiner, previously reproached by Vitelleschi for involving himself in affairs that were inappropriate to his station,⁸⁵ took it upon himself to procure a new agent for Leopold. The chosen man, Abbot Camillo Cattaneo, was, according to Scheiner, Cameral secretary to Urban VIII, very devoted to the house of Austria and an admirer of the Jesuit order.⁸⁶ Previously, on Vitelleschi's personal recommendation, he had been the agent of Scheiner's deceased patron Archduke Karl.

When Scheiner was preparing to depart from Rome he sent most of his belongings ahead to Innsbruck in fourteen wooden boxes. The boxes were well stuffed with hay and straw, bound twice, and with a waxen fabric to protect them against 'any injuries from heaven and earth'.⁸⁷ A final covering was marked with a picture of a bottle [*signum Flasconis*] to warn the bearers of the fragility of the objects within. In numerous letters to Archduke Leopold, Scheiner emphasised the preciousness of the objects contained in these boxes, and the terrible dangers that might arise if they were opened before his arrival in Innsbruck.[88] The contents of these boxes, then, provide an important clue to the system of values within which Scheiner's career was embedded.

Scheiner explained to the Archduke[89] that about five of the boxes were intended for Leopold himself. The contents of three or four of the boxes were for the Emperor and the remaining boxes contained Scheiner's personal affairs. As Scheiner reveals in a later letter,[90] when he fears that some of the boxes may be opened before his arrival, their prized contents included a total of around ten 'bodies' – saintly relics procured by Scheiner 'not through the generosity of the Pope, but through [my] private endeavours'.[91] The boxes also contained a huge number of *Agni Dei* – oval wax tablets blessed by the pope on Holy Saturday and marked with the image of the Paschal lamb, the words '*Ecce Agnus Dei qui tollit peccata mundi*', the name of the current pope and the year of his pontificate. For Leopold alone there were twelve thousand of these wax tablets in Scheiner's boxes, probably intended to be distributed within the regions under his control.

As Pope Paul II had written in a Bull of 21 March 1470,[92] the *Agnus Dei* was invested with remarkable powers. It effaced sins, incited Christians to praise God, and protected the bearer against fire, shipwreck, hurricanes, lightning, hailstones and evil influences. Pregnant women who wore an *Agnus Dei* were guaranteed a safe delivery.[93] Given the supernatural powers of these consecrated tablets, it was not surprising that they were in such high demand that forgery became a serious problem in the fifteenth century. In fact it was to deal with this very problem that Paul II issued his bull, *Immoderata perversorum cupiditas* which gave details of the punishments that would be incurred by those who continued to manufacture or sell illicitly the *Agni Dei* which it was his unique privilege to distribute. Offenders would be excommunicated, incarcerated and, if clerical, stripped of all ecclesiastical benefits and offices.[94] Paul II also ordered that those who were currently in possession of forged *Agni Dei* should bring them, within eight days, to the bishop of Lesina, who would exchange them for an equal quantity of wax or equivalent compensation.[95]

The 'face value' of an *Agnus Dei* wax tablet clearly exceeded its 'intrinsic value' as a piece of wax by a great deal in the systems of exchange operating in the worlds inhabited by Paul II and Christoph Scheiner.

The ten 'bodies (*corpora*)' sent by Scheiner to Innsbruck were mostly the ashes of saints. No amount of inspection of such ashes could establish their intrinsic sanctity. Their value was established by the marks on their containers and, perhaps most importantly, by the words written by Christoph Scheiner in his frequent letters to Archduke Leopold from Rome.[96] After the rediscovery of the Roman catacombs in 1578, there was a dramatic growth in the distribution of early Christian relics through clerical channels, particularly to the German lands where Jesuits, Capuchins and Praemonstratensians used them as a powerful means of combatting heresy. The arrival of such relics in Austria and Southern Germany, sometimes accompanied by miracles, provoked pilgrimages to their new places of rest.

The system of values revealed by two of the main components of Scheiner's luggage is, then, crucially extrinsic – it is founded on chains of belief associated with inscriptions which connect the items to privileged sites and conditions of production. In this respect, as the words of Copernicus and Paul II suggest, it is structurally isomorphic with the system of values associated with both the standards laboratory – the standard metre bar in Paris for example[97] – and the mint. Bellarmine's distinction discussed earlier in this chapter can be usefully applied in this context – even if certain scriptural passages are not articles of faith <u>intrinsically</u> (*de fide ex parte obiecti*) they are <u>extrinsically</u> (*ex parte dicentis*) beyond doubt, as the credentials of the speaker – the Holy Spirit – are unchallengeable.

The exceptionally laudatory *imprimatur* of Galileo's *Assayer* written by Niccolò Riccardi, later Master of the Holy Palace during the 1633 trial, relates philosophical innovation to metrological analysis very explicitly. Deviating radically from the formalised norms of the genre, Riccardi concludes:

> I count myself lucky to have been born when the gold of truth is no longer weighed in bulk and with the steelyard, but is assayed with so fine a balance[98]

Galileo's text presses the analogy further:

> But since it seemed to me that [Grassi] used too crude a steelyard in his weighing of Sig. Guiducci's propositions, I have elected to employ an assayer's balance precise enough to detect less than the sixtieth part of one grain.[99]

In answering Orazio Grassi, Galileo is proposing to test the truth-content of Grassi's philosophical coin intrinsically, by assaying it with a fine balance. He is thus ignoring the claims of privilege – noble witnesses, Grassi's own high reputation, the authority of the Collegio Romano – that Grassi used widely in his *Libra Astronomica*[100] to support the foundation of his astronomical and philosophical claims. By questioning the skill, authority and honesty of Jesuits like Grassi and Scheiner, Galileo threatened to undermine public confidence in the value of their philosophical coin. The threat became especially acute when Galileo's chosen audience included powerful patrons of the Society, such as Archduke Leopold of Austria.

In 1618 Leopold, whose sister, Maria Magdalena, was married to the Grand Duke of Tuscany, had met Galileo in Florence. Galileo, despite a long period of illness, attempted to cultivate Leopold as a patron by sending him telescopes, a copy of the *Sunspot Letters* (written in answer to Scheiner's *Tres Epistolae de maculis solaribus* and *Accuratior Disquisitio*) and his unpublished treatise on the tides. In return for his gifts, Galileo sought the Archduke's opinion on his tidal proof of the Copernican theory, which he prudently described as 'a poem, or a dream' in view of the 1616 Decree of the Index.[101] He also hoped to use Leopold to increase his favour with the Grand Duchess in order to further secure his position in the Florentine court.[102] Mario Guiducci's 1619 *Discorso delle Comete* was dedicated to Leopold and Galileo, the real author of Guiducci's discourse, did not miss an opportunity to attack Scheiner caustically in one of the opening paragraphs of the work:

> May I be granted the ability to explain them to you vividly, for I esteem more highly the praise of having been a good imitator than I do that other kind which is usurped by those who have attempted to make themselves the inventors of views that are really [Galileo's], pretending themselves to be Apelleses,[103] when with poorly coloured and worse designed pictures they have aspired to be artists, though they could not compare in skill with even the most mediocre painters.[104]

Leopold's astronomer, Johannes Remus Quietanus, summarised the contents of the book for his master (prevented from reading it immediately by pressing affairs of state) and wrote to Galileo to describe the Archduke's reaction:

> He was very pleased by the work and was most grateful. He sent it immediately to Father Scheiner, who answered that he will pay your Lordship back in the very same coin.[105]

The attacks on Jesuit natural philosophy in the *Assayer*, which Galileo also sent to Leopold,[106] thus posed a clear threat to the patronage niche carefully cultivated by Jesuits such as Scheiner in the Empire, precisely the patronage niche attacked by Inchofer and Scotti in the *Monarchia solipsorum*.

In return for substantial financial contributions from Leopold to Jesuit activities, Christoph Scheiner provided him with an intelligence agent in Rome, holy relics, *Agni Dei* and astronomical expertise. The value of all of these commodities rested entirely on Scheiner's credibility. In his letters to Archduke

Leopold Scheiner repeatedly slips from Latin into a childish vernacular to assure Leopold of his loyalty, truthfulness and right conduct: 'Ich sey so unschuldig als wie ein Khind',[107] 'Ich bin halt der alte Scheiner, und einfaltige Schwaab',[108] 'Ich bin noch der alte Scheiner, Euer Hochfürstliche Durchlaut mag mir wol trawen'.[109]

The contrary projects of Galileo and Melchior Inchofer both implied that these claims, and the various types of information that they sought to validate, were worth decidedly less than the paper on which they were written.

From the Monarchia solipsorum (1645) to the Ordinatio pro studijs superioribus (1651)

In 1645, just after Vitelleschi's death, a vitriolic satire of the Jesuit order was published in Venice. This work, entitled *Monarchia solipsorum*[110] ('The Monarchy of the Solipsists'), was dedicated to Leone Allacci, and was generally assumed by contemporaries to be the work of Giulio Clemente Scotti,[111] previously a classmate of Sforza Pallavicino in the Collegio Romano. Evidence has come to light that points to Inchofer's involvement in the composition of this work.[112] On 3 January 1648 Inchofer's room in the German College was searched under the orders of General Vincenzo Carafa, and manuscripts were found linking him to Scotti. These included letters from Scotti and manuscript versions of works criticising the Jesuits.[113] Scotti, no longer a Jesuit, was beyond the disciplinary measures of the order but Inchofer had no such escape-route and was forcibly taken to Tivoli the following morning, where, after a long disciplinary process, he was condemned to a month's penance 'according to the rules of St. Ignatius'[114] in the monastery of the Holy Trinity. He died only a few months later. Inchofer's involvement in this episode may be seen as part of a more general desire to reform what he perceived as the corruptions of the Society of Jesus of his day. The principal departure from the Ignatian origins of the Society criticised by in the 1645 *Monarchia Solipsorum* and also in Inchofer's unpublished, *Historia Octavae Congregationis Generalis* was the monarchical system of government which conferred lifelong, absolute power on the General. 'The word Monarchy was either unknown to Ignatius or rejected by him', Inchofer insisted in his *Historia*.[115] Bad government, epitomised by Vitelleschi's tyrannical rule, also permitted exotic theological and philosophical opinions to spread unchecked amongst Jesuit professors.[116] By arguing against the concentration of power in a single individual, the General, the *Monarchia* attempted, as Giorgio Spini points out, to enable provincial Jesuit colleges to challenge the authority of the General if his position appeared to be in conflict with the Ignatian *Rules* and *Constitutions*.[117] The words of Ignatius in the foundational documents would thus acquire a degree of legislative authority that was explicitly rejected by Ignatius himself on many occasions in his advocation of a flexible, jurisprudential approach to the government of the order and '*moderatio*' in the interpretation of rules. Obedience to one's superiors, and accommodation to local traditions are generally emphasised above obedience to any textual authority in Ignatius's own writings.[118]

The criticisms of the order contained in the *Monarchia solipsorum* seem to have been brought before the Inquisition through the mediation of Leone Allacci, according to Inchofer's own unpublished history of the Ninth General Congregation, apparently one of the documents discovered when his rooms were searched.[119] The newly elected pope wrote a memorandum to the Fathers of the 8th General Congregation, emphasizing a number of points on which the Jesuit order urgently required reform. Innocent X insisted on the reform of the absolute power of the General, which had come in for particular criticism in the *Monarchia*:

> If it is not decided that the perpetuity of the Generalate should be abolished, then a way should be sought to moderate the absolute authority of the General. General Congregations should also take place every eight years without fail, and it should not be possible for them to be prevented or postponed either by the General or by the entire Society, as has occured in the past.[120]

Echoing Inchofer/Scotti's criticisms of theological and philosophical exoticism in the Order, the Pope also required that

> No doctrine should be taught or professed other than that of St. Thomas, or those commonly accepted by the Church Fathers[121]

An examination of the proceedings of the Eighth General Congregation reveals that the Pope's memorandum set the agenda. As soon as Carafa had been elected, six deputations were set up to deal with the different problems pointed out by Innocent X. The fifth deputation was destined for 'the promotion of the study of letters, the examination and revision of the books of the *Ratio Studiorum*, especially the parts relating to the selection of opinions, so that any excessive license or novelty of opinions is curbed, especially as these points are most strongly recommended now by the Pope.'[122] It was also decided to divide the deputation for studies into two parts, a first for *literae humaniores* and a second for theology and philosophy.[123] The link between the doctrinaire Thomism advocated by Innocent X and a return to doctrinaire Aristotelianism in philosophy, including natural philosophy or *physica*, was emphasised by a lengthy document on the reform of studies composed shortly after the Congregation by the Prefect of Studies of the *Collegio Romano*, Leone Santi.[124]

> Scholastic theology signifies none other than that which supposes Aristotelian philosophy. If, therefore, our authors commonly depart from Aristotle, they are not transmitting not scholastic theology, but, as some would say, fantastic theology, for each individual forges his own with great confusion and perturbation to the Church. But how much less can someone defend and explain the theology of Saint Thomas in his theological conclusions [...] if in his philosophy he departs from the principles of Aristotle and the entire Peripatetic school? For unless minds are contained within certain limits their excursions into exotic and new doctrines will then be infinite, as will their ways of talking, with constant danger lest we should be brought before the Holy Tribunal of the Inquisition.[125]

When this document was composed Santi was waging a war with the unorthodox theological and philosophical opinions of Sforza Pallavicino within the Collegio Romano itself.[126] The Ninth General Congregation in 1649 ensured that the *Revisores* of the Collegio Romano would play a central role in establishing the reponse to the problem of doctrinal indiscipline. Even before the Eighth General Congregation, in 1645, the *Revisores* could criticise a series of philosophical 'paradoxes', possibly due to Honoré Fabri in Lyon, with the words

> We judge that these theses should in no way have been permitted to be defended, much less to be printed, because their author is seen to pursue novelties studiously and to abhor common opinion. The doctrine and method of Aristotle are not followed and in the name of paradoxes many things are inverted.[127]

Clearly in this case the response of the *Revisores* came too late – the conclusions had been defended publicly and printed in 1643. The response to the practical problems of centralised policing of opinions and printed works was to compose and distribute lists of prohibited propositions. The *Revisores* had been ammassing dangerous philosophical and theological propositions from the Provinces with increasing vigour under Vitelleschi and Carafa. The *Revisores* submitted to the Ninth Congregation that Carafa had embarked on a very useful practice in insisting that both philosophical and theological propositions that were suspected of exoticism should be sent from the provinces to Rome in order to be judged by them.[128] The deputation for studies submitted a list of propositions not to be taught in Jesuit colleges.[129] The secretary of the Congregation, Pierre Cazré, himself a mathematical practitioner,[130] objected immediately to the suggested prohibition of the Cartesian proposition that matter was indistinguishable from quantity that

'The 23rd proposition is most common, and in many places more common than the other, and although it has already been said to be prohibited, the prohibition has not been effective, and if this is done again, I hardly think that [the prohibition] will be observed in the future'.[131] In spite of Cazré's objections, the list of prohibited propositions were sent unaltered to the *Revisores* of the *Romano*[132] and were published unchanged in 1651 as the *Ordinatio pro studiis superioribus*. Even after the publication of the *Ordinatio*, objections continued to be voiced. Roderigo de Arriaga postulated to the Tenth General Congregation that

> [A]s there are great quarrels everywhere concerning the number and quality of the opinions rejected by Rev. Fr. Piccolomini of good memory, the Congregation is asked if it would be worth asking many Universities to send their judgements on those opinions to our Rev. Fr. General, so that once these are considered he would judge which must still remain prohibited and which not. This is especially so because some sentences are rejected which are defended by most approved authors of the Society and of other orders, which do not concern the faith or good morals, and cannot cause any offense, even by their appearance. It appears to me that our faculty of discoursing about these [opinions] should not be removed in any way.[133]

Arriaga's objections went unheeded and the 1651 *Ordinatio pro studiis superioribus* retained its hold over the two public theatres of Jesuit erudition – print and classroom. Although principally directed against the public teaching of novel doctrines or 'discussing useless subtleties' in Jesuit colleges, the terms of the *Ordinatio* made it clear that it was also directed to printed works:

> For this reason this catalogue has been communicated to all of our professors of Philosophy and Theology, as well as the Prefects of the higher faculties, so that they can ensure that opinions of this kind are not propagated either in theses or in disputations, and that they are not taught in our schools. It has also been sent to all of the *Revisores*, so that if it happens that they are found inserted in books which they are examining, they will not permit them to be published.[134]

The agents of this reform of Jesuit teaching would be the Rectors of each college, the Provincials carrying out their routine *visitationes* to the colleges in their province, and the local Prefect of Studies. A college that was especially sound in doctrine would serve as a model for the other colleges in its province. In the Collegio Romano, 'totius Societatis facile primo', the prohibitions would be observed most carefully.[135]

The *Ordinatio* had significant effects on the way the Jesuit 'physico-mathematician' could manifest himself in print. In criticising works such as Athanasius Kircher's

Itinerarium Extaticum[136] or Francesco Maria Grimaldi's *Physico-mathesis de lumine*,[137] the Jesuit *Revisores* could now refer to the propositions of the *Ordinatio* directly, as grounds for refusing publication. Reactions to this increased disciplinary vigour were varied. Honoré Fabri attempted to avoid the scrutiny of the *Revisores* altogether by publishing many of his works under the name of his student Pierre Mousnier.[138] Orazio Grassi hinted that he planned to destroy a work in a letter sent to Baliani in 1652 that 'I see that I will not be able to publish my study of colours because of the rigorous orders made [...] in these last General Congregations, in which ours are forbidden to teach many opinions, some of which are the substance of my treatise, and they claim to prohibit them not because they consider them to be bad or false, but because they are new and not ordinary. It will thus be necessary for me to sacrifice them to Holy Obedience, by which I will undoubtedly gain more than I would be publishing them'.[139] As late as 1674 Fabri felt it necessary to demonstrate that he had never defended the propositions prohibited by the *Ordinatio* in his third *Epistola* to Ignatius Pardies, to which he appended a list of Aristotelian propositions to which he subscribed.[140]

Clearly, Piccolomini's *Ordinatio* was particularly intended to eradicate the exoticism in theology and philosophy ridiculed by the *Monarchy of the Solipsists* and frowned upon by Innocent X. As well as adopting pseudonyms, destroying manuscripts or adopting the 'doctrinal duplicity' manifested in Grimaldi's *De Lumine*,[141] a significant number of Jesuit authors thus responded to the new rigour by presenting their work as technique rather than physics – *ars* rather than *philosophia naturalis*. Thus, perhaps somewhat paradoxically, the period of enforced disciplining of exoticism in natural philosophy within the Jesuit order coincided with an explosion of works composed by Jesuit mathematicians dealing with artificial magic, exemplified by the works of Athanasius Kircher, Kaspar Schott and Francesco Lana Terzi. The new, magical space in which many Jesuit mathematicians carried out their work after the 1640s nonetheless compromised the authority with which they could make pronouncements about the operation of causes in the natural world.

Notes

1 Parts of this chapter were included in Michael John Gorman, 'A Matter of Faith? Christoph Scheiner, Jesuit censorship and the Trial of Galileo', *Perspectives on Science*. 1996; 4(3): 283–320. I am grateful to Mordechai Feingold and The MIT Press for permission to reproduce here.

2   Vitelleschi to Guldin, Rome, 12 July 1631, ARSI Austr. 4¹, pp.484: 'cum horologio rotata inter instrumenta mathematica compari non soleant, et si horologio aliquando opus fuerit dubiter cum eidem Domino Praefecto facendum erit, ipsi proculdubio non sit defuturus quo instruere. Utatur Qoare RV. precor non moleste fera sibi non concedi quod Societas religiosa paupertati consuit minime convenire', Vitelleschi to Henricus Philippus, ARSI Austr. 4¹, pp. 484–85: 'Usum horologi rotati R.V. cupidissime concederem, nisi viderem me eo concesso rogendum ad idem pluribus, qui non minus illo quas R.V. reputant indigere, concedendum. Quare ne suo exemplo permittere cogar pluribuas talis machinulae usum quam Societas puritati religiosae paupertatis adversari censuit, rogo ut pro sua religiosa paupertatis amore, nolit se alijs exemplum fieri quo liberus peta ut se lege illa paupertatis solui'.

3   Nicholas Fabri de Peiresc to Athanasius Kircher, Aix, 17 August 1633, APUG 568, ff.198r–199v, on f. 198r-v.

4   On the sunspots polemic, see especially Feldhay, *Galileo and the Church*, cit., pp. 256–291, A. Favaro, *Oppositori di Galileo, III. Cristoforo Scheiner*. Venezia, 1919 and W. R. Shea, *Galileo, Scheiner and the Interpretation of sunspots*, Isis, 61; 1970: 498–519,

5   On the relations between Peiresc and Galileo see A. Favaro, *Amici e Corrispondenti di Galileo*, ed. Paolo Galluzzi, Firenze: Libreria Editrice Salimbeni; 1983, pp. 1535–1582 (*s.v.* Niccolò Fabri di Peiresc).

6   Peiresc wrote to Gassendi that 'M. Naudé m'escript que le P. Scheyner escrivoit dez lors ex professo contre le pauvre Galilée, qu'il travailloit puissamment et avec grandissime animosité, à ce qu'on leur mandoit de Rome; dont les effects n'ont que trop paru à mon grand regret et peult estre au dezadvantage des arts liberaulx', Peiresc to Pierre Gassendi, 25 June 1633, OG XV pp.164–165.

7   OG XV pp. 87–88. See also Pietro Redondi, *Galileo Heretic*, transl. Raymond Rosenthal, Princeton: Princeton University Press; 1987 (orig. publ. 1983), p. 34.

8   Scheiner to Kircher, Rome; 16 July 1633; in Kircher to Peiresc, Avignon, 9th August 1633, BN Fonds Français, no. 9538, f. 227r. See OG XV p.184 for an excerpt from Scheiner's letter. Peiresc's interpretation of this letter is clear from his letter to Gassendi of 12 August 1633: 'Vous aurez aussy une lettre que m'a escripte le bon P. Athanaze Kircher, où il en a transcrit une aultre par lui reçeüe du P. Scheiner de Rome, ou vous serez bien aise de voir à quel poinct monte l'estime qu'il faict de vous, mais bien mortifié aussy de voir ce qu'il y dict du pauvre Sʳ Galilée, que je plains grandement; ce que je seroys bien d'advis de ne pas divulguer, si vous m'en croyez, pour bons respects, puisque la chose avoit esté tenüe dans Rome si secrette jusques à present. Si cela se doibt publier, il vauldra mieux qu'il vienne d'aultre main que de la nostre' OG XV p. 219

9   'Je me suis laissé dire que les Jésuites avoient aidé à la condamnation de Galilée', Descartes, *Oeuvres*, Correspondance I, pp. 281–282. On Descartes relationship with

the Jesuits see also Roger Ariew, 'Descartes and the Jesuits: Doubt, Novelty and the Eucharist', in Feingold, *Jesuit Science and the Republic of Letters*, pp. 157–194.

10 Galileo to Elia Diodati, Florence, 25 July 1634, OG XVI, p.117

11 On Scheiner see the Lamalle biographical card in ARSI, the works cited in note 5 above, Franz Daxecker, *Briefe des Naturwissenchaftlers Christoph Scheiner SJ an Erzherzog Leopold V von Österreich Tirol 1620–1632*, Innsbruck: Publikationsstelle der Universität Innsbruck, 1995, A. von Braunmühl, *Christoph Scheiner als Mathematiker, Physiker und Astronom*, Bamberg, 1891, A. Müller, *Der Galileo-Prozess (1632-1633) nach Ursprung, Verlauf und Folgen dargestellt*. Freiburg im Breisgau: Herdersche Verlagshandlung; 1909, Corrado Dollo, *Tanquam nodi in tabula- tanquam pisces in aqua. Le Innovazioni della cosmologia nella Rosa Ursina di Christoph Scheiner*, in *Christoph Clavius e l'attività scientifica dei Gesuiti nell'età di Galileo*, edited by Ugo Baldini. Rome: Bulzoni; 1995, pp. 133–158 and Steve J. Harris, *Les chaires de mathématiques*, in L. Giard, ed., *Les jésuites à la Renaissance. Système educatif et production du savoir*, Paris: Presses Universitaires de France; 1995, pp. 251–261.

12 Benedetto Castelli to Galileo, 19 June 1632, in OG XIV p. 360 'Il Padre Scheiner, ritrovandosi in una libraria dove un tal padre Olivetano [Vincenzo Renieri] venuto di Siena a' giorni passati, si ritrovava; e sentendo che il Padre Olivetano dava le meritate lodi ai Dialoghi, celebrandoli per il maggior libro che fusse mai uscito in luce, si commosse tutto con mutatione di colore in viso, e con un tremore grandissimo nella vista et nelle mani, in modi che il libraio, quale mi ha raccontata l'istoria, restò maravigliato: e mi disse di più che il detto Padre Scheiner haveva detto, che havrebbe pagato un di quei libri dieci scudi d'oro per poter rispondere subbito subbito'

13 Scheiner to Gassendi, Rome; 23 February 1633; BN Fonds Français, no. 9531 c. 201r (cited in OG XV p. 47): 'Ego pro me et veritate defensionem paro'.

14 In this report, Inchofer argued that Galileo's principle aim in writing the Dialogo was to attack Scheiner's *Rosa Ursina*, which was anti-Copernican, thus rendering Galileo's work Copernican. See Favaro, *Cristoforo Scheiner*, cit., pp. 93–94, and W. R. Shea, *Melchior Inchofer's 'Tractatus Syllepticus': A Consultor of the Holy Office answers Galileo*, in P. Galluzzi, ed., *Novità Celesti e Crisi del Sapere*, Florence: Giunti Barbèra; 1983, pp. 283–92. Inchofer's damning report is published in Maurice A. Finocchiaro, *The Galileo Affair*, cit., pp. 262–270. Pio Paschini says on the subject in his biography of Galileo: 'Del padre Inchofer, del quale conosciamo il voto nel processo del 1633, non si sa altro in riguardo al Galilei. Egli incidentalmente fa anche il nome del padre Scheiner; ma se questo abbia esercitato qualche influsso, almeno nel campo dottrinale nel'esito del processo, non siamo in grado di stabilire', Pio Paschini, *Vita e Opere di Galileo Galilei*, Rome: Herder, 1965, p. 587

15 The precise date of Scheiner's departure from Rome for Vienna is difficult to establish. Favaro 1919 argues from Raffaello Magiotti's letter to Galileo from Rome

of 14 October 1633 that Scheiner was still in Rome at this date, preparing his
Prodromus (OG XV pp. 300-301). The letter from Kircher to Peiresc cited below
confirms that he was still in Rome on 1 December. On 3 January 1634 Wilhelm
Weilhamer wrote from Parma to Giannantonio Rocca to say that 'Expecto in dies
P. Scheinerum, qui etiam forte apud vos transiturus est' (*Lettere d'uomini illustri del
secolo XVII a Giannantonio Rocca*, Modena: Società Tipografica, 1785 pp. 7-9),
suggesting that Scheiner left Rome at the end of December or in the first days of
January.

16 For a restatement of the possibility of Scheiner's authorship of a denunciation of
Galileo see Mario D'Addio, *Il caso Galilei: Processo/Scienza/Verità*, Roma: Edizioni
Studium; 1993, p. 147 note 52: 'Le testimonianze conservateci dall'epistolario
galileiano concordano nell'individuare il padre Scheiner come il promotore delle
iniziative intese a sottoporre il Dialogo e il suo autore al giudizio del S. Ufficio'. For a
similar position see F. V. Ferrone and M. Firpo, *Galileo tra inquisitori e microstorici*,
Rivista Storica Italiana, 97, 1985, pp. 177-238, 957-68, on pp. 511-513.

17 Kircher to Peiresc, Rome, 1 December 1633, BN Fonds Français 9538, ff.234r-v.

18 For bibliographical information on Kircher see especially John Fletcher, ed.,
Athanasius Kircher und seine Beziehungen zum gelehrten Europa seiner Zeit,
Wiesbaden: Harrassowitz, 1988 and M. Casciato, M. G. Ianniello and M. Vitale, eds.,
*Enciclopedismo in Roma barocca: Athanasius Kircher e il museo del Collegio Romano
tra Wunderkammer e museo scientifico*. Venice: Marsilio, 1986. For biographical
details see especially Carl Brischar, *Athanasius Kircher: Ein Lebensbild*, Würzburg:
Katholische Studien, Jahrgang 3, Heft 5, 1877 and Conor Reilly, *Athanasius Kircher:
A Master of a Hundred Arts, 1602-1680*, Wiesbaden: Edizioni del Mondo; 1974.

19 In fact, it was Kircher who was originally invited to the Imperial court, shortly after
Kepler's death in 1631. However, Pereisc petitioned Urban VIII and Cardinal
Francesco Barberini to have Kircher transferred to Rome, to continue his work on
hieroglyphics, instead of devoting himself to less important mathematical matters.
The petition was initially unsuccessful, so Peiresc was delighted to hear from Kircher
that Scheiner too had been invited to the Imperial court and expressed the hope in a
letter to Kircher that this might bode well for Kircher's future. According to Kircher's
autobiography, he set off for Vienna nonetheless and, after a number of adventures at
sea, his ship from Genoa was blown off course and landed at Civita Vecchia instead
of Livorno. He then, we are told, made his way to Rome on foot, only to learn on his
arrival (erroneously dated as 1634) that General Vitelleschi had obtained an order
from Urban VIII, calling him to Rome, and had been sending letters to him while
he was at sea. This fortuitous episode ('ex hoc capite Divinam Providentiam satis
mirari non potuerim'), described in Kircher's autobiography (*Vita*, Augsburg:
S. Utzschneider; 1684, pp.41-54) is flatly contradicted by further letters from Peiresc
sent to Cassiano dal Pozzo and Claude Saumaise during the Autumn and Winter of

1633. These make it clear that he fully planned a stay in Rome even before leaving France. e.g. Pereisc to Saumaise, 14 November 1633: 'Ce jesuite eut commandement de s'en aller à Vienne en Austriche et passa par icy, ayant prins sa routte du costé de Rome, où j'estime qu'il soit encores à présent, ayant eu de ses lettres de Genes, et m'ayant fort solennellement promis de m'escripre de Rome et de Vienne' (Peiresc, *Lettres à Cassiano dal Pozzo (1626–1637)*, ed. J.-F. Lhote and D. Joyal, Clermont-Ferrand: Adosa, 1992, pp. 27–50) and Peiresc to Cassiano dal Pozzo, 10th September 1633 (ibid. pp. 111–112). See also ARSI Lugd. 14, f.263. Kircher subsequently travelled to Malta, in the company of the famous convert, Landgrave Ernst of Hessen-Darmstadt (ARSI Ital. 6, ff.366, 373) and Sicily (ARSI Sic. 156, f.53v), before returning to Rome in 1639 (ARSI Rom. 57, f.153, n.14) and taking up the position of mathematics professor in the Collegio Romano, where he remained for the rest of his life.

20 'Ego quoque Illustrissimae V[estr]ae Dom[ination]i salutem humilli[ssi]ma et officiosa servitia offero. Christophorus Scheiner', Kircher to Peiresc, Rome, 1 December 1633, cit.

21 'Ego spero me tandem etiam in Germaniam rediturum, ad quam sensim aspero ... Die Teütsche Sprach hab ich noch nit vergessen, dan ich khan die Welsche nit. No io niente parlare Italiano', Scheiner to Archduke Leopold of Austria, Rome, 8 May 1627, in Franz Daxecker, ed., *Briefe des Naturwissenschaftlers Christoph Scheiner SJ an Erzherzog Leopold V von Osterreich Tirol 1620–1632,* Innsbruck: Publikationsstelle der Universität Innsbruck, 1995, p. 127.

22 'Consarcinaui iam pleraque in Germaniam euehenda', Scheiner to Archduke Leopold of Austria, Rome, 2 June 1628, in Daxecker, *Briefe des Naturwissenchaftlers Christoph Scheiner SJ*, cit., p. 137.

23 'Aus all den veröffentlichten Dokumenten und selbst der uns zugänglichen noch unveröffentlichen vertraulichen Korrispondenz P. Scheiners mit seinem Ordensgeneral ergibt sich auch nicht der leiseste Anhaltspunkt für irgend eine tatsächliche Beteilung Scheiners am Galilei-Prozess', A. Müller, *Der Galileo-Prozess,* cit., p. 131. Admittedly Müller might merely have been making a negative claim as to the lack of evidence for Scheiner's part in the trial, but this will not be fully clear until Scheiner's correspondence with Vitelleschi is published in its entirety.

24 In fact, from 1628 they were both living in the same building – the *Domus Professa* beside the Church of the Gesù (previously Scheiner lived in the *Collegio Romano*). See Daxecker, *Briefe*, p. 138.

25 E.g. Pietro Redondi, *Galileo Eretico: Anatema*, Rivista Storica Italiana, XCVII (1); 1985: 934–56, on p. 949 'Sono state dichiarate altre prove scagionanti nella «corrispondenza inedita e confidenziale di Scheiner con il padre generale del suo ordine, accessibile ai padri» [...] Rivolgo qui, pubblicamente, un appello al padre Edmond Lamalle e ai suoi collaboratori dell'Archivio storico della Compagnia di

Gesù perché sia fatta luce su questa segnalazione che né Favaro né altri studiosi hanno potuto accertare'.

26 Some letters from Vitelleschi to Scheiner before Scheiner's arrival in Rome can be found in ARSI Austria 3¹ pp. 171, 181, 184, 207, 213, 242, 246, 248, 270, 287, 312, 340, 411, 425. Letters sent to Scheiner after his departure from Rome are in ARSI Austria 5¹ pp. 18, 38, 53, 65, 74, 106. The only exchange of letters between Scheiner and Vitelleschi during Scheiner's time in Rome that I have been able to find took place in 1626, when Scheiner was still in the Collegio Romano (see Daxecker, *Briefe*, pp. 171–2).

27 Although, as Ugo Baldini points out, 'Le censure erano scritte in forma che potrebbe dirsi epistolare, indirizzate al Generale e consegnate direttamente a lui o al suo segretario privato' Baldini, *Legem impone subactis*, cit., p.89

28 Melchio Inchofer, *Tractatus Syllepticus,* Rome: Lodovico Grignani, 1633. A manuscript version of this work is in the Biblioteca Casanatense in Rome (Ms. 1331 ff. 147r–213r)

29 Melchior Inchofer was born to a noble Lutheran family in Kőszeg, Hungary (not Vienna as is sometimes supposed) around 1585. From 1605 he studied in the German College in Rome, before entering the Roman Jesuit Novitiate of Sant'Andrea al Quirinale on 26 March 1607 (ARSI Rom. 172, f. 108 no. 608). He seems to have spent the following years studying in the *Collegio Romano* (ARSI Rom. 110 f. 44, ARSI Rom. 54 f. 256v, n.56). During the years 1617–1629 he lived in Messina, where he taught metaphysics, physics, theology and mathematics in the Jesuit college (ARSI Sic. 155, f.38, f.45v), and professed the four vows on 4th June 1623. His book *Epistola B. Mariae Virg. ad Messanenses Veritas Vindicata*, first published in Messina in 1629, was prohibited *donec corrigatur* by the Congregation of the Index. Inchofer went to Rome in 1630 to negotiate with the Cardinals of the Congregation (ARSI FG 675 ff.213–215v and 220–222v), and managed to publish a second edition of the book, which authenticated a letter reputed to have been written to the people of Messina by the Virgin Mary by making some very minor changes, the most important of which was a substitution of *Conjectatio* for *Veritas Vindicata* in the title. This was the reason that Inchofer happened to be in Rome in 1632 when Riccardi, Master of the Holy Palace, was setting up a committee to evaluate Galileo's Dialogo. Inchofer died in Milan on 28th September 1648 (ARSI Hist. Soc. 47 f. 50v). For biographical information, see the Lamalle biographical *scheda* in ARSI, Dezsö Dümmerth, *Les combats et la tragédie du Père Melchior Inchofer S. J. à Rome (1641-48)*, Annales Universitatis scientiarum Budapestinensis. Sectio Historica, 17: 81–112, 1976, Shea, W. R. Shea, *Melchior Inchofer's 'Tractatus Syllepticus'*, cit., Oudin's biography in R. P. Nicéron, *Mémoires pour servir à l'histoire des Hommes Illustres dans la République des lettes,* Paris: Briasson; 1736, t. 35 pp. 322–346 and Sommervogel *s. v.*. Inchofer's dealings with the Cardinals of the Congregation of the

Index do not seem to have ended with Galileo's condemnation. A 1680 inventory of the Jesuit archives (ARSI Miscel. 8) gives details of one compartment (Armarius EE. Capsula n° XXX) entirely devoted to the Holy Congregation of the Index. This apparently contained an *Epistola P. Melchioris Inchofer de libris prohibitis*, 1642. (The inventory also gives details of a large collection of manuscripts pertaining to the Sacred Congregation of the Holy Office (EE Capsula no. XXXII), but it is difficult to ascertain where these documents might be found in the present archives, if indeed they have even survived). Furthermore, Inchofer signs his *imprimatur* for Athanasius Kircher's *Ars Magna Lucis et Umbrae*, dated Rome, 21 December 1644, as Melchior Inchofer S.C.I.C. which must surely stand for *Sacrae Congregationis Indicis Consultor*.

30 Finocchiaro, *The Galileo Affair*, cit., pp. 262–270

31 The work seems to have been begun before 13 February 1633, the date of the *imprimatur* of Allacci's *Apes Urbanae*, which mentions a work by Inchofer with the title of *An sit de fide terram esse immobilem, ubi affirmativa multis ostenditur. Tractatus* as forthcoming. This is almost certainly the *Tractatus Syllepticus*, and the original title, which emphasises the stylistic similarities between Inchofer's work and an article of Aquinas's *Summa Theologiae*, acquires special relevance in light of Scheiner's censure, discussed below. Coincidentally 13th February was also the date on which Galileo arrived in Rome to await trial. The reports on Galileo's *Dialogo* by Melchior Inchofer, Zaccaria Pasqualigo and Agostino Oriego were presented to the Holy Office on 17 April 1633.

32 For an account of the context of publication and content of Inchofer's book see Shea, *Melchior Inchofer's 'Tractatus Syllepticus'*, cit. For background information on Inchofer and the Jesuit College in Messina, see Rosario Moscheo, *Melchior Inchofer (1585-1648) ed un suo inedito corso messinese di logica dell'anno 1617*, in Quaderni dell'Istituto Galvano della Volpe, 3: 1982: 181–94.

33 See Shea, *Melchior Inchofer's 'Tractatus Syllepticus'*, p. 290

34 Baldini suggests that Inchofer's *Tractatus* was directly influenced by Bellarmine's approach to biblical exegesis. See Baldini, *Legem impone subactis*, p. 297

35 The second decree of the fourth session (8 April 1546) of the Council of Trent allowed for a certain amount of ambiguity, exploited by the Galileians, by stating that 'nemo [...] *in rebus fidei et morum*, [...] sacram scripturam ad suos sensus contorquens, contra eum sensum, quem tenuit et tenet sancta mater ecclesia, [...] aut etiam contra unanimem consensum patrum ipsam scripturam sacram interpretari audeat, etiamsi huiusmodi interpretationes nullo unquam tempore in lucem edendae forent' (S. Ehses, ed., *Concilii Tridentini Actorum pars altera*, Freiburg im Breisgau, 1911, p. 92, emphasis added). Bellarmino's position, as Baldini points out, nonetheless insists that none of Scripture is external to *rebus fidei et morum*: '[I]n Scriptura non solum sententiae, sed etiam verba omnia, et singula ad fidem

pertinent. Credimus enim nullum esse verbum in Scriptura frustra, aut non recte positum', *Roberti Cardinalis Bellarmini opera omnia*, Naples, 1856, Tom. 1. II, cap. XII. See Baldini, *Legem impone subactis,* p. 338 and Blackwell, *Galileo, Bellarmine, and the Bible*, cit. .

36 Robert Bellarmine to Paolo Antonio Foscarini, Roma, 12 April 1615, in OG XII, pp. 171-172. On Bellarmine's famous letter see Blackwell, *Galileo, Bellarmine, and the Bible*, cit., pp. 103-108.

37 ARSI Fondo Gesuitico 661 f.194r-196r

38 In the following sketch of the practice of censorship in the Jesuit order I am greatly indebted to the description given by Ugo Baldini in Baldini, *Legem impone subactis*, pp. 75-119. On the issue of *uniformitas doctrinæ*, see also Anita Mancia, *Il concetto di 'dottrina' fra gli esercizi spirituali (1539) e la Ratio Studiorum (1599)*, Archivum Historicum Societatis Iesu, LXI, 1992: 3-68.

39 ARSI Instit. 46, f.61r-v. These are published in Baldini, *Legem impone subactis*, p. 85

40 Vitelleschi, elected General of the Society in 1615, was born in Rome on 2nd December 1563 and died there on 9th February 1645, making him a close contemporary of Galileo. Before he began to be involved in the administration of the Society, he taught philosophy and theology in the *Collegio Romano*. At least three of his courses in natural philosophy from this period have survived, *Lectiones R.P. Mutii Vitelleschi in octo libros physicorum et quatuor de coelo, Romae, Annis 1589 et 90. In Collegio Romano Societatis Iesu,*. 4° ff. 389 (in the library of Bamberg), *Commentarius in libros de coelo et mundo*, APUG Fondo Curia 392. and his *In libros meteorologicorum, 1590,* BNR FG 747, 2876. A theology course, *De actibus humanis: In Prima 2ª D. Thoma 1602*, is conserved in ARSI Opp. NN 5. Documents pertaining to Vitelleschi's life, the controversy surrounding his election to the position of General and his death are in ARSI Vitae 127. See also Sommervogel *s.v.*.

41 ARSI Congr. XXI, f. 224r, cit. in Baldini, *Legem impone subactis,* pp. 75-119. The practice continued unhindered despite this prohibition, even in Vitelleschi's full knowledge, as attested by a letter he sent to the Rector of the College of Paris in 1627 concerning the *imprimatur* of Orazio Grassi's pseudonymous *Ratio Ponderum*: 'Nunc aliud quoddam est quod R.V. admonitum cupio. P. Grassi librum fecit sub aliene nomine de Ratione ponderum, isque liber nuper vulgatis est opera Domini Cramoisy Bibliopolae Parisiensis, qui etiam aliquot exemplaria ad Autorem misit. Verum animadversum est nullam in eo positam fuisse de more approbationem, quam ob causam hic videri non posset, quin subito a Magistro Sac. palat. supprimeritur', Vitelleschi to Ignace Armand, Rome, 23 March 1627, ARSI Franc. 4, f. 247v.

42 ARSI, Rom. 3¹, f.85r, cit. in Baldini, *Legem impone subactis,* pp. 75-119

43 This letter, subsequently distributed to the other provinces, gives a very clear idea of Vitelleschi's conception of the role of the *Revisores*: 'Sa molto bene V.R. quanti,

e gravi, travagli in diversi Regni, e Provincie ci sono per l'indietro avvenuti con occasione de' libri stampati da alcuni della Comp[agni]a, et ancora non siamo sicuri, che non ne sopravengano de' nuove. Questo negotio mi è stato sempre grandemente a cuore, e sono stato di continuo avvertito, perche non uscisse cosa alcuna, che potesse offendere, e pareva che si fusse e con regole, e con avvisi particolari provisto a bastanza: ma perche veggo, che non ha conseguito affatto l'intento, per soddisfare al mio debito, ho giudicato necessario di mandare a V.R. alcuni punti, che dovranno essattamente osservare tanto gli Autori, come li Revisori per il ben commune, e pace della Compagnia.

1. Che li Revisori nelle Provincie si deputino dal Generale, come si nominano gl'essaminatori per la professione, e siano obligati a leggere attentamente tutto quello, che sara loro dato a rivedere.

2. Che alle Regole de' P[ad]ri Revisori tanto universali, come delle Provincie s'aggiunga, che non passino cosa alcuna in offesa de Principi qualunque si siano, ne di loro sudditi: ne questi, o quelli si nominino da gl'autori per dispreggio, o con poco rispetto o siano vivi, o morti.

3. Lascino gl'Autori affatto le questioni, che in questi tempi sono state cagione di tanti rumori.

4. Procurino le Superiori, che le Revisori su generali, come particolari habbino, e sappino distintamente queste cose, e leggano le regole de Revisori generali, e particolari. Se queste cose s'osservaranno essattamente, come spero, che sara col divino favore, e con la vigilanza di VR. cesserà il mio timore, e sollecitudine, e s'evitaranno affatto gli disgusti, che sin' hora hanno dato tanto trauaglio alla Compagnia', Vitelleschi to P. Provinciale Romano, Rome 31 January 1632; ARSI Rom 3^l ff.173r-v

44 The *Ordinatio pro studiis superioribus* of 1651 is published in Pachtler 1887–1894 Tom. III: pp. 235–249. On the *Ordinatio*, see Costantini, *Baliani e i Gesuiti*, cit., pp. 95–109, and also Marcus Hellyer, 'Because the authority of my superior's commands': Censorship, physics and the German Jesuits. Early Modern Science and Medicine. 1996; 1(3): 319–354.

45 See Ugo Baldini, *La conoscenza dell'astronomia nell'Italia meridionale anterioramente al Sidereus Nuncius*, in *Atti del Convegno Il Meridione e le scienze (secoli XVI–XIX), Palermo 14–16 maggio1985*, ed. P. Nastasi, Palermo and Naples: Istituto Gramsci Siciliano, Istituto Italiano per gli studi filosofici; 1988, pp. 127–68, on pp.162–163 and Michel-Pierre Lerner, L'entrée de Tycho Brahe chez les Jésuites, ou le chant du cygne de Clavius, in *Les jésuites à La Renaissance: Système educatif et production du savoir*, ed. Luce Giard, Paris: Presses Universitaires de France, 1995, p. 175. The thesis of celestial fluidity was strongly supported, partly on scriptural grounds, by none other than Cardinal Bellarmino, who combined a tough line on scriptural exegesis with significant departures from Aristotelian teachings in natural philosophy. See Baldini, *Legem impone subactis*, pp. 293–296.

46 Vitelleschi to Claude Bonyol, Rome, 21 March 1631, ARSI Lugd. 5, f. 608v, copy in ARSI Gall. 117 p. 142.
47 The theologian and philosopher Giovanni Battista Rossi (1576–1656), who worked as Revisor for a period of twenty-two years, approved the book as follows 'Perlegi libellum cui est titulus Tractatus Syllepticus de terrae coelique motu vel statione etc. et nihil inueni contrarium fidei, vel bonis moribus, neque doctrinae sanae, et uideo posse utiliter typis mandari. In Collegio Romano Societatis Iesu XII Aug. 1633. Io: Bapt. Rubeus' ARSI FG 661 f.195r. The other Revisor Ioannes Rho (1590–1662), reputed for his oratorical skills (Sommervogel *s.v.*), approved the book in similar terms but added 'Sunt in eo aliqua mathematica de Theorijs planetarum quam minus esse cultus me esse ingenue fateor, quippe nihil de illis censeo. Opus totum valde placuit, cui si adeat integra tractatio ex principijs philosophicis et Mathematicis, confectum dabit negotium de stabilitate terrae. Ita censeo Io. Rho Romae Nonis Augusti 1633.'
48 ARSI FG 661 f.194r. The whole document is published in M. J. Gorman, *A Matter of Faith? Christoph Scheiner, Jesuit censorship and the Trial of Galileo*. Perspectives on Science. 1996; 4(3): 283–320, on pp. 314–6. Page 34 of Inchofer's *Tractatus Syllepticus* contains the phrase 'Quare Terram stare non solum per se est de Fide, sed etiam quatenus immediatè deducitur ex alia Propositione de Fide, quae est, Solem moveri circulariter, in qua proprie virtualiter continetur. Id fortasse eo sit certius, si etiam de Fide sit Terram esse Centrum Universi, quod an dici queat, infra in loco videbimus'. Of course, it is unlikely that the pagination of the printed work should coincide with that of the manuscript examined by Scheiner and the other *Revisores*.
49 Lothario Sarsi [Orazio Grassi], *Ratio ponderum Librae et Simbellae*, Paris: Cramoisy, 1626, published in OG VI, pp. 485–490, quoted in Redondi, *Galileo Heretic*, p. 336 [emphasis added].
50 'Cum tractatum syllepticum, de terrae, solisque motu, vel statione, &c.P. Melchioris Inchofer, nostrae Societatis Theologi; Tres eiusdem Societatis Theologi, quibus id commissimus, recognoverint; ac in lucem edi posse probauerint; facultatem concedimus, ut Typis mandetur [...] Romae XVIII. Augusti MDCXXXIII', Inchofer, *Tractatus Syllepticus*, cit. (*imprimatur*).
51 Magni to Giovan Battista Barsotti, Warsaw; 8 March 1647, BAV Vat. Lat. 13512, c.78r, cited in Massimo Bucciantini, *Valeriano Magni e la discussione sul vuoto in Italia*, Giornale Critico della Filosofia Italiana, Serie VI, Volume XIV (Anno LXXIII [LXXV]): 73–91.
52 See Chapter 6 below.
53 Cf. Giambattista Riccioli's claim that 'The Holy Congregation of Cardinals, separated from the Pope, cannot make propositions de fide, either by defining them as de fide or by declaring the contrary propositions as heretical; thus, since no pastoral letter from a Pope or a Council directed or approved by him has yet appeared, it is not yet

an article of faith that the Sun moves and the Earth is at rest [...] In spite of this, we Catholics, through both prudence and obedience, are obliged to accept that which has been decreed by this Congregation, or at least not to teach anything to the contrary', G.B. Riccioli, *Almagestum Novum astronomiam veterem novamque complectens*, Bologna: Ex Typographia Haeredis Victorij Benatij, 1651, Pars Prior, p. 52.

54 Christoph Scheiner to Athanasius Kircher, Rome, 16 July 1633, OG XV p. 184. On the very same day Scheiner wrote a letter to Gassendi in which he openly attacked Galileo as 'mearum Inventionum Invasorem', OG XV p. 683.

55 Christoph Scheiner, *Rosa ursina, sive, Sol*, Bracciani: Apud Andream Phaeum Typographum Ducalem; 1630.

56 Scheiner's work was certainly discussed in a very heated, and generally abusive manner in Galileian circles. See Pio Paschini, *Vita e Opere di Galileo Galilei*, cit., pp. 465–467

57 Galileo Galilei, *Dialogue Concerning the Two Chief World Systems – Ptolemaic and Copernican*, translated by Stillman Drake, Brekeley and Los Angeles: University of California Press, 1967, p. 346. Further attacks on Scheiner (*Apelles*) are to be found elsewhere in the *Dialogo* (e.g. pp. 357, 367)

58 The notion of 'common astronomy' deployed by Scheiner points to a high degree of disciplinary inertia. As Westman paraphrases Kuhn, 'what is at stake [in a revolution] is the overturning of a whole way of scientific life, not the abstract and transcendental deliberations of some scientific jury using a calculus of relative problem-solving capabilities'. Robert S. Westman, *Two Cultures or One? A Second Look at Kuhn's* The Copernican Revolution, Isis; 85: 79–115, on p. 82.

59 See note 13 above.

60 The 'silencing' of Galileo in 1633 did not, of course, remove the need for an astronomical and physical refutation of Copernicanism, as demonstrated by Riccioli's monumental *Almagestum Novum* (cit.). However, what I am suggesting here is that it is difficult to reconcile the act of silencing involved in a denunciation with the visibility-enhancing dynamic of the seventeenth-century astronomical duel. Riccioli's enquiry should, perhaps, be read not as a vindication of Scheiner, but, in so far as it relates to the trial at all, as a judicial post-mortem, an enquiry into the justness of the causes of the disputants. On scientific duels see Biagioli, *Galileo Courtier*, cit., pp. 60–73.

61 ARSI FG 655 ff.198r-200v *Judicium Revisorum Collegij Rom. de Vindicijs P. Melchioris Inchoferi 29 Jan. 1636*. The report of the *Revisores*, Jakob Bidermann (1578–1639), better known for his theatrical works than his work as Revisor (see above Chapter 2), Giovanni Battista Rossi (1576–1656) and Ioannes Alvarado,is followed by a reply by Inchofer [f.199r], which is in turn followed by the final critical judgment of the *Revisores* on f. 202v: 'Atque liber hic R. P. Melchioris Inchoveri talis

omnino est, ut argumentum gravissimum, idque magis Theologicum quam mere mathematicum contineat. Debebat igitur & hic eius liber, ut edi iubeatur, lectorisque exspectationi satisfacturus putatur, esse multo solidior, gravior & efficacior, quam si de quocunque minusculo solum argumento tractaret'. The censorship of this work by Inchofer, a manuscript of which is amongst Sforza Pallavicino's papers in the Biblioteca Casanatense in Rome (Ms. 182), is discussed in Baldini, *Legem impone subactis*, pp. 297–298.

62 'Respondeo titulum libri hunc esse. Vindiciarum S. Sedis Apostolicae, Sacreorum tribunalium et Authoritatem ac libri duo. Si hic titulus nimis magnificus videtur, facile emendari potest, hoc, aut alio modo, Vindiciarum Sacrearum Auctoritatem' Melchior Inchofer, *Responsio ad ea quae Patris Censores opponunt in meo libro*, ARSI FG 655 ff.199r

63 See Gorman, *A Matter of Faith?* cit., p. 316. I have not found a corresponding passage in the *Tractatus Syllepticus*, perhaps suggesting that on this point Scheiner's criticisms were taken into consideration.

64 Scheiner, *Prodromus pro sole mobili, et terra stabili, contra Galilaeum a Galilaeis, qui nunc primum in publicam lucem prodit*, Pragae: Gosvinus Nickel, 1651 [Copy consulted: Österreichische Nationalbibliothek, Vienna, 72.D.54]. On Scheiner's *Prodromus* see Favaro, *Cristoforo Scheiner*, cit., pp. 98–107

65 See above note 63.

66 See Wilhelm Weilhamer to Giannantonio Rocca, Mantua; 23 April 1642, published in Rocca 1785 p. 301 'De Scheineri libro nondum scio aliquid certi quo pretio vendatur; suam quamprimum jam ipsi Authori scriptum fuit: Et liber editus in folio (sunt 30. folia cum figuris intermediis aeneis) contra Galilaeum libri tres, Inscriptio est Prodromus de terra stabili, & Coelo, seu Sole mobili; non habet indicem, nec praefationem: reservavit Author ista ad pleniorem impressionem: interea Lectori ista praegustanda proponit.'

67 L. Allatius, *Apes Urbanae, seu de viris illustribus qui ab anno 1630 per totum 1632 Romae adfuerunt, ac typis aliquid evulgarunt*, Rome: Ludovicus Grignanus, 1633. It is noteworthy that Allacci removed the entry on Galileo's Dialogo from his work before publication. He compromised by adding an entry on Pierre Gassendi, in the hope that this would be reflected by a parallel substitution in the Barberini entourage. In spite of Allacci's plans to mould the real theatrum mundi according to his paper pantheon, Galileo's place was filled by Kircher instead of Gassendi. See Allatius, *Apes Urbanae,* cit., pp. 70–71, and Gabriel Naudé's letter to Gassendi in OG XV p. 88. On Kircher's move to Rome see also above note 20.

68 'Prodromus pro Stabilitate terrae contra eundem Dialogistam, in quo compendiose afferentur Galilaei errores Logici, errores Physici, errores Mathematici, errores Ethici, errores Theologici, atque sacri: adeoque ex omnibus his constabit detracta larua doctrinam hactenus mentita imperitia', Allatius, *Apes Urbanae*, pp. 68–71

69 The work does seem to have suffered some modifications before publication, as is borne out by the statement on the title page that a larger version of the work was composed *ante annos 20*.

70 OG XVII pp. 276–7.

71 See ARSI FG 1368/8/12

72 On Galileo's programme, see W. R. Shea, *La contrariforma e l'esegesi biblica di Galileo Galilei*, in A. Baboli, ed., *Problemi religiosi e filosofia*, Padua: La Garangola, 1975, pp. 37–62). His principle of exegesis was expressed in his 1615 *Lettera a Madama Cristina di Lorena, Granduchessa di Toscana*, in which he defended Copernicanism from the accusations of heresy that had been flowing from Florentine pulpits. In this work, he argued that 'I will say here that which I have heard from an extremely eminent ecclesiastical figure [i.e Cardinal Baronius] which is that the intention of the Holy Spirit is to teach us how to go to heaven, and not how heaven goes', OG V p.319. In his 1992 speech to the Pontifical Academy of Sciences, Pope John Paul II used exactly this passage from Baronius to define his own position on scriptural hermeneutics, which he related explicitly to that of Galileo's *Lettera a Madama Cristina di Lorena* (*Discours du Saint-Père à l'Académie pontificale des Sciences, 31 October 1992*, published in Festa 1995 pp. 389–406 on p. 402). Galileo drew further support for his views from another theological work that was to be placed on the Index in 1616 donec corrigatur, Diego a Zuñiga's commentary on Job (see OG V p. 336). Galileo's exegetical principles were strongly influenced by those put forward by Benito Pereira in his *Commentariorum et disputationum in Genesim*, Rome, 1599, and thus highly consonant with Jesuit theological sensibilities of his time. See Baldini, *Legem impone subactis*, pp. 296–297

73 Indeed, a crucial role in preventing the complete prohibition of Copernicus in 1616 was played by Cardinal Maffeo Barberini, before he became Urban VIII, as we are shown by the diary of Cardinal Giovanfrancesco Buonamici for 2 May 1633: 'In tempo di Paolo V⁰ fu contrariata questa opinione, come erronea et contraria a molti luoghi della Sacra Scrittura; perciò Paolo V⁰ fu di parere di dichiarla contraria alla Fede: ma opponendosi li SS^ri Cardinali Bonifatio Gaetano et Maffeo Barberino, hoggi Urbano 8o, fu fermato il Papa di testa, per le buone ragione addotte da loro Eminenze...', OG XV, p.111. Buonamici reaffirms this statement in his *Relazione* of July 1633, in OG XIX p.410–11. Urban VIII's biographer, Herrera, adds that the future pope was motivated to prevent the prohibition of De Revolutionibus because of its utility in the Gregorian reform of the calendar, an argument that was then taken up by Ingoli in his suggested corrections to Copernicus's work (see below note 93). Herrera also states that 'Cardinal Gaetani judged the same and Bellarmine, who consulted the geometers (*geometri*), approved greatly'. Presumably the 'geometers' in question are the Jesuit mathematicians of the *Collegio Romano*, headed at that time by Christoph Grienberger and Orazio Grassi. Herrera's biography is

cited in D'Addio, *Il caso Galilei,* cit., p. 48. See also Massimo Bucciantini, *Contro Galileo: Alle origini dell'Affaire,* Florence: Olschki, 1995, p. 154.

74 The term is familiar from discussions of Duhem's study, 'ΣOZEIN TA ΦAINOMENA' (Pierre Duhem, *To Save the Phenomena,* transl. E. Doland and C. Maschler, Chicago: University of Chicago press, 1969 (orig. publ. 1908)). For an account of the many faces of astronomical pragmatism during the sixteenth and early seventeenth centuries, see N. Jardine, *The Birth of History and Philosophy of Science. Kepler's 'A Defence of Tycho against Ursus', with Essays on its Provenance and Significance,* Cambridge: Cambridge University Press, 1988 pp. 225-257.

75 'Dico che mi pare che V.P. et il Sigr. Galileo facciano prudentemente a contentarsi di Parlare ex suppositione e non assolutamente, come io ho sempre creduto che habbia parlato il Copernico. Perchè il dire, che supposto che la terra si muova et il sole stia fermo si salvano tutte l'apparenze meglio che con porre gli eccentrici et epicicli, è benissimo detto, e non ha pericolo nessuno; e questo basta al mathematico: ma volere affermare che realmente il sole stia nel centro del mondo, e solo si rivolti in sè stesso senza correre dall'oriente all'occidente, e che la terra stia nel 3o cielo e giri con somma velocità intorno al sole, è cosa molto pericolosa non solo d'irritare tutti i filosofi e theologi scholastici, ma anco di nuocere alla Santa Fede con rendere false le Scritture Sante', Bellarmino to Foscarini; 12 April 1615 in OG XII, pp.171-172. On Bellarmine's conception of astronomy see Baldini, *Legem impone subactis,* pp. 285-303

76 Sforza Pallavicino to Carlo Roberti; c. 1665. Rome, Biblioteca Casanatense Ms. 4983 ff. 38r-v.

77 OG XIX pp. 400-401. In her study of the trial, Rivka Feldhay uses the 1620 decree to construct a 'Dominican voice' in the Galileo affair, based on the fact that it is signed by the Dominican friar Franciscus Magdalenus Capiferreus: 'Thus, the decree of the Index of 1620, which attempted to control any further investigation of the motion of the earth, may be seen as a reflection not of scriptural fundamentalism but of the theology of the Inquisitors with which it was entirely in keeping', Feldhay, *Galileo and the Church,* p. 212. However, Ingoli, despite working as a Consultor of the Congregation of the Index, was no Dominican. More importantly, the Jesuit mathematicians to whom the Congregation submitted his text for evaluation 'all approved and praised the wishes of the said Lord Francesco, and judged it to be wholly profitable that it should be permitted that the work be amended and corrected in accordance with his corrections' (W. Brandmüller and E.J. Greipl, eds., *Copernico, Galileo e la Chiesa. Fine della controversia (1820). Gli atti del Sant'Uffizio.,* Florence: Olschki, 1992, pp. 444-445, cited in Bucciantini, *Contro Galileo,* cit., p. 87). Feldhay's use of the document to argue for a fundamental opposition between Jesuit and Dominican 'celestial hermeneutics', central to her re-interpretation of the 1632-33 trial, is therefore inappropriate. On the genesis of the 1620 decree see Bucciantini, *Contro Galileo,* pp. 141-147.

78 BAV Barb. Lat. 3151 ff.58r-61v: 'I state that it is possible for these amendments to be made without danger to truth or Holy Scripture: because as the science which is treated by Copernicus is astronomy, whose proper method is to use false and imaginary principles in order to save celestial appearances and phenomena, as seen from the Epicycles, eccentrics, equants, apogees and perigees of the Ancients, if the places in which Copernicus does not treat the motion of the earth hypothetically are rendered hypothetical, they will challenge neither truth nor Holy Writ. On the contrary in a way they will fit together with the latter because it is customary for the science of Astronomy in particular to make use of false suppositions'. The task of 'correcting' *De Revolutionibus* was originally assigned to Bonifacio Cardinal Gaetani. However his death in June 1617 led Ingoli to take over the job. See Bucciantini, *Contro Galileo*, pp. 141–147 (the document is published on pp. 207–209). Surprisingly, the document does not appear in any of the collections of documents relating to Galileo's trial. An English translation of the document is in O. Gingerich, *The Censorship of Copernicus's De Revolutionibus*, Annali dell'Istituto e Museo di Storia della Scienza di Firenze, Anno VI: 45–61.

79 This is not to deny the existence of a certain degree of flirtation with the Copernican theory by Jesuits in unpublished works and letters. Much of the evidence for Copernican sympathies among the Jesuits is based on hearsay, e.g. Pietro Dini's letter to Galileo of 16 March 1615 'I understand that many Jesuits are secretly of the same opinion, although they keep quiet about it' (OG XII p.181) Interestingly, there have even been suggestions that Scheiner himself was a closet Copernican: As late as 1626 Francesco Stelluti wrote to Galileo that Scheiner 'is in agreement with you about the system of the world' (OG XIII p. 300). Other Jesuits reputed to be inclined towards the Copernican theory include Orazio Grassi, Niccolò Cabeo, Wenceslas Kirwitzer and Wilhelm Weilhamer. On Copernican Jesuits see Favaro, *Cristoforo Scheiner*, cit., pp. 105–106, J. L. Russell, *Catholic Astronomers and the Copernican System after the Condemnation of Galileo*, Annals of Science, 46 (4); 1989: 365–86 and J. Lattis, *Between Copernicus and Galileo*, pp. 202–205. On Kirwitzer's declaration of Copernicanism in his letter to Christoph Grienberger of 7 June 1615 (in APUG 534 ff. 90r-91v) see Baldini, *Legem impone subactis*, pp. 215–216 note 35 and Lattis, *Between Copernicus and Galileo*, p. 205.

80 See Scheiner, *Pantographice seu Ars delineandi res quaslibet per parallelogrammum lineare seu cauum, mechanicum mobile*, Romae; 1631.

81 For details of Karl's monetary donations to Scheiner for the foundation of the College of Neisse see ARSI 1368/8/7. On Scheiner's ill-fated church for the College of Innsbruck, which collapsed in 1626, see *Memoriale P. Scheineri pro fabrica 1621. Fabrica templi*, Copy in Museum Ferdinandeum, Innsbruck, Ms. FB 51838. The original, dated 16 October 1621, is conserved in the archives of the Jesuit college of Innsbruck.

82 Daxecker, *Briefe des Naturwissenchaftlers Christoph Scheiner SJ*, cit., p. 15
83 Among other things Scheiner provided Leopold with detailed information about the private affairs and political intentions of his brother, Archduke Karl in 1621, around the time when Karl asked Scheiner to be his confessor. On Karl's future intentions towards the Jesuits Scheiner wrote that he would be easily won over to the Society 'from which he has sucked milk since infancy'. Daxecker, *Briefe*, p. 45
84 Scheiner's list of reproaches of Michael Will included his lack of affection for the Jesuits, his association with a prostitute of uncertain gender, and the infestation of his house with terrifying ghosts the night after his death. Christoph Scheiner to Archduke Leopold of Austria, Rome, 8 January 1627, in Daxecker, *Briefe*, pp. 114–15
85 Daxecker, *Briefe*, p. 68
86 Anticipating Leopold's distrust of a non-German agent, Scheiner lapses into a macaronic German that he seemed to feel would reassure the Archduke: 'Es sind nit alzeit alle Teütschen Teütsch, und auch nit allezeitt alle Welschen [i.e. Italians] Welsch: das Blättlein kheret sich zue Zeitten umb. Ego scio hunc Virum uere Germanum [a play on words between 'sincere' and 'German'], et si non pro certo scirem [...] nequaquam ita expresse scriberem'. Christoph Scheiner to Archduke Leopold of Austria, Rome, 8 December 1626, in Daxecker, *Briefe*, p. 107
87 Christoph Scheiner to Archduke Leopold of Austria, Rome, 30 March 1630, in Daxecker, *Briefe,* p. 144
88 'Rogo si Oenipontum peruenerint, ut Vestra Serenitas in meum aduentum ipsas clausas retineat, cuius postulati caussas grauissimas habeo; cum enim res Caesaris sint ubique intermixtae p fieret chaos horrendum, et facile possent aliquid pati ...', ibid.
89 Scheiner to Archduke Leopold, Rome, 2 June 1628, in Daxecker, *Briefe*, p. 137
90 Christoph Scheiner to Archduke Leopold, Rome, 27 July 1630, in Daxecker, *Briefe*, p. 148
91 'Corpora sacra afferam minimum 8, verum non ex liberalitate Pontificia, sed industria privata, quam auxit studium gratificandi Serenitati Vestrae et Collegio Nissensi benefaciendi amor', Scheiner to Archduke Leopold of Austria, Rome, 2 June 1628, in Daxecker, *Briefe*, p. 137
92 *Bullarium Romanum* Tom. V pp. 199–200
93 '[I]nter cetera, invitentur ad Dei laudes, ab incendio atque naufragio liberentur; procella quoque turbinum, fulgura, grandines, tempestates et omne malignum molimen procul ab eis pellantur; praegnantes absque partus periculo conserventur', ibid.
94 'Si quis autem contra inhibitionem nostram huiusmodi temere venire praesumpserit sententiam excommunicationis incurrat; et si clericus fuerit, omnibus, quae obtinuerit, beneficiis et officiis ecclesiasticis priuatus existat; et nihilominus laicus, quoties id fecerit, per annum carceri mancipetur ob tanti facinoris ultionem', ibid.

95 'Volumus autem, quod ii, qui cereas imagines sive Agnos Dei confectos huiusmodi apud se habuerint, infra octo dierum spatium a prohibitione praesentium computandum, illos venerabili fratri nostro Nicolao episcopo Pharensi consignare teneantur, recepturi pro cera aequivalentem compensam vel satisfactionem condignam, alioquin sententias et poenas praedictas, ipso termino elapso, incurrant', ibid.

96 E.g. 'Item de Corpore S. Iulii Cineres, et similiter de Corpore S. Saturnini, quos hic Romae pro me singulariter exoraui, per quos tamen non excludo partem tertiam corporis', Christoph Scheiner to Archduke Leopold, Rome, 27 July 1630, in Daxecker, *Briefe*, p. 148

97 On the difference between the metrology of intrinsic standards and the metrology of artefact [extrinsic] standards see Joseph O'Connell, *Metrology: The Creation of Universality by the circulation of Particulars*, Social Studies of Science, 23, 1993: 129–73. O'Connell uses the theological differences between Calvinism and Catholicism as a metaphor for this difference (see e.g. p. 154). However, I am arguing here that the metaphor should be taken seriously – the conflict of social models discussed by O'Connell is not limited to the field of metrology.

98 OG VI p. 200, translated in Stillman Drake and C. D. O'Malley, *The Controversy on the Comets of 1618*, Philadelphia: University of Pennsylvania Press, 1960, p. 152.

99 OG VI p. 220, Drake and O'Malley, *The Controversy on the Comets*, p. 171

100 Drake and O'Malley, *The Controversy on the Comets*, p. 69: 'But why was it so readily believed that this Gregoriana of ours, renowned for the many interests of its academicians should be considered as, among other things, the eyes of all, and that it ought especially to be consulted and its answers awaited?', ibid. p. 111: 'I have no few witnesses to the fact that I say this not more surely than truly; first, many fathers of the Collegio Romano – however, many others were willing to recognise this on the authority of my teacher – and many others as well'.

101 '[R]eputo questa presente scrittura che gli mando, come quella che è fondata sopra la mobilità della terra overo che è uno degli argumenti fisici che io producevo in confermazione di essa mobilità, la reputo, dico, come una poesia overo un sogno, e per tale la riceva l'A[ltezza] V[ostra]' Galileo to Archduke Leopold, Florence; 23 May 1618, OG XII 389–392.

102 See Biagioli, *Galileo Courtier*, cit., p. 219 note 41

103 Scheiner's early works on sunspots were written under the cumbersome pseudonym of *Apelles post tabulam latens*, referring to the story of Apelles, the favourite painter of Alexander the Great, who hid behind one of his paintings so that he could hear genuine appraisals of his work. Galileo's suggestion seems to be that Scheiner, as a plagiarist and unskilled draftsman is unworthy of the generous patronage that he has received on behalf of his order from the house of Austria.

104 OG VI pp. 47–48, Drake and O'Malley, *The Controversy on the Comets* p. 24.

105 'L'ha piacuto assai l'opera et l'è stata gratissima, e subito l'ha mandato al P. Scheiner il quale rispose che pagarà V.S. con la medesima moneta' Johannes Remus to Galileo Galilei, Vienna; 24 August 1619, OG XII pp. 488–489.
106 Leopold's letter of thanks for the *Assayer* (OG XIII p. 162, 26 December 1623) is short, formal and in Latin, unlike his previous letters to Galileo, which are in Italian and relatively informal. The fact that Leopold sent no further letters to Galileo suggests that, although he compliments the book, Galileo's ridicule of Grassi and, to a lesser extent, Scheiner, in the *Assayer* may not have been altogether to his taste.
107 'I am as innocent as a child', Daxecker, *Briefe,* p. 125
108 'I remain the old Scheiner and simple-minded Schwab' (Scheiner was originally from Wald, a village near Mindelheim in Schwaben), ibid., p. 127
109 'I am still the old Scheiner, Your Most Royal Majesty can truly trust me', ibid., p. 144
110 *Lucii Cornelii Europaei Monarchia solipsorum*, Venice, 1645.
111 On Scotti, see the biography by P. Oudin in R. P. Nicéron, *Mémoires pour servir à l'histoire des Hommes Illustres dans la République des lettes,* Paris: Briasson; 1736, t. 39, pp. 65–85
112 ARSI Hist. Soc. 166, *Causa P. Melchioris Inchofer*. Inchofer's confession is on ff. 65r-74v. These documents form the basis of the account of the last year of Inchofer's life given in Dümmerth, *Les combats et la tragédie du Père Melchior Inchofer S. J.*, cit.
113 Many of these documents are contained in the volume ARSI Congr. 20^d, *Melchioris Inchofer scripta de VIII Congregatione generali, aliaque Epistolae ad eum datae.*
114 Although Dümmerth suggests that Inchofer would have been subject to regular flagellation during this period, there is nothing to suggest this in ARSI Hist. Soc. 166, and physical punishment was generally discouraged by the Society. In reality, the documents contained in this codex suggest that Sforza Pallavicino played an important role in mediating between Inchofer and the General to allow Inchofer to have his books so he could continue his studies during his period of incarceration.
115 'Ignatio vocabulum Monarchia aut ignotum aut explosum fuit', Inchofer, *Historia 8ae congregationis generalis*, ARSI Congr. 20^d, f. 56r.
116 'Philosophica apud eos fere huiusmodi sunt. An Scarabeus paradigmaticè stercora voluat in orbem, An si Mus in mare mingat timendum naufragium. An puncta mathematica sint receptacula spirituum. An hiatus ventris, sit spiraculum animae. An canum latratus lunam reddat maculosam; & plura id genus, quae pari contentione dictantur, & excipiuntur' *Monarchia solipsorum*, p. 29.
117 See Giorgio Spini, *Ricerca dei Libertini: La Teoria dell'Impostura delle Religioni nel Seicento Italiano*, Roma: La Nuova Italia; 1983 (2nd edition), p. 244. On the *Monarchia Solipsorum* in general, see ibid., pp. 233–246. Although Spini was unaware of the documents relating to Inchofer's secret trial in ARSI, he judges that

Inchofer is the most likely author of the Monarchia on grounds of style and content, as well as contemporary rumours in Jansenist circles.

118 See O'Malley, *The First Jesuits,* pp. 335–345, *Constitutions,* #822.

119 'Sub id tempus prodijt libellus Lucij Cornelij Europaei de Monarchia solipsorum. Qui quantum novitate placuit, tantum obscuritate absterruis, et nihilominus curiosa ingenia exercuit. Leo Allatius cui inspeciem dicatus est, primus ipse apud Sacrum Tribunal nomen deposuit', Inchofer, *Historia 8ae congregationis generalis,* ARSI Congr. 20^d f. 8r.

120 Innocent X to 8th General Congregation, ARSI Congr. 1, ff. 199r-v (two further copies inside the same codex, inside front cover and ff. 201r-202v).

121 'Che non si possa leggere ne professare altra dottrina che quella di S. Thomaso, et altre che communamente sono state abbraciate dai Santi Padri', ARSI Congr. 1, f. 199r.

122 ARSI Congr. 1, f. 226r: 'Ad promovenda l[itter]arum studia, examinando et recognoscendo libros de Ratione studiorum, in illis maxime quae pertinent ad opinionum delectum, ut refrenetur nimia quaedam opinionum licentia et novitas, praesertim cum iam aliquot Summi Pontifice hoc valde Societati commendarint'.

123 ibid., f. 227r: 'De promovendis studijs Deputatio partienda in duas: de humanioribus litteris unam; de Theologicis, et Philosophicis alteram'.

124 [Leone Santi], *Dubia et Postulata Praefecti Studiorum Collegii Romanii de formandis Decretis circa Studia. 1646,* ARSI Rom. 150^l, ff. 268r-271v

125 Ibid.

126 See ARSI FG 657 ff. 431–454.

127 Ibid.

128 *Revisores Generales* to Ninth General Congregation, Rome, 28 December 1649, ARSI Congr. 20e ff. 49r-v.

129 *Propositiones quas Patres Deputati pro studijs censuerunt non a praeceptoribus nostris docendas. 1649.* ARSI Congr. 20e f. 226r

130 Cazré was professor of mathematics in Pont-à-Mousson (1622–23) and later Reims (1623–24, 28–29). His works include a criticism of Galileo's science of motion, *Physica demonstratio qua ratio, mensura, modus, ac potentia, accelerationis motu in natuali descensu gravium determinantur. Adversus nuper excogitatam a Galileo Galilei,* Paris: Jacques du Breuil, 1645.

131 ARSI Congr. 20e f. 234r

132 *Instructio pro studiis superioribus Iussu nona Congregationis Generali confecta et missa a R.P.N. Francisco Piccolomineo Praeposito Generali,* Rome, 3 October 1650, ARSI FG 657 pp.641–667.

133 ARSI Congr. 20^e f. 413r

134 *Ordinatio,* cit., p. 7.

135 *Ordinatio,* cit., p. 98.

136 Athanasius Kircher, *Itinerarium Extaticum*, Rome: Vitalis Mascardi; 1656. See Franciscus Le Roy, *Responsio ad quandam censuram in qua redarguuntur propositiones sex, ex Itinerario Extatico P. Athanasii Kircheri desumptae*, ARSI, Fondo Gesuitico 675, ff.247-248. On the censorship of this work see Carlos Ziller Camenietzki, *L'Extase interplanetaire d'Athanasius Kircher: Philosophie, Cosmologie et discipline dans la Compagnie de Jésus au XVIIe siècle.* Nuncius. 1995; X(1): 3-32.

137 Francesco Maria Grimaldi, *Physico-mathesis de lumine, coloribus, et iride*, Bologna, 1665. See ARSI FG 670 ff. 53r-55r, published in Baldini, Legem impone subactis, pp. 102-3, e.g. f. 53r: 'Propositione 42a significat colores, etiam qui appellantur permanentes, non distingui reipsa a lumine, nec proinde esse qualitates. Et propositione 14a docet operationem magneticam consistere in effluvio substantiali omnia corpora pervadente: cumque propositione 15a dicat se ex occasione ostensurum quis sit verus conceptus rarefactionis et condensationis, verisimile est eum itidem explicaturum esse per ingressum corpusculorum, minus consentanee ad propositionem 37a in Ordinatione pro Studiis'. The 27th proposition of the *Ordinatio* was: 'Elementa non transmutantur invicem, sed unius particulae in alio delitescunt incorruptae, quarum ingressus rarefactionis et condensationis est ratio', *Ordinatio*, cit., p. 92.

138 On Fabri see E. Caruso,*Honoré Fabri, gesuità e scienzato. Miscellanea secentesca. Saggi su Descartes, Fabri, White*, Milano: Università di Milano; 1987: pp. 85-126. A. Boehm, *L'aristotélisme d'Honoré Fabri (1607-1688)*. Revue des sciences réligieuses. 1965; XXXIX: pp.305-360, E. A. Fellman, *Die Mathematischen Werke von Honoratus Fabry*. Physis. 1959; 1: pp.6-25, 69-102 and David C. Lukens, *An Aristotelian Response to Galileo: Honore Fabri S.J. (1608-1688) on the causal analysis of motion*. Dissertation: University of Toronto; 1980.

139 Orazio Grassi to G. B. Baliani, Savona, 25 August 1652, Biblioteca Braidense, Milan, AF XIII, 13, 4, f. 233r, in Serge Moscovici, *L'Experience du mouvement. Jean-Baptiste Baliani, disciple et critique de Galilée*, Paris: Hermann, 1967, pp. 251-2, on p. 252.

140 *Honorati Fabri Societatis Iesu ad P. Ignatium Gastonem Pardesium ejusdem Societatis Iesu Epistolae Tres de sua Hypothesi Philosophica*, Mainz: Apud Joan. Petrum Zubrodt, 1674, p. 116: 'Unum restat mihi faciendum, ac respondendum circa ea (humanissime Pardesi) quorum per te certior factus sum: scripseras enim, me a multis pro Cartesiano, aut Democritico, seu Atomista, ut vocant, traduci, & nonnulla in mea Hypothesi Philosophica, vel adstrui, vel contineri, que per Societatis leges, aut Decreta, a nostris doceri vetantur: in duabus prioribus literis, quas ad te scripsi, primum ni fallor, evici, nimirum me, nec Cartesianum, nec Gassendistam, nec demum Atomistam, aut Democriticum esse; alterum in hisce ultimis praestare conabor, ostendamque, mea Placita iis Propositionibus omnino adversari, quas R.P.N. Franciscus Piccolomineus, in sua Ordinatione pro studiis

superioribus, juxta deputationem, quae de illis habita est, in Congregatione nona Generali, ad Provincias missa, Anno 1651. ex scholis nostris proscripsit: quamvis enim nulla Censura illas affecerit: eas tamen in nostris gymnasiis doceri, prohibuit; quod ut penitus evincam, singulas juxta numerorum seriem, in qua extant, recensendas esse, duxi, & cum meis placitis conferendas'.

141 The title of Grimaldi's *Physico-Mathesis* speaks eloquently in this respect. An approximate translation is 'Two books of Physico-mathematics of light, colours and the rainbow, and other connected matters, in the first of which new experiments are put forward, along with the reasons deducted from them in favour of the substantiality of light. In the second, however, the arguments put forward in the first book are dissolved, and it is taught that the Peripatetic doctrine of the accidental nature of light can be held as probable', Grimaldi, *Physico-mathesis*, cit.

4

The Jesuits and the Vacuum Debate[1]

Your Most Illustrious Lordship should not affirm the existence of the vacuum so strongly because of that famous experiment; as now even the disciples of Galileo deny it, and would even like to claim that it is impossible, even by divine power, which I do not believe. It is certain that one experiences that small animals can live and sound can be produced within that space in the vessel that they thought remained empty, so therefore there is body.

Sforza Pallavicino[2]

In July 1660, the Jesuit natural philosopher Honoré Fabri wrote from Rome to Lorenzo Magalotti, secretary of the Accademia del Cimento, to congratulate him on the ongoing experimental investigations of the Florentine experimental academy. While approving in enthusiastic terms of the advancement of natural philosophy through the systematic accumulation of experimental data by the academy,[3] Fabri warned Magalotti that the experimental enterprise itself was fraught with dangers:

> I frequently say, and perhaps you might have heard me say it on some occasion, an experiment is like a very good sword, in that if an agile man uses it for the defence of prince and realm it achieves many glorious deeds but if, however, it is used by someone driven by fury, nothing is to be expected but terrible slaughter and patricide.[4]

One of the politically dangerous experiments that Fabri undoubtedly had in mind was the experiment first carried out by Evangelista Torricelli sixteen years previously to demonstrate the existence of the vacuum and the weight of air. From the late 1640s to the 1660s a large number of works were published by Jesuit natural philosophers, theologians and mathematicians in Rome, Bologna, Vilnius, Paris, Lyon and elsewhere which attempted to refute the anti-Aristotelian conclusions derived by Torricelli and the other vacuists from the experiment. Although the reaction was by no means completely uniform, and Jesuit writers employed radically different hypotheses in their efforts to provide alternative explanations for the anomalous phenomenon of the suspension of the mercury in the Torricellian tube, the large-scale nature of the Jesuit reaction suggests that in attempting to understand the positions adopted by Jesuit participants in the debate we should look to non-local institutional factors. The debate had significance for the order that went far beyond the bounds of pure intellectual curiosity.

This chapter will attempt to trace some of the different political and theological factors that combined to give the Torricellian experiment such a highly charged meaning for Jesuit experimental practitioners in the middle of the seventeenth century. By documenting the relationship between the participation of Jesuits in specific replications of the void experiment and discussions of its significance with broader goals of the order in the complex political context of the time around the Peace of Westphalia, I hope that a picture will emerge of the way the work of the individual Jesuit experimenter was given meaning by the goals of the order as a whole on the turbulent political and theological stage of Europe at the end of the Thirty Years' War.

The genesis of an experimental fact

Although the empty space left by the mercury at the top of the Torricellian tube became invested with a huge variety of theological and political meanings by the end of the 1640s, the experimental work that led to Torricelli's elated exclamation in 1644 that 'We live submerged at the bottom of a lake of elementary air, which we know to have weight by indubitable experiments'[5] was ostensibly driven by a problem in civil engineering.[6] In July 1630 Giovanni Battista Baliani, a high dignitary and port-official in Genoa, was attempting to build a siphon in order to transport water over a hill of 84 *palmi di Genova* in height.[7] He wrote to Galileo to give a detailed explanation of the behaviour of the siphon. 'When it is opened, while the upper part remains closed, the water escapes everywhere, and if one side is kept closed and the other is opened, [water] still comes out on the open side.'[8] In particular Baliani informed Galileo that the water descended to a certain fixed point on the opened siphon, before stopping, a phenomenon that left him 'stupefied'.[9] Galileo wrote back to Baliani, suggesting that the expense of constructing the copper siphon would have been spared had Baliani first thought to ask him for advice, as the question was one which he had considered long before. Although Baliani, like Torricelli later on, wished to account for the phenomenon in terms of air-pressure, Galileo's interpretation, later expanded in his *Discorsi* consisted in advocating a theory of matter as being held together by an infinite number of infinitestimal vacua. Just as a rope of a given thickness will break under its own weight if it exceeds a certain length, the water in a siphon of more than 18 *braccie* will 'break' under its own weight, when the limited resistance of the interstitial vacua is overcome. In the *Discorsi*, Galileo linked his discussion to a much-contested demonstration using the 'Aristotle's wheel' problem that the mathematical continuum was actually composed of an infinite number of indivisible points, separated by interstitial vacua.[10] The type of atomist approach used by Galileo in this explanation was closely related to the revived interest in Hero's *Pneumatica* in late sixteenth-century Italy,[11] and Galileo's Venetian friend Paolo Sarpi also leaned towards a Heronian conception of matter in his private writings.[12]

Readers of the *Discorsi*

Although much of the work had been written in 1631–34, the *Discorsi* were only finally published in Leiden in 1638.[13] The discussions of siphons inspired a number of different groups to conduct further experiments along the lines by Galileo. In the presence of the Minim mathematician Emmanuel Maignan and the Jesuits Athanasius Kircher and Niccolò Zucchi, Gasparo Berti conducted a spectacular, but equivocal experiment outside his Roman palace at some time between 1641 and 1643. The experiment, initially little more than an urban version of Baliani's troublesome siphon, involved attaching a lead pipe of around 22 *braccie* in height to the side of Berti's house. The pipe was filled with water through an open valve from one of the bedrooms in the house with the valve at the base of the pipe remaining closed. When the upper valve was closed the lower valve was opened allowing the water to fall into a tub in the courtyard.[14] A glass tube fitted to the top of the pipe allowed the water level to be observed to fall only to a certain point before stopping, leaving an apparently empty space above. A sophistication of the experiment suggested to Berti by Kircher incorporated a bell, which was made to ring in the empty space by means of a magnet.[15] The various reports of the experiment are contradictory, in both their accounts of the phenomena observed and their interpretation. Years later Raffaello Magiotti remembered the small bubbles that he had seen rising through the water in the glass portion of the siphon, which he suspected of going to replace the air in the top of the tube. 'Those bubbles [*pulighe*]', he wrote to

Mersenne in 1648, 'have always been on my mind'.[16] The sound of the bell in the space convinced Athanasius Kircher that it contained air,[17] although Maignan countered this argument with the possibility that the sound of the bell might be transmitted along the string attaching it to the tube.[18] Magiotti wrote to Torricelli describing the experiment, and suggesting that seawater might come to rest lower in the tube than freshwater. This, according to Magiotti, inspired Torricelli and his collaborator Michelangelo Ricci to conduct experiments with heavier and heavier fluids, culminating in the use of mercury in the famous Florentine experiment of 1644.[19]

Berti was not alone in being inspired by Galileo's work to begin experiments on siphons. During the negotiations to publish the *Discorsi* outside Italy, an attempt had been made in 1636 to have the work published in Moravia, under the patronage of Cardinal Dietrichstein, who died before this could be carried out. One of the prime movers behind this effort was the Capuchin friar from Milan, Valeriano Magni.[20] Magni was a vehement opponent to the hold of Aristotelian philosophy over the universities of Europe and was later to dismiss Aristotle in scathing terms in his *De Atheismo Aristotelis*, published in 1647,[21] writing that '[Aristotle] is a tyrant, who has burdened the human race, more pernicious than any heresiarch, or any man brought forth by any age'.[22] During the 1630s and 40s he had attempted to develop a new philosophical system of his own, drawing on Platonic and other sources, to constitute an alternative to the Christianised Aristotelianism sanctioned by the decrees of the Council of Trent and by the Jesuit hegemony over education in Catholic Europe. Magni's first published philosophical work, a popular treatise expounding a Platonic Christian philosophy influenced by Bonaventure's *Itinerarium mentis in Deum*, entitled *De luce mentium*, and published in 1642, was submitted to numerous theologians for approval.[23] Athanasius Kircher attacked the work in scathing terms in a report that may have been prepared at the request of the Congregation of the Index, writing that 'The treatise *De luce mentium* contains many dogmas extracted from the workshops of the Pythagoreans and Platonists'.[24]

Magni's next philosophical publication was his 1647 *Demonstratio ocularis*,[25] in which he described his experiments on the vacuum.

As he wrote in 1647, his reading of Galileo's *Discorsi*, perhaps the manuscript that Pieroni had brought to Bohemia, had encouraged him to conduct his own investigations on the behaviour of liquids in tubes:

> After I had recognised from a certain little work by Galileo Galilei, that it is not possible to raise water in a tube or a pipe above eighteen *braccie*, I realised, on

this account, that it would not be possible to raise mercury up to a height of two *braccie*, on account of the difference between its weight and that of water. For this reason, I was beset with an enormous desire to prepare glass tubes in the same way, until I had brought out to the eye a famous, disputed question hidden from the world. I did this and saw with my own eyes: Place without a located [body], the successive motion of a body in the vacuum, [and] light inhering in no body.[26]

Magni performed this experiment in the court of the Polish King Wladislaus IV Vasa in July 1647, as he narrates in the *facti historia* of his *Demonstratio*. To the suggestions of members of the Mersenne circle, especially Gilles Personne de Roberval, that he had plagiarised the experiment from Torricelli and others during a visit to Italy in 1644–1645,[27] he replied, in a subsequent *Narratio apologetica*, appended to the fourth 1647 edition of the *Demonstratio ocularis*[28] that

When I was in Rome I never saw, or even heard the name of Angelo Ricci. In Florence I never saw Evangelista Torricelli, not because these men are not famous, but because I am unknown to them. In Florence I had frequent discussions, even with Prince Leopold, of the obstinacy with which the Peripatetics hold onto the sentences of Aristotle, even against what is seen and felt. Even when I was asked there if there might be a plan for my philosophy to be published in that city, under the auspices of the Most Serene Grand Duke of Tuscany, nobody mentioned the vacuum to me in that city. In Rome in 1645 the Reverend Father Mersenne did not share any [information] with me regarding that experiment.[29]

Although fully acquiescing in the priority of Michelangelo Ricci and Evangelista Torricelli in carrying out the experiment, Magni reiterated, in the same letter, that he had elaborated his own 'plan to overcome the impossibility of the vacuum' from Galileo's work, and from the use of 'an Archimedean balance' (i.e. a hydrostatic balance) received as a gift in 1644, from which he had discovered that the proportion between the weights of water and mercury was approximately 1 to 13.[30]

The fact that others had performed the experiment before Magni was no reason for him to be unhappy:

By these [tubes] I exhibited the experiments of the vacuum that were published not in order to be able to teach the possibility of the vacuum, but so that from this I could argue for the falsity of the first principle of Aristotelian Physics. For this reason I am delighted for these experiments to be [known] earlier amongst you, and to be confirmed by your authority. I seek life from truth, not praise.[31]

Although the *Demonstratio Ocularis* had only made cursory reference to the path by which Magni came to attempt the mercury experiment, his *Narratio Prima* in response to Roberval gave further details of the labours that had preceded his performances in the Warsaw court of Wladislaus IV:

> It was eight years ago [i.e. 1639–40], when obstructed by various matters, I attempted in vain to have a tube made over eighteen cubits in height, and four years ago (that is, since using the Archimedean balance) when I began to seek a glass tube of three cubits in height. Two years elapsed since my first effort: The glass-makers in Cracow neither wanted to make nor could make a tube suitable for the work at hand. I produced a wooden one, which I still have now, and which has been seen by, among others, the noble man Girolamo Pinocci, an outstanding cultivator of the liberal arts. However, in this the mercury came to rest far below the correct [*debita*] height, and this varied on successive occasions, because of the different quantities of air drawn through the pores of the wood'.[32]

Magni's troubles with wood were ended by the arrival of the expert Venetian glass-blower Gasparo Brunori in Warsaw:

> From this most courteous person I received many tubes of different heights and diameters, including some with a bulbous end.[33]

Glass was to be crucial to Magni's public 'ocular demonstration' in the Warsaw court. Magni could satisfy himself as to the emptiness or otherwise of the space above the mercury in his wooden tubes, most probably by tapping the sides and 'listening' for the meniscus, and could place his finger over the end of a shorter wooden tube to feel, in private, the pull of the 'prohibitive virtue of the vacuum'.[34] In order to render the experiment, and in particular his conclusion concerning the nature of light, public to the successive gatherings of theologians, nobles and *literati* in the Warsaw court, the vacuum had to be transparently open to public inspection.

Magni paraded the various pieces of evidence for the independence of his experimental research before Mersenne's interlocutor in the Polish court, the Queen's secretary Pierre Desnoyers. These included his copy of Galileo's *Discorsi*, the hydrostatic balance, the wooden tubes used in his unsuccessful earlier trials, and two letters from eminent members of the Capuchin and Dominican orders in Rome, confirming that before the appearance of the *Demonstratio* nobody had heard anything about a vacuum experiment in Rome.[35] His trump-card was yet to come, however, and he continued:

> Two years ago the Reverend Father Giovanni Battista Andriani of the Society of Jesus left Rome, where he taught Rhetoric,[36] and came to Warsaw, where he

witnessed [*spectavit*] the vacuum experiments, as a complete novelty. Therefore, it is possible for a Capuchin to be unaware of what is new in philosophy in Rome, if even the Collegio Romano of the Society of Jesus knew nothing of it'[37]

If the Jesuits in the Collegio Romano knew nothing of the mercury experiments in 1645 or 1646,[38] they made strenuous efforts to remedy this after the publication of Magni's *Demonstratio Ocularis*.

Jesuit reactions

1. Niccolò Zucchi

In early 1648 an anonymous nineteen-page pamphlet was published in Rome, with the title *Magno Amico nonnemo ex Collegio Rom[ano] S[ocietatis] I[esu] S.D. Experimenta vulgata non vacuum probare, sed et antiperistasim stabilire*. The author of the pamphlet, written in the form of a letter to the French Jesuit mathematician Jacques Grandamy,[39] was the rector of the Collegio Romano, Niccolò Zucchi.[40]

'You have interrogated a Davus', Zucchi began his response to Grandamy, 'but one who often converses with Oedipi.[41] He has lived with so many of them that perhaps something of their erudition will have stuck to me [sic], by which I might satisfy what has been proposed'.[42] One of the domestic 'Oedipi' to whom Zucchi was alluding was undoubtedly Athanasius Kircher, who, although he had yet to publish his enormous study of Egyptian culture, the *Oedipus Aegyptiacus*,[43] had been brought back to Rome from Malta through the intercession of Francesco Barberini in order to continue his studies of hieroglyphics. Zucchi's treatment of the vacuum experiments was consciously non-technical, in order not to breach the protocols of the epistolary genre. 'As a guardian, it will be permitted for me to have carried out experiments of my own, which I have frequently repeated, in addition to those that have been exhibited by others', he writes, and

> but I do not add the measurements of the glass tubes, of the mercury poured into them, nor of the parts of these left empty by [the mercury] when it descends, both in the vibrations and when at rest, nor do I add the number of the same vibrations, or similar things, as they would be of little help for the solution of those matters that are proposed to me, and my desire is to retain the brevity of the letter, as much as is possible, without paying the price of vigilant truth.[44]

Zucchi describes four experiments: the mercury experiment as performed by Magni, the water experiment carried out by Gasparo Berti, the experiment carried out by Zucchi himself on antiperistasis with a thermoscope in a cave, and an experiment carried out on the freezing of water in glass containers.

To justify his inclusion of such a large number of experiments he writes

> It has pleased me to multiply knots, so that by untangling them it will appear more clearly that it is not necessary to forge new dogmas every day from the difficulties that arise, or to recast the antiquated comments of the ancients, but that it is better to curb intemperate minds by cooling them down, than to heat them up, so that while some attack the received philosophy with a sincere heart, others will not accomplish greater destruction by the study of novelty.[45]

Zucchi's treatment of the Berti experiment, later echoed by Kircher in his *Musurgia Universalis*,[46] used the fact that the bell in the space above the water produced a sound to argue that, as no accident could exist without a subject, the space was full.[47]

Zucchi's 'solution' to mercurial suspension was the introduction of a 'subtle body', or 'spirit', which was extremely distended and filled the upper part of the Torricellian tube.[48] When water was reintroduced to the tube, and displaced the mercury, this spirit, which had the property of being extremely abhorrent to water, but extremely attractive to mercury, was reabsorbed violently into the mercury, and hence the space disappeared.

Very little evidence has survived to document Zucchi's entry into the debate on the vacuum. A large collection of his papers, consisting for the most part of an enormous spiritual diary, contains only a single-leaf, inserted inside the back cover, on which there is a fragment of what appear to be notes made by Zucchi on Magni's work prior to composing his reply.[49]

Zucchi's biographer, Daniello Bartoli, recounts how nature's abhorrence of a vacuum entered Zucchi's spiritual advice on humility to those in his charge, to such an extent that he adopted the rhyming nickname 'Turabuchi' or 'stop-gap'. Zucchi, according to Bartoli, used to advise future Jesuit preachers that

> As nature performs violent acts, that appear to be miracles against the order of nature itself in order to avoid the vacuum, such as when heavy bodies rise and light ones descend [...] thus, so that there is no Vacuum in the pulpit you must make yourselves ascend to it, not for the ability that you have, but because of the extrinsic virtue of obedience.[50]

Zucchi's pamphlet did not go unnoticed outside Rome. From Münster, Fabio Chigi, the papal nuncio, wrote to Athanasius Kircher in Rome to describe the reception of the pamphlet.[51]

I passed Father Zucchi's diatribe on to the experts [*periti*] and it [was] disputed zealously with four members of the Society for two hours, with many reasons and experiments, just as he relates. I hope that your letter might please Father Maderson, whose mind is oppressed by continuous worries, and afflicted by such a long effort to obtain peace for the city, wavering between the various vicissitudes of war.[52]

Before travelling to Münster, as the lengthy peace-negotiations drew to a close, Chigi had been in Cologne. From there the future pope had written to Kircher to criticise the vacuum experiments performed by Magni in no uncertain terms: 'I strongly doubt the vacuum of the other grey-bearded innovator; and I fear lest one should discover that he is complaining in his Hood, having prattled on so often in vain'.[53] Kircher's relationship with Chigi dated back to Kircher's time as confessor to the famous convert Cardinal Frederick, Landgrave of Hessen-Darmstadt in Malta, which coincided with Chigi's period as apostolic delegate.[54] Chigi's links to the Jesuits became ever closer as he became involved in the anti-Jansenist debate that followed the 1640 publication of Jansenius' *Augustinus*.[55] Most probably Kircher sent him Zucchi's 'diatribe' in response to his earlier letter criticising Magni.

2. Paolo Casati: The Eucharist enters the debate

Another recipient of Zucchi's treatise, this time from the author himself, was Paolo Casati, professor of mathematics at the Jesuit college in Bologna.[56] For Casati, neither Zucchi's answer nor the treatise written by the Bolognese physician Giovanni Fantuzzi sufficiently confuted the claims of the Capuchin friar. 'Many things are intact', Casati wrote to Giannantonio Rocca.[57] 'I have people who are vigorously encouraging me to write an answer, more adequate and distinct than Fr. Zucchi's one'.[58]

Casati's lengthy response, the *Vacuum Proscriptum*, was published in Genoa the following year, after being strongly endorsed by the Jesuit censors.[59] As well as including a huge number of new experiments and physical arguments on the vacuum, Casati introduced, very early on in his treatise, an element that had been absent from Zucchi's pamphlet. Responding to Magni's claim to have demonstrated that light inheres in no substance, he wrote:

> But here who does not hear the heterodox miracle-haters shouting out and mocking the carelessness of the orthodox, who have been so negligent in the investigation of nature, that they have taken an accident not inhering in a

substance in the most sacred mystery of the Eucharist for a miracle? [...] Indeed, what do the holy operations of the priest confer on the accidents of the bread if they are now separated from the bread by the work of nature? Is it that which the Tridentine Synod defined by Transubstantiation? But, so that in carrying out the mysteries the minister does not differ from Christ, it is required that before transubstantiation, the substance remaining can be truly said to be bread. This is very well, but who defines whether the sensible accidents are separated, or conjoined to the substance? For however much it is conceded that this combination of accidents inheres in the substance of bread, with the ineluctable bond between substance and accidents dissolved, whenever no body stands under an accident (as Fr. Magni affirms about the light pervading the empty glass tube) who will not equally be uncertain, whether the whiteness, flavour and the other attributes of the bread remain abstracted from its body by the powers of nature?[60]

Casati was alluding to the famous thirteenth session of the Council of Trent of 11 October 1551, in which the Thomist account of transubstantiation had been re-affirmed against the deviant theories propounded by Luther and Melanchthon.[61] Magni had feared such a reaction, which was also hinted at in the contemporaneous works of Pallavicino[62] and Kircher,[63] since at least July 1647. On 3 August of the same year he wrote to Giovanni Barsotti in Rome, to tell him that:

I was visited on the last day of July by the Auditor of the Most Illustrious Mons. Nuncio, while I was with two Piarist fathers. When we spoke about my demonstration of the vacuum, the objection of some theologians was mentioned, that with such a demonstration the miracles due to transubstantiation in the Holy Sacrament are removed. This gave me occasion to say what follows. I said, that St. Augustine and St. Thomas Aquinas in particular, disapproved of the authorities of the Faith becoming involved in philosophical questions easily, as very dangerous to the Faith itself, whence I added that this was the current situation, that is, the involvement of the truth of the aforementioned miracles with the most violent ocular demonstration of the vacuum.[64]

Magni was apprehensive that 'some people who are not well-disposed towards me', might cause difficulties with the nuncio, and that the latter would find himself abandoned by theologians, none of whom desired to be the author of the proposition that the vacuum removed the miracles from the Eucharist.[65] Fearing that his projected philosophical reform might prejudice his chances for the Cardinal's purple, despite the recommendations of Wladislaus IV,[66] Magni completed his letter with a bid for recognition of his services to the Faith, and a plea that 'Your Lordship will not show my [letter] to anyone living, except for the brother of the Most Illustrious Monsig. Nuncio, when you know it to be necessary to do so'.[67]

3. Nature on trial: Étienne Noël's response to Pascal

Blaise Pascal's *Expériences Nouvelles touchant la vide* were published in October 1647, three months after the appearance of Magni's *Demonstratio*.

The work occasioned a rapid response from the Jesuit rector of the Collège de Clermont in Paris, Étienne Noël, who had recently published a resumé of the Aristotelian physics courses that he gave at the college of La Flèche.[68] The Collège de Clermont seems to have been well provided with the material prerequisites for experimental investigation and other artificialia in this period, if we are to judge from the notes made by Samuel Hartlib in his *Ephemerides*, admittedly written eight years later:

> The Iesuits in Collegio Claremontano kept an 1. Hydraulicum. 2. Opticum et 3. Mechanicum Collegium. They are very communicative and are trying all manner of Optical. etc. conclusions having many choice Rarities [...] They have made in one roome an exact Representation of systema Copernicanum Mobile. And they are going about to make an other systema Tycho Brahaicum Mobile [...] The Claremont-Iesuits have in their Church a Lampe with a Concave Glasse which gives as much light as a hundred candles.[69]

Noël's first letter to Pascal complemented the *Expériences Nouvelles*, on being '*fort belles et ingénieuses*' but gently rejected the existence of the vacuum on physical grounds.[70] Glass is porous – we know this from the fact that light passes through it and from various other phenomena, such as the fact that a sealed bottle placed on hot ashes will not break.[71] The apparent vacuum is another body, which enters through the pores of the glass and follows the retreating mercury down the Torricellian tube.[72] Pascal's reply to Noël attempted to display the incoherence of his Aristotelian terminology, such as his definition of light as '*un mouvement luminaire de rayons composés de corps lucides, c'est-à-dire lumineux*', and magnanimously complimented him that 'one cannot deny you the glory of having upheld the peripatetic physics as well as it is possible to do so'.[73]

Unlike his initial letter to Pascal, Noël's response brought the theological problems of the vacuum to the foreground. 'If you say to me that the species of the Holy Sacrament have parts, separated from each other, but are nonetheless not bodies, I will answer that firstly, by the composite of parts separated from each other we understand that which we normally call long, wide and deep, and secondly, one can explain the doctrine of the Roman Catholic Church concerning the species of the Holy Sacrament, by saying that the small bodies that remain in the species are not the substance of the bread. This is the reason for which the Council of Trent never uses the word accident when speaking of the Holy Sacrament, despite that in effect these little bodies are really the accidents of the bread, according to the definition of accidents accepted by the whole world, which does not destroy the subject, whether present or absent. Thirdly, [I will say that] without a miracle, every composite of parts separated from each other is a body, and I think that, in order to decide on the question of the vacuum, it is not necessary to have recourse to miracles, seeing as we are presupposing that all of your experiments contain nothing beyond the forces of nature'.[74] Noël's explanation of the Eucharist, equating the quantitative and qualitative accidents of the Eucharist with 'little bodies' was heterodox, and ran against the directives of the Jesuit college of revisors in Rome. A document composed by the *Revisores* in May 1632 made this clear for the case of accidents 'which have a contrary', i.e. qualities such as whiteness, while censoring the proposition that 'both material substantial forms and accidental forms which have a contrary, have a matter [*materia*] of which they are constituted: it is otherwise for immaterial substantial forms and material accidental forms having no contrary'.[75]

The *Revisores* of the Collegio Romano assessed this proposition as follows:

> Of the second proposition we say that it, similarly, is improbable and must be retracted, all the more so because certain absurdities, contrary to the Mystery of the Eucharist, can be deduced from it; e.g. that the whiteness of the Host (as a material accidental form having a contrary) is constituted from matter [*materia*]; or rather that the thing itself is completely composed of matter and form. This opinion, if defended openly by [the] author, would not pertain to our censure but to that of the Inquisition.[76]

Despite Noël's heterodox opinion of the nature of the Eucharistic accidents, the tone of his second letter, and the publication of his *Le Plein du Vide* in 1648 make it clear that he regarded Pascal's short treatise as a major threat to the Aristotelian structure around which his career had been built. The dedication of Noël's work, which appears to have been published at the end of January 1648, proclaims that Noël will demonstrate the 'integrity' of Nature, accused of the void, and show the falsehood of the deeds [*faits*] with which She is charged, and the impostures of the witnesses that are opposed to Her. If She was known to everyone as well as She is known to Your Highness, to whom she has revealed all her secrets, She would have never been accused by anyone, and one would have never dared to try Her on the basis of false depositions, and experiments that are poorly understood and even more poorly confirmed.[77]

In Warsaw, Magni too was quick to see the significance of Pascal's work for his anti-Aristotelian agenda. Desnoyer's informed Roberval that 'Our Capuchin wants to translate the little treatise of Monsr. Pascal into Latin, and print it, as he is making a shield of everything [against Aristotle]. His principal aim does not end with the vacuum, as *he doesn't just want Aristotle to vacate the universities, but to be excommunicated*'.[78] Just as Pascal was later to recruit the anti-Jesuit invective of Magni's *Commentarius . . . De homine infami*[79] in the *Provinciales*,[80] Magni attempted to recruit Pascal to multiply his lines of attack on Aristotelian physics, although the promised translation never appeared. He was persuaded to remove his *De Atheismo Aristotelis* from later editions of his *Demonstratio Ocularis*, on the advice of friends who persuaded him that he would prejudice his chances of the Cardinalate.[81] Although Noël's *Le Plein du Vide* addressed, principally, the experiments conducted by Pascal, as reported in the *Expériences Nouvelles*, Noël made it clear to Pascal privately that his most bitter invective was aimed at Magni.[82]

Paris, Rome and Bologna were not the only centres from which the Jesuits attacked Magni's ocular demonstration. In distant Lithuania, two members of the Jesuit college in Vilnius responded to the *Demonstratio*. In 1648 the prolific philosophy professor and theologian Albertus Kojalowicz-Wijuk[83] composed his *Oculus ratione correctus*[84]- 'The eye corrected by Reason', a 104-page treatise attacking Magni. Kojalowicz-Wijuk's older colleague, the mathematics professor at Vilnius, Oswald Krüger[85] published his *Dissertationes de vacuo*[86] in the same year.

Defending Jesuit educational hegemony in the empire

The massive scale of the Jesuit reaction to Magni's version of the Torricellian experiment demands explanation. Vilnius is a case in point. The *Annual Letters* of the *Collegium Vilnensis* for the years 1646–1648 paint a harrowing picture of daily life:

> Two Jews were brought towards the sacred font, and well-disposed [for conversion]. In the areas withdrawn from the frequency of parishes, about twenty clandestine marriages were ratified. Many people were turned away from bestial habits, and shameful acts of sodomy, many more were called from the earnings of prostitution and disgracefully licentious lives back to a life worthy of a Christian [...] A certain woman, who was for a time pestered by the incessant and impudent molestations of an *incubus*, lost almost all her strength.[87]

In the midst of this quotidian chaos, Krüger and Kojalowicz-Wijuk replicated and disputed Magni's experiments, and printed their refutations at the expense of the College. The situation in Münster is perhaps even more extraordinary – Fabio Chigi, papal nuncio and one of the key players in the Westphalia settlement, is spending valuable time writing letters about the consequences of Magni's experiment, and discussing it for hours in the company of the Jesuit fathers of the college in which he is lodged. In Bologna, shortly before he is sent on the crucial mission to Stockholm to convert Queen Christina, Paolo Casati is being 'vigorously encouraged' to answer Magni's experiments. Zucchi, having failed to convert Kepler on a diplomatic visit to the court of Ferdinand II, is willing to take time off from his multiple duties as Rector of the Collegio Romano in order to conduct experiments with thermoscopes in caves, while preparing his refutation of Magni.

To say that these Jesuits were concerned to defend Aristotelian Physics against Magni's onslaughts is undeniably true, but does not account for the scale or violence of the reaction. The censures of philosophical propositions conserved in the Roman Jesuit archives allow one to document the attempts to police internal deviance from Aristotelian physics by the Roman College of *Revisores*. These reveal that propositions denying the 'ineluctable bond' between substance and accident described by Casati and propounding the composition of the continuum of indivisibles, both foregrounded in Magni's interpretation of the vacuum experiment, were regarded with increasing suspicion by the Jesuit *Revisores* during the 1630s and 40s. In 1633, General Muzio Vitelleschi wrote to Ignace Cappon in Dole to emphasize his strong feelings on the issue:

> As regards the opinion on quantity made up of indivisibles, I have already written to the Provinces many times that it is in no way approved by me and up to now I have allowed nobody to propose it or defend it. If it has ever been explained or defended, it was done without my knowledge. Rather, I demonstrated clearly to Cardinal Giovanni de Lugo himself that I did not wish our members to treat or disseminate that opinion.[88]

De Lugo, who was a a significant opponent to Jesuit 'Zenonism' – the term used to describe those who attempted to define the continuum intrinsically in terms of an infinity of indivisible points, and in opposition to Aristotle – had been Casati's old teacher at the Collegio Romano.[89] It was widely feared that such a theory of the continuum, as espoused by Galileo and later Cavalieri, was close to Wyclif's theory of the Eucharist as a *'corpus mathematicum'*, and had been condemned by the Council of Constance in 1415. In February 1642, Mersenne worried that the Cartesian theory of body might be prone to such a theological attack, and forwarded Descartes a letter describing the propositions condemned at the Council of Constance.[90]

Ten years earlier, the Jesuit philosopher and theologian Roderigo de Arriaga had published his *Cursus Philosophicus*,[91] which advocated a theory of the continuum as composed of indivisibles. Arriaga felt it necessary to distance himself from Wyclif explicitly, and included a chapter in which he denied that his theory of the continuum was among the condemned theses of Wyclif.[92] Although his theory contravened Vitelleschi's wishes, and went against the position adopted on the continuum by the *Revisores* in their censures of philosophical positions, successive editions of the work still contained the unorthodox treatment. In the 1659 edition, Arriaga included a preface which suggested a reason for his deviance, claiming that he had been permitted to publish his anti-Aristotelian opinions 'in part because they are completely accepted here at the university of Prague'.[93]

The Jesuits and the Carolinum: Roots of the conflict with Magni

Behind Arriaga's defence lies a deep tension. Since long before the 1618 rebellion, the University of Prague had been a 'center of heresy, infecting all of Bohemia through the many lawyers, notaries, teachers and government officials it graduated'.[94] Jesuit control in the University dated from the negotiations of 1622–23, when Ferdinand II, advised by his Jesuit confessor Lamormaini, agreed to give the Jesuits responsibility for the teaching of philosophy and theology in the University, and direction of the new university, which was to be an amalgamation of the old Carolinum and the Jesuit Collegium Clementinum.[95] With the defeat of the Protestant forces at the Battle of White Mountain, Ferdinand II revoked the brief of Rudolph II granting tolerance to Bohemian Protestants. The handing-over of the Carolinum to Jesuit control consolidated the new aggressively anti-Protestant imperial policy. Ferdinand's decision was not without its opponents, and one of the loudest critical voices was that of none other than Valeriano Magni, then the guardian of the Capuchin convent on the Hradschin and provincial superior of the Capuchins in Bohemia. Magni was the spokesman for the archbishop of Prague, Cardinal Ernst von Harrach, and began a zealous campaign to 'liberate' the Carolinum from Jesuit control, and restore the autonomy that it had enjoyed since the bull of Boniface IX of 1397.

As Robert Bireley recounts, after 1622 the university was intended to become a 'center for orthodoxy for all Bohemia', and a model for the Empire as a whole. The consequences included the banning of all non-Jesuits from the theological faculty of the university and Jesuit responsibility for the censorship of all books published anywhere in the kingdom. Lamormaini, the imperial confessor, took a hard-line on restoring the Catholic faith in the empire, and chastised even other Jesuits for moderacy. He wrote to General Vitelleschi to insist that 'piety can only be restored in Bohemia by a powerful authority',[96] a position that was vigorously countered by Magni, who, with Von Harrach urged an independence of Church and Empire. Magni's conflicts with the Jesuits, which took many forms in the successive decades, date from this time.[97] While Lamormaini urged both Ferdinand II and Muzio Vitelleschi to extremes of religious intolerance, Magni preached accommodation and ecumenism, and attempted to curb Jesuit power by every means possible, including a letter to Muzio Vitelleschi to plead for the reform of the order. By the 1630s relations between Magni and the order were soured beyond repair. Arriaga wrote to Vitelleschi in 1628 to denounce Magni as a mendacious slanderer, a near-apostate who was attempting to blacken the reputation of the Jesuits with Urban VIII.[98] Vitelleschi advised Lamormaini that Magni was 'the chief author and inciter of every controversy and difficulty' between the Society and the Archbishop of Prague.[99] In 1631, Paolo Anastagi wrote from Prague to complain to Vitelleschi that 'whoever wishes to be helped and promoted by Father Magni only needs to show himself to be disgusted with the Jesuits. I can say with good conscience that all of those here who have just left the Society have been promoted beyond their merits and capacities by Father Magni'.[100]

By attempting to temper the new Jesuit hegemony over the oldest university of the Empire with the philosophical traditions of the Carolinum, Arriaga was running the risk of being accused of perverting his teachings with the Wyclifian and Hussite theses that had been embraced so warmly in the earlier history of the university.[101] Small wonder, then, that he should distance himself explicitly from Wyclif in his discussion of the composition of the continuum.

After the death of Ferdinand II in 1637, the alliance between the Jesuits and the emperor was substantially weakened. Although the new emperor Ferdinand III had himself been taught by Arriaga as a child, he did not nurture the strong political and ecclesiastical links with the Jesuits cultivated by his father. Lamormaini's projected coërcion of Bohemia into orthodoxy through a Jesuit order empowered by the Emperor lost force considerably with the Peace of Prague in 1635, and the terms of the Peace of Westphalia ended all hopes of religious unity within the empire, just a few months after Lamormaini's death. It was in this context that Magni, removed from the political power that he had enjoyed as Von Harrach's interlocutor, struck at the philosophical principles on which the legitimacy of Jesuit hegemony over education in the empire was based with the new metaphysics, expounded in the *De luce mentium* and the experimental physics of the *Demonstratio ocularis* and *Experimenta de incorruptibilitate aquae*.

Domesticating the Torricellian experiment

In contrast to the telescope, in dealing with the Torricellian experiment, and, later, Guericke's *antlia pneumatica*, Jesuit authors repeatedly insisted that you could not believe your eyes. In attempting to disable the device as a philosophical instrument, they recruited a host of invisible agents. These ranged from the sound of the bell discussed by Kircher (in an experiment that all but defied replication) and the sense of touch emphasised by Linus in his *Tractatus de corporum inseparabilitate*, to the minute pores in the glass of the tube invoked by Noël and later Schott,[102] the subtle mercurial spirit of Zucchi, and the real presence in the Eucharist deployed by Casati.[103] The Jesuit theologian and natural philosopher Melchior Cornaeus expressed this position eloquently in his discussion of the new vacuum experiments of Magni and Guericke:

> And since the Vacuists appeal so earnestly to the judgement of the eyes in this business, why do they not see that this thing has itself been confectioned for the judgement of the eyes? Why will they not finally acknowledge that subtle air, drawn out by an occult fear, takes the place of the extracted denser air, or water or smoke and prevents a vacuum?[104]

A refusal to allow the instrument to produce new natural philosophy did not put an end to Jesuit discussions of hydraulics. Instead, the device was removed from circulation in the philosophical domain and relocated within the context of the *Wunderkammer*. Kaspar Schott's *Mechanica-Hydraulico Pneumatica* largely an account of the hydraulic devices present in Kircher's museum in the Collegio Romano,[105] includes both the Torricelli/Magni experiment and the Berti experiment in a section entitled *De machinis hydraulicis variis*, where it is surrounded by a ball made to spin in the air, a perforated flask for carrying wine known as the 'Sieve of the Vestal Virgin', and a 'phial for cooling tobacco smoke'. Unhealthy philosophical readings of *Machina VI* (the Torricelli and Berti tubes) are dismissed by Schott, echoing his former teacher Kircher, as the writings of '*Neotherici Philosophastri*' and 'insolent and unmannerly braggarts proclaiming a triumph before victory'.[106] To situate the Torricellian device in the context of trick fountains and water-vomiting seats was to insulate it from the Aristotelian philosophy taught in the classrooms of Jesuit colleges that Magni wished to 'excommunicate'. In the Prague context, a work on springs and fountains published by Jakob Dobrzensky de Nigro Ponte in 1657 included various refutations of the possibility of the vacuum by Jesuits and close supporters of the intellectual programme of the order including Joannes Marcus Marci[107] and Godefridius Aloysius Kinner.[108] The work also contained numerous hydraulic machines and clocks designed by Prague-based Jesuits including Theodore Moretus and Valentin Stansel[109] and other Jesuit exponents of the hydraulic arts such as Niccolò Cabeo,[110] Mario Bettini and Kircher. Later, Francesco Lana Terzi, another disciple of Kircher, suggested the use of globes evacuated by the Berti technique to make an airship, while sidestepping the issue of whether the globes were truly empty.[111]

The story of the Torricellian experiment is the story of a shrinking instrument. From Baliani's large-scale copper siphon, to Berti's unwieldy outdoor device to Magni's wooden tubes and Torricelli's glass tubes, a messy, unstable, opaque, ambiguous phenomenon became reduced to a manageable size. In the hands of Magni and Pascal this mobile, transparent device was coupled to a militant anti-Aristotelian agenda that threatened to weaken the Jesuit grip on philosophical education in Catholic Europe, especially during a period in which Jesuit authors were being disciplined more strenuously than ever for departures from Aristotle in natural philosophy.[112]

Magni himself saw no clear boundary between his political dispute with the Jesuits and the vacuum debate. Writing to Mersenne in 1648 to describe his disgust at Albertus Kojalowicz-Wijuk's *Oculus ratione correctus* he added 'this is not going to excite my love towards that Society'.[113] He dismissed the theological arguments adduced against the vacuum, writing of his accusers that 'if they lack reasons, they adduce truths from theology, not revealed by God but commented to them in private, above those that we commonly believe'.[114]

Much later, in his *Principia et specimen Philosophiae axiomata*, published in 1661 Magni was to return to the problem of reconciling his theory of the vacuum with the mystery of the Eucharist. His method consisted in placing the Eucharistic wafer itself inside the Torricellian tube:

> If bread is placed in the empty part of the tube, and light [*lumen*] shines on it, the light [*lux*] inhering in the bread will be affected by the size, shape and colour of the bread in such a way that this light will be distinguished to the eye from the remainder of the light shining through the tube, empty of any body. If God was to annihilate the substance of the bread, and the light that was in the bread was not annihilated, this light would remain, in the presence of the light source, without a miracle, but freed from the imperfections that are gathered together in the body of the bread. That is, the light would not bring back [*refere*] the size, shape and colour of the bread, and neither could that light conserve the species of size, shape and colour of the same bread without the bread as subject. By this I have explained in passing, that my opinion on light in the vacuum detracts neither from the creation by God nor from the miracle of the Most Holy Sacrament. Everything, however, is subject to ecclesiastical censure.[115]

Unfortunately for Magni, in the same year that this attempted marriage of the philosophy of vacuum with eucharistic dogma was published he was incarcerated in Vienna on the orders of Alexander VII because of his polemical writings against the Jesuits.[116] As the Jesuits confined the Torricellian tube to the *Wunderkammer*, their long-term ally Fabio Chigi confined its most outspoken exponent to a prison-cell.

The antlia pneumatica

Shortly after the Torricellian experiment had been domesticated, Jesuit plenism came under renewed threat. The manuscript for Gaspar Schott's *Mechanica Hydraulico-Pneumatica* was submitted to the Roman *Revisores* by January 1654.[117] Schott had completed the body of the work in Rome, while still a '*Socius*' of Athanasius Kircher.[118] On Schott's return to Mainz,[119] however, he heard word of a 'new instrument to show that a vacuum can exist, or which wanted to show me this',[120] which he witnessed at first hand in Würzburg the following year in the residence of the Archbishop of Mainz, Johann Philipp von Schönborn. Although the *imprimatur* for his *Mechanica* had been issued by General Goswin Nickel on 23 January 1655, in view of the new experiments of Guericke, and sanctioned by both the written approval of Kircher[121] and a *'privilegi[um] Sacrae Cesarae Majestatis'*,[122] Schott later added an appendix dealing with the new Magdeburg experiments that by-passed the Roman censors.

Guericke's principal devices worked by extracting water from glass or copper receivers, a laborious process that required robust experimental assistants. Early trials with wooden casks had been inconclusive – after water was extracted from a cask by vigorous pumping, a noise was heard 'like vigorously boiling water',[123] and the keg was later found to be filled with air. By the time of the Imperial Diet of 1653–54, Guericke, who had previously worked as an engineer for the Swedish army during the Thirty Years' War, had ironed out his initial problems and produced a workable pump. Towards the end of the *Reichstag* Guericke demonstrated his experiments to various dignitaries, including Ferdinand III and Schönborn. Magni was also present at the Diet, and displayed his mercury experiments to Guericke, to whom he also presented a copy of his *Demonstratio ocularis* [**See Figure 4.1**] Schönborn, whose period as Elector of Mainz represented the first real break with Viennese political hegemony in the Empire,[124] was particularly impressed by Guericke's equipment and purchased the pump from him. When the device was reassembled in Schönborn's residence in Würzburg, the Jesuits from the local college, including Schott, had an opportunity to examine it carefully.

Schott sent a report of these examinations and further experiments to the German Assistant of the Jesuits, resident in Rome, in which he asked for the opinions of Kircher and Niccolò Zucchi. He published their replies, and a discussion of the experiment by Melchior Cornaeus, the local theology professor in the Jesuit college in Würzburg, in the appendix to his *Mechanica*, along with letters from Guericke giving his own interpretation of the experiment and answering Schott's queries. Schott writes:

> There are some who attempt in every way by means of this engine [*machinamentum*] to affirm vacuum (that was until now a phantasm either to think of or to hope for, being resisted by the plenitude of nature, invulnerable from even an angel). Others, on the other hand, affirm that [the possibility of a vacuum] cannot be eliminated by anything more effectively than by this experiment.[125]

Describing Guericke's machine in words and pictures, Schott allows his reader to be the judge between the opposing positions, as long as he will not be guided by his preconceptions.[126] Despite this posture of impartiality, it is telling that three of the 'experts' to whom he refers the machine[127] are Jesuits, whereas the only voice in favour of the vacuum is that of Guericke himself, whose interpretation

Figure 4.1 Otto von Guericke's antlia pneumatica, from Kaspar Schott, *Mechanica hydraulico-pneumatica*, Würzburg, 1657

remains irredeemably tied to his person. Kircher's response to Schott's enquiry emphasises the labour involved in the experiment:

> If there is a vacuum there, I ask, what should make such difficulty? Certainly not air, as it has been extracted, so therefore that 'nothing' that has been left after the extraction of the air. But who can conceive of nothing causing resistance, or has ever heard of this in Philosophy?[128]

Zucchi's response refers to the virtue in bodies preserving unity and contiguity in the universe that he had described in his letter to Grandamy.[129] Cornaeus refused to admit that such an experiment was sufficient to refute the authority of antiquity and described the collapse of the sides of a well built in the Jesuit college in Paderborn, demonstrating the triumph of nature over all human efforts to produce a vacuum.[130]

Although Guericke approved of Schott's wishes to publish the first account of his experiments and their various interpretations,[131] the way in which the experiments were published arguably reinforced the status of both the Collegio Romano and the Jesuit college in Würzburg as centres of expertise in matters both philosophical and hydraulic. Schott dedicated the appendix of his work to Kircher, praising his immense erudition at some length, and thus lending further authority to his opinion of the experiment as presented in the body of the text. Schott anticipates Kircher's reaction to the appendix in the letter of dedication:

> But do you offer me a vacuum? – you ask – On the contrary, not a vacuum at all, even by your own calculation.[132]

Kircher's letter reinforced his familiarity with hydraulic techniques, stating that he had experienced nature's enormous resistance to rarefaction 'hundreds of times in similar machines',[133] but regretted that he 'didn't have more time to refute the whole contraption [*machinatione*] from first principles'.[134] Kircher also sends Schott greetings from his new '*socius*', the Jesuit Valentin Stansel, 'also most versed in Hydrostatics'.[135] Zucchi's letter thanks Schott for his letter seeking opinions on Guericke's device 'both from me and from other more expert people'.[136]

In the case of the Guericke experiment, we see the poles of Würzburg and Rome functioning as a split laboratory. The experiments performed in Würzburg do not need to be repeated in Rome, due to the highly conductive channels of information between the two colleges. Although a diagram of the Guericke device is present among Niccolò Zucchi's papers in Rome,[137] the fact that he suggests that the Würzburg Jesuits attempt the experiment in a darkened room to see if any glowing rays or flashes of light appear in the empty space confirm that he did not undertake any replication of the experiment.[138] The faithful reports of the Würzburg experiments undertaken by Schott and his fellow Jesuits were deemed to be sufficient grounds for his expert opinion.

Expert knowledge of matters other than religion had come to be highly expedient to the apostolic goals of the Society of Jesus during the course of the seventeenth century. Jesuits who attained privileged status through various types of consultancy were in strong positions for encouraging piety and advancing the wider interests of the order. As Orazio Grassi suggested in 1619[139] and Fabio Chigi confirmed in 1648,[140] Jesuit colleges were perceived as sites to which a curious elite could travel in order to find answers to arcane questions, whether concerning comets or the Torricellian tube.

The tight epistolary links between colleges, so clearly evinced by the appendix to Schott's *Mechanica* meant that even if the requisite expert knowledge was unavailable locally, distant specialists could be mobilised with little difficulty. Such a presentation of the Jesuit college could come under threat from both within and without.

Individuals or groups outside the order, who laid claim to privileged access to arcane knowledge, could undermine Jesuit authority, and attacks on Paracelsus, Van Helmont, the Rosicrucians, Jansenists, alchemists, astrologers, Galileo and Magni by Jesuit authors must in part be read as attempts to defend the authority of the Jesuit order over erudite questions regarding the natural world.[141] As Sforza Pallavicino said of philosophy, 'the monopoly on that precious commodity is not conceded to certain people who, in the guise of necromancers, with certain horrendous and obscure words, make the common people venerate them for their singular wisdom'.[142] The material corollaries of expertise included mathematical instruments and machines.

One resource for individuals and groups competing with the Jesuits as consultants in different arcane areas was to claim that the supposedly objective information provided by Jesuit 'experts' was not the product of disinterested enquiry, but was instead highly coloured by the religious goals of the order.[143] In the case of the vacuum, Boyle's response to Linus accounted for the popularity of an explanation of the Torricellian experiment in terms of some rarefied substance among Jesuits in terms of 'perhaps its congruity to some articles of their religion'.[144] The series of moves to enforce adherence to Aristotle in the order on many fundamental physical matters during the 1630s and 40s, culminating in the 1651 *Ordinatio pro studiis superioribus*, made it increasingly difficult for Jesuit natural philosophers to strike a disinterested pose when such points were at stake, hence the very real danger represented by opponents such as Magni, to the intellectual reputation of the Jesuits.

Notes

1 This chapter expands significantly on the discussion of the role of the Jesuits in the vacuum debate in Michael John Gorman, *Jesuit explorations of the Torricellian space: carp-bladders and sulphurous fumes*. Mélanges de l'Ecole Française de Rome. Italie et Méditerranée. 1994; tome 106(fasc. 2): 7–32.

2 Sforza Pallavicino to Carlo Roberti, n.p., n.d., Biblioteca Casanatense Ms. 4983 ff. 38r-v in Pallavicino, *Opere,* Tom. II pp. 19–20.

3 'Laudo, probo, et ut voce tua utor, osculor sanctissimas illas Academiae vestrae Leges; Libertatem illam dicendi, arcanum illud, altercationum Scholasticarum proscriptionem approbo, eamque experimentorum Sylvam in re Physica maximi facio, quae nisi praemittatur, nihil unquam in illa facultate assequi possumus', Honoré Fabri to Lorenzo Magalotti, Rome; 31 July 1660; BNCF, Ms. Gal. 283 ff. 76r-77v, on f. 76v.

4 Ibid. Magalotti seems to have encountered Fabri while a student at the *Collegio Romano*. See Eric Cochrane, *Florence in the forgotten centuries 1527-1800*. Chicago and London: The University of Chicago Press; 1973, pp. 231-4. On Fabri's troubled relationship with the academicians of Prince Leopold during the 1660s see also Albert van Helden, *The Accademia del Cimento and Saturn's ring*. Physis. 1973; 15: 237-259 and John L. Heilbron, *Honoré Fabri, S.J. and the Accademia del Cimento*. Actes XII Cong. Int. Hist. Sci.; 1968.; 1971: 3b, pp. 45-49.

5 Evangelista Torricelli to Michelangelo Ricci, Florence, 11 June 1644, in Torricelli, *Opere*, vol. III, p. 216.

6 On the relationship between the science of hydrostatics and practical problems of water management in seventeenth century Italy see Cesare S. Maffioli, *Out of Galileo: the Science of Waters, 1628-1718*, Rotterdam: Erasmus; 1994.

7 'Ci conviene far che un'acqua di due oncie di diametro in circa traversi un monte, e, per farlo, conviene che l'acqua salisca a piombo 84 palmi di Genova, che son circa 70 piedi geometrici', Gio. Battista Baliani to Galileo Galilei, Genova; 17 July 1630, OG XIV, 124-125

8 'Però questo sifone non fa l'effetto desiderato; anzi aperto, ancorchè chiuso dal di sopra, l'acqua esce da tutte le parti, e se si tien chiuso da una parte, aprendo dall'altra, ad ogni modo da questa esce l'acqua', ibid.

9 'Avviene un'altra cosa che mi fa stupire; et è, che aprendosi la bocca A, esce l'acqua sin che dalla parte D sia scesa per la metà in circa, ciò è sin a F, e poi si ferma', ibid.

10 Galileo Galilei, *Two New Sciences*, translated by Stillman Drake, Madison; 1989, p. 33.

11 See William R. Shea, *Galileo's Atomic Hypothesis*, Ambix. 1970; 17: pp.13-27, Marie Boas, *Hero's Pneumatica: a study of its transmission and influence*. Isis. 1949; 40: pp. 38-48.

12 See Paolo Sarpi, *Pensieri Naturali, Matematici e Metafisici*, a cura di Luisa Cozzi e Libero Sossi. Milan and Naples: Riccardo Ricciardi; 1996, pp. 238-9, 333, 419, also Gaetano Cozzi, *Paolo Sarpi tra Venezia e l'Europa*. Torino: Einaudi; 1979, and Libero Sosio, *Galileo Galilei e Paolo Sarpi*, in *Atti del Convegno Galileo Galilei e la Cultura Veneziana, Venice, 18 June 1992*, Venice: Istituto Veneto di Scienze, Lettere ed Arti; 1995: 269-311.

13 Stillman Drake, *Galileo at work*. Chicago, London: University of Chicago Press; 1978, p. 386.

14 On the Berti experiment see Athanasius Kircher, *Musurgia universalis, sive Ars magna consoni e dissoni*, Rome: Francesco Corbelleti, 1650 p.11 ff, Emmanuel

Maignan, *Cursus Philosophicus concinnatus ex notissimis cuique principiis*, Toulouse: Apud Raymundum Bosc.; 1653 Tom. IV esp. p.1849, Kaspar Schott, *Mechanica Hydraulico-Pneumatica*, Würzburg; 1657, p. 306 ff, idem., *Technica Curiosa sive mirabilia artis, Libris XII. comprehensa.* Würzburg: Jobus Hertz; 1687 [1664] pp. 202-204. The genesis of the experiment is discussed in Frank D. Prager, *Berti's Devices and Torricelli's Barometer from 1641 to 1643.* Annali del Istituto e Museo di Storia della Scienza di Firenze. 1980; Anno 5(Fasc. 2): pp. 35-53, and Cornelius de Waard, *L'Experience barometrique: ses antécédents et ses explications.* Thouars: J. Gamon; 1936. On Berti's career as a 'mechanikos', see Joseph Connors, *Virtuoso Architecture in Cassiano's Rome*, in *Cassiano Dal Pozzo's Paper Museum*, London, 1992, vol. II (Quaderni Puteani 3), pp. 23-40 on pp. 27-8.

15 'Intra vero phialam me suggerente, campanulam una cum malleolo lateribus phialæ ea dexteritate inferuit, ut malleolus ferreus magnete ab extra attractus elevatusque mox a magnete liber proprio pondere campanulae illisus sonum faceret', Kircher, *Musurgia*, loc. cit.

16 'Infatti, quelle pulighe mi sono restate sempre nella mente', Raffaello Magiotti to Marin Mersenne, Rome; 12 March 1648, in Mersenne, *Correspondance*, XVI, 168-171.

17 'Concludimus itaque, quod tametsi vacuum in natura rerum possibile foret, sonus tamen in eo contingere minimè posset. Nam cum sonus sit affectio aeris, imò aër sit materialis causa soni, illo deficiente, sonum quoque deficere necesse est; & contra ex proposito experimento clare ostendimus, vacuum in natura rerum minime assignari posse', Kircher, *Musurgia*, loc. cit.

18 This is also suggested by Honoré Fabri. See Pierre Mousnier [Honoré Fabri], *Metaphysica demonstrativa*, Lyons: 1648, p. 579: 'cum affixum sit superiori basi, licet in ea cavitate vacuum esset, sonus tamen adhuc audiri posset; quia cum tremulo motu tintinnabuli in ipsam fistulae basim traducitur'.

19 The standard study of the experiment remains Cornelius de Waard, *L'Experience barometrique: ses antécédents et ses explications.* Thouars: J. Gamon; 1936. See also W. E. Knowles-Middleton, *The History of the Barometer*, Baltimore: Johns Hopkins University Press; 1964, pp. 10 ff., S. Moscovici, *L'Experience du mouvement. Jean-Baptiste Baliani, disciple et critique de Galilée*. Paris; 1967, C. Costantini, *Baliani e i Gesuiti*. Florence: Olschki; 1969 and Massimo Bucciantini, *Valeriano Magni e la discussione sul vuoto in Italia*. Giornale Critico della Filosofia Italiana, 1994; Serie VI, Volume XIV (Anno LXXIII (LXXV)): pp. 73-91.

20 See Jerzy Cygan, *Das Verhältnis Valerian Magnis zu Galileo Galilei und seinen wissenschaftlichen Ansichten.* Collectanea Franciscana. 1968; 38(n. 1-2): 135-166.

21 Jerzy Cygan, *Valerianus Magni (1586-1661). 'Vita Prima', operum recensio et bibliographia.* Rome: Istituto Storico dei Cappuccini; 1989, pp. 315-6 #8.

22 'Tyrannus est, qui premit genus humanum perniciosius ulla heresiarcha, ullove hominum quos tulerit aetas ulla', Valeriano Magni, *De Atheismo Aristotelis*, Warsaw, 1647

23 For a detailed study of the *De Luce Mentium* see A. Boehm, *Deux essais de renouvellement de la scolastique au XVIIe siècle. I. L'Augustinisme de Valerien Magni (1586-1661)*. Revue des Sciences Réligieuses. 1965; 39: pp.230–267. Magni's philosophical system has been the subject of a monograph, Stanislav Sousedík, *Valerianus Magni 1586-1661. Versuch einer Erneuerung der christlichen Philosophie im 17. Jahrhundert*. Sankt Augustin: Verlag Hans Richarz; 1982. The tone of this study is somewhat coloured by the author's aim to demonstrate that Magni's philosophical system was a precursor of Kantian transcendental idealism. A study that relates Magni's anti-Aristotelianism to his ecumenism is Cesare Vasoli, *Note sulle idee filosofiche di Valeriano Magni* in Vittore Branca and Sante Graciotti, eds. *Italia, Venezia e Polonia tra medio evo e età moderna*, Florence; 1980; Studi 35: pp. 79–112.

24 'Censura libri de luce mentium. Tractatus de luce mentium multa continet ex Pythagoraeorum Platonicorumque officina deprompta dogmata: quorum principem locum oblivet Lux mentium quam Author toto passim opusculo demonstrare conatur increatum esse, atque identificari cum Numine assistente et illuminante; iuxta illud priscorum philosophorum, quos sectatur epiphomena', Athanasius Kircher, *[Censura libri de Luce Mentium]*, APUG 561 f.101r.

25 Valeriano Magni, *Demonstratio ocularis loci sine locato: corporis successive moti in vacuo: luminis nulli corpori inhaerentis*. Warsaw: In officina Petri Elert; 1647.

26 Valeriano Magni, *Demonstratio ocularis* p. 6.

27 Gilles Personne de Roberval to Pierre Desnoyers (in Warsaw), Paris, 20 September 1647; Mersenne, *Correspondance*, Vol. 15, pp. 427–441, on p. 429: 'at ibidem praecipue vero Romae atque Florentiae celeberrimas inter eruditos de ea re viguisse controversias quas non potuit ignorare Valerianus qui circa eadem tempora in regionibus degebat et cum doctis illis conuertebatur'.

28 See Jerzy Cygan, *Valerianus Magni (1586-1661). 'Vita Prima', operum recensio et bibliographia*. Rome: Istituto Storico dei Cappuccini; 1989, p. 314, n. 6

29 Valeriano Magni, Dedicatory letter to Gilles Personne de Roberval of *De inventione artis exhibendi vacuum Narratio apologetica*, Warsaw, 5 November 1647, in Mersenne, *Correspondance*, Vol. 15, pp. 527–531, on p. 528.

30 'Consilium ergo de superanda impossibilitate Vacui, incidit mihi apud Galilaeum, quod aqua nequeat per attractionem ascendere in fistula ultra cubitum decimum octavum et ab usu librae Archimedis, quam Cracoviae anno 1644. dono accepi a Tito Livio Buratino, viro erudito in Mathematicis: qua occasione cognovi proportionem gravitatis inter aquam, et mercurium esse 1 ad 13 proxime', ibid. .

31 Ibid., pp. 529–30.

32 Ibid., p. 529.

33 'Demum, cum Gaspar Brunorius Venetus, qui apud Reges Angliae, Daniae, et Sueciae suam in fabrica vitri celebravit artem evocatus Dantisco a Serenissimo *Rege*

Poloniae, venit Warsaviam, ab illo humanissime accepi plures diversae altitudinis et diametri tubos, ex quibus aliquos, quorum altera extremitatum protuberat'.

34 'Virt[us] prohibitiva vacui', ibid., p. 529.

35 'Dominus de Noyers vidit allegata documenta, scilicet librum Galilaei, libram Archimedis, tubos ligneos, epistolas duas' ibid., pp. 530–1.

36 Andriani was professor of rhetoric in the Collegio Romano from 1638–1646, so his departure from Rome may have been later than Magni suggests. Villoslada, R. G., *Storia del Collegio Romano dal suo inizio all soppressione della Compagnia di Gesù*. Rome; 1954, p. 335.

37 'Biennio ab hinc R.P. Joannes Baptista Adrianus Societatis Jesu discessit Roma, ubi docuerat Rhetoricam, venitque Warsaviam, ibique spectavit experimenta Vacui, velut inaudita: Ergo Capuccinum nescivisse quid Romae innovatus in Philosophia, est possibile: siquidem id ignoravit Collegium Romanum Societatis Jesu'. ibid. p. 530.

38 It is worth noting that Kircher had developed a mercury thermoscope to measure the differences of the winds, which would probably have been physically highly similar to the Torricellian device, by 1641: 'Hac arte ego machinas argento vivo animatas alias me construxisse memini, quibus omnes ventorum differentiae propè verum cognoscebantur, vis et qualitas elementaris uniuscuiusque rei iuxta gradumsuum certa applicatione dispiciebatur, quae omnia perfectius naturaliusque in Sphaera, cuius spirales Solis Cycli, e vitreis Syphonibus constructi sint repræsentari possunt; cuius arcanas rationes libenter hic ostenderem, nisi eas nostrae Meteorologicæ arti reseruassam.' Athanasius Kircher, *Magnes, sive de magnetica arte libri tres*, Rome, 1654 [1641], pp.410–411

39 Cygan (op. cit., p. 387) suggests that the *Magnus Amicus* of Zucchi's title was Magni himself, on the basis of a letter written by Fabio Chigi, the future pope Alexander VII, to Franciscus Van der Veetren on June 26 1648 (BAV Chigi a I 46, f.283) which says of Zucchi's pamphlet that 'Auctor adiuncta epistolae est Pater Zucchi Rector Collegi Romani, is cui scribet est Pr. Valerianus Magni Cappuccinque'. However, Zucchi himself, in a letter written to Kaspar Schott, mentions that the pamphlet was addressed to Grandamy. See Schott, *Mechanica Hydraulico-Pneumatica*, cit., p. 464.

40 On Zucchi, see Baldini, Ugo. *Una lettera inedita del Torricelli ed altri dei gesuiti R. Prodranelli, J.C della Faille, A. Tacquet, P. Bourdin e F.M. Grimaldi*. Annali dell'Istituto e Museo di Storia della Scienza di Firenze. 1980; V (no.1): pp.14–36.

41 The reference is to the Terentian disclaimer, 'Davus sum, non Oedipus', i.e. 'I am no solver of enigmas'.

42 'Davum interrogasti, sed versari solitum cum Oedipis, quibus conuiuere tanti est, ut mihi ex illorum eruditione aliquid fortè adhaeserit, quo propositis satisficiam', Niccolo Zucchi, *Magno amico nonnemo ex Collegio romano S.I. experimenta vulgata non plenum sed vacuum et antiperistasim stabilire*, Rome: L. Grignani;1648, Sig. A recto.

43 *Oedipus Aegyptiacus hoc est universalis hieroglyphicae veterum doctrinae temporum iniuria abolitae instauratio*, Rome: Vitalis Mascardi, 1652–4. 3 tom, 4 vols. Kircher seems initially to have harboured hopes that Ladislaus IV, Magni's patron, might be willing to sponsor the publication of this work (Kircher to Cyprian Kinner, Rome, 20 February 1648, Hartlib Papers 1/33/31B). However Ladislaus IV died on 20 May 1648 prompting Kinner to remark to Hartlib that 'Kyrcherus a Poloniae Rege nihil sperare habet: quia nullus nunc est Rex' (Kinner to Hartlib, Danzig, 5 August 1648 Hartlib Papers 1/33/44B). Although Kinner mentions Magni's experiments in the same letter, I have been unable to find any direct connection between this episode and the vacuum experiments of the previous year (by this time Magni had left Warsaw for Danzig). On Kircher's *Oedipus Aegyptiacus* see Daniel Stolzenberg, Egyptian *Oedipus; Athanasius Kircher and the Secrets of Antiquity*, Chicago: University of Chicago Press, 2013

44 'Ut praestem, licet praeter exhibita ab alijs, propria, & saepius iterata experimenta adhibuerim; non apponam tamen mensuras Tuborum vitreorum, aut infusi argenti vivi, aut partium in illis ab eo descendente tam in vibrationibus, quam in quiete derelictarum, nec vibrationum ipsarum numerum, aut similia; quia ad solutionem eorum, quae mihi proposita sunt, minus faciunt, & Epistolae brevitatem quantum fieri possit, sinè veritatis intentae dispendio, retinere animus est', Zucchi, *Magno amico*, loc. cit..

45 'Placuit multiplicare nodos; ut ex illorum solutione clarius constet, non oportere ex difficultatibus occurrentibus nova quotidie dogmata cudere, aut antiquata veterum commenta recoquere; sed satius esse quorundam ingeniorum intemperem iniecta frigida compescere, quam fovere; ne studio novitatis, dum animo sincero aliquid Philosophiam iam receptam impetunt, ad meliora labefactanda alij progrediantur.' ibid., Sig. A3 r – v.

46 'Nos vero ut falsitatem eorum opinionis, vel ipsa auriculari experientia demonstraremus, arreptem magnetem phialæ vitreæ è regione malleoli ferrei foris applicuimus, qui mox attractum malleolum eleuauit, abstracto vero magnete malleolus pondere proprio illisus campanulæ limpidissimum sonum edidit', Athanasius Kircher, *Musurgia universalis, sive Ars magna consoni e dissoni*. Rome; 1650, p.11.

47 'Porrò in superiori parte, quae huiusmodi expiratione repletur, appulsu malleli ad campanulam donum edi, nullo ibi existente vacuo, nullo accidente sine subiecto; & evidenter convinci, ibi fuisse corpus violenter dilatum ex eo' ibid., Sig. A4 *verso*.

48 'Est igitur corpus tenue, seu spiritus, qui superiores Tubi partes maxime distensus replet', ibid. Sig. [A6] *verso*.

49 Nicolò Zucchi, *[Nicolaus Zucchius Manuscripta Diversa]*, APUG Fondo Curia 1595, inside back-cover (crossed-out): '[…]vivum, elevatur, et sustinetur aqua inter dictos Tubos, non esse aliquid extra illos tubos, sed intra illos, dum tale grave elevatum, vel

manens suspensum (patet alioqui deorsum tubo) tali sua determinata gravitate non potest ex sua, vel contigui corporis substantia ulterius distendere quod repleat spatium suo descensu derelinquendum. Ille ut secretus à Vulgo [*mutil.*] ita inter Patricios [*mutil.*] calculo recensendus qui nulla habet difficultatem dim.. [*mutil.*] etiam pro maiori spatio vacuum inter maiora corpora, sicut inter atomos, quibus illa componit spatia vacua passim constituit; Asserit sustineri ex vi medij extrinseci gravitantis, et sua gravitate aequilibrantis illud liquidi spissoris, quod elevatur, ex sustinetur inter vitreos, iuxta experimentum primum'.

50 Daniello Bartoli, *Della vita del padre Niccolo Zucchi*, Rome: presso il Varese; 1682, p. 35. Curiously, the 'estrinseca virtù' sounds more similar to Bartoli's own interpretation of the mercury experiment, in terms of pressure, than to Zucchi's, which Bartoli criticises strongly in his late work on the Torricellian experiment: 'La cagione di questo natural sintoma dell'argentovivo la trove attribuita da un valente huomo [i.e. Zucchi] alla necessità di multiplicare spiriti con che poter riempire quel vano del cannello che starà sopra l'argento: e ogni secesa che fà è come una strappata che si dà alle viscere di quell'infelice mercurio, accioche così agitato, scomosso, e premuto, fumichi e svapori in maggiore abbondanza', Daniello Bartoli, *La tensione, e la pressione disputanti qual di loro sostegna l'argento vivo ne' cannelli, dopo fattone il vuoto*, Rome; 1677.

51 'Communicavi diatribam P. Zucchi peritis, et cum quatuor Societatis post mensam ad duas horas multis rationibus atque experimentis acuiter disputatum, prout refferet, spero Pater Madeson oppressum animum assiduis occupationibus tractatum, afflictum civito tam diu conatu pacis assequendeque, ancipitem inter bellorum successus varios, recreant literes Paternitatis Vestris, aliosque amicos, quos nominat, quosque plurimum nomine me veluti salvere iubeat, expectans interim ipsos literas ac si quid eorum gratia possim, ut mandent libere confidenteque. Ita faciat P[aternit]a V[estr]a ac mei memor luius in specibus ac sacris esse velit', Fabio Chigi to Athanasius Kircher, Münster; 29 May 1648; APUG 556, f.27r.

52 'Communicavi diatribam P. Zucchi peritis, et cum quatuor Societatis post mensam ad duas horas multis rationibus atque experimentis acuiter disputatum, prout refferet; Spero Pater Maderson oppressum animum assiduis occupationibus tractatum, afflictum civito tam diu conatu pacis assequendeque, ancipitem inter bellorum successus varios, recreant literae Paternitatis Vestrae', Fabio Chigi to Athanasius Kircher, Münster; 29 May 1648, APUG 556 f.27r.

53 'De vacuo alterius seniculi iam Novatoris valde dubito; et vereor ne nimis vere se deprehendisse conqueratur in suo Galero, toties et frustra decantato', Fabio Chigi to Kircher, Cologne; 14 February 1648, APUG 556, f.18r. Copy in BAV Chigi a. I 45, ff. 164r-165v.

54 See Vincent Borg, *Fabio Chigi, Apostolic delegate in Malta (1634–1639). An edition of his official correspondence*, Vatican City: Biblioteca Apostolica Vaticana; 1967, pp. 272–3, 312, 318, 328, 330 and *passim*. In a letter to Cardinal Francesco Barberini

written on 1st February 1638, after Kircher's departure from Malta, Chigi expresses his high esteem of Kircher to the papal nephew: 'E qui per fine, le fo humilissima riverenze, non lassando di attestare il godimento mio grandissimo della litteratura del padre Atanasio, il quale hora meritamente sotto la protettione di V. Em.a, viene a fatigare per benefitio universale in teatro più proportionato al suo ingegno', Borg, op. cit., p. 352.

55 See Marcel Albert, *Nuntius Fabio Chigi und die Anfänge des Jansenismus 1639–1651. Ein römischer Diplomat in theologischen Auseinandersetzungen*, Rome, Freiburg and Vienna: Herder; 1988.

56 See Paolo Casati to Giannantonio Rocca, Bologna; 1 June 1648, in *Lettere d'uomini illustri del secolo XVII a Giannantonio Rocca*, Modena: Società Tipografica; 1785, pp. 386–389, 'Godo che alle mani di V.S. sia capitato l'opusculo di quel Cappuccino Polacco, di cui pure avevo inteso parlare, ed avevo letto una risposta scritta dal P. Niccolò Zucchi in Francia a chi lo richiedeva del suo parere, intorno a certa esperienza fatta con l'argento vivo per provare il vacuo, & egli dalla stessa pretende si provi l'opposto: detta risposta è stata stampata in Roma, e me ne fu mandato una copia dallo stesso Padre'.

57 'Stavo quasi per scrivere, che non mandasse detto piego, pensandomi, che solo vi si contenessero le cose del Cappuccino, il cui opusculo con quelle due aggiuntarelle della disputa datta in Varsavia e quell'altra de possibilitate vacui, ho ultimanente veduto, essendosi stampato detto Opusculo nel fine d'un Libretto scritto contro di esso dal Dottore Fantucci. Ma la risposta è assai tenue, e molte cose sono intatte', Paolo Casati to Giannantonio Rocca, Bologna; 20 July 1648, in *Lettere d'uomini illustri del secolo XVII a Giannantonio Rocca*. Modena: Società Tipografica pp. 394–395.

58 'Ho chi mi sollecita gagliardamente a scrivere una risposta adeguata, e distinta più che quella del P. Zucchi' ibid.

59 For the remarks made by the Roman *Revisores* on Casati's work see ARSI FG 662, ff. 473r-475r., e.g. the censure by Casati's friend Mario Bettini, which suggests the importance which the Revisors attached to the defence of Aristotle from the attacks of Magni (f. 474r): 'Ho scorso il Vacuum proscriptum, etc. (della quale disputa havevo ancora havenda qualche congnitione con occasioni du qualche congresso con l'autore) et mi pare opera dotta, ben fondata, seriosa, utile, anzi necessaria per stabilire la verita della dottrina Aristotelica contro il vacuo. Mario Bettino'.

60 Paolo Casati, *Vacuum proscriptum. Disputatio physica*, Genvae: Ioannes Dominicus Peri; 1649, p. 5

61 See *Conciliorum Oecomenicorum decreta*, Bologna 1973, pp. 693–695, Session XIII, 11 October 1551, esp. ch. I, III, IV and canons 1–4. On the question of physical explanations of the Eucharist during the sixteenth and seventeenth centuries see Redondi, *Galileo heretic*, cit., pp. 203–226 and *passim*. Redondi's principal claim, that the incompatibility of an atomist conception of matter with the Tridentine account

of transubstantiation was the prime motive behind the Galileo trial of 1632–33 has been widely disputed. See especially F. V. Ferrone and M. Firpo, *Galileo tra inquisitori e microstorici*, Rivista Storica Italiana, 97, 1985, pp. 177–238, 957–68. Other useful treatments of the Eucharistic question include J.R. Armogathe, *Theologia Cartesiana: L'explication physique de l'Eucharistie chez Descartes et dom Desgabets*, La Haye: Nijhoff, 1977, with a particularly helpful bibliography. For an important introductory treatment see the article by F. X. Jansen, 'Eucharistiques (accidents)' in *Dictionnaire de théologie catholique*, vol. V, cols. 1360–1452. For a more recent discussion of the physics of the Eucharist, see Hellyer, 2005, pp. 90–113.

62 'Paucis hisce annis Thomas Campanella dominicanus, vir qui omnia legerat, omnia meminerat, praevalidi ingenij, sed indomabilis, quid non ausus est aut contra Aristotelem in philosophia, aut contra Divum Thomam adeoque Scholasticos universos in Theologia? Neque illi multum absimilis Valerianus Magnus franciscanus, pius utique ac doctus, nec minus ad actionem, quam ad contemplationem natus, idemque vel in sacco regibus carus, haereticis formidolosus. Is enim, & libellum vulgavit ubi assiduam quamquam aenigmatis indigentem; et aliam dissertatiunculam nuper dedit in qua non tantum inane solidis permixtum adversus Aristotelem se demonstrasse gloriatur, verum etiam ibi vacuo stantes asseverat; adeoque praecipua Theologorum dogmata super ineffabilis Eucharistiae mysterio convellit? Nec tamen haec qualiscumque novandi libido ad Ordines illos praeclarissimos manat. Alia enim longe sunt haec duo: praefervidis quibusdam ingenijs laxiores habenas permittere ne constricta violentius erumpant; ac, eisdem ceu inculpatis, laudandisque favere, patrocinari, subscrivere. Praetereo, inquam, haec omnia; & id unum noto a nostris censoribus vel inobservatum, vel dissimulatum' Sforza Pallavicino, *Vindicationes Societatis Iesu, quibus multorum accusationes in eius institutum, leges, gymnasia, mores refelluntur.* Romae: typis Dominici Manelphi; 1649. pp. 223–4.

63 Kircher does not mention the Eucharist explicitly in his attack on Magni in the *Musurgia Universalis*, but hints at the dangers posed by Magni's interpretation to religious orthodoxy: 'Hinc argumentantur; spacium RH. in superiori tubi parte relictum vere & proprie vacuum esse, cum fieri non possit, ut interim aliud corpus in abeuntis mercurii locum substitui potuerit. Hinc veluti insolentes & importuni iactatores triumphum ante victoriam canentes multa sane essutire *non tantum naturalium rerum principiis repugnantia, sed & in orthodoxa fide periculosa*; ut dum locatum sine loco, accidentia sine subiecto naturaliter subsistere subtilissimi hoc experimento se demonstrare posse imprudentius iactitant.' Kircher, *Musurgia universalis*, cit., pp. 11–13, emphasis added.

64 Valeriano Magni to Giovan Battista Barsotti, Warsaw; 3 August 1647; BAV Vat. Lat. 13512, f.78r.

65 'Lo pregai, che, come da se, consigliasse a sottoscritto, et da loro sottoscritto, temendo io, che alcuni, a me non bene affetti, mi movessero difficoltà con l'auttorità

di Mons. Illmo. Nuntio, et che poi, al stringere della causa, egli si trovasse abbandonato da Theologi, de' quali niuno volesse essere auttore, che il Vacuo tolga i miracoli al S[antissi]mo Sacramento', ibid., f. 78v. On the complex relationship between theories of vacuum, atomism and the Eucharist in the seventeenth century, see Redondi, *Galileo Heretic*, cit.

66 [Wladislaus IV Wasa], *Epistola ad sanctissimum dominum nostrum Urbanum papam VIII. scripta ab Vladislao IV Poloniae et Sueciae rege serenissimo.*; Gedani; 1636 (Manuscript of Urban VIII's copy, dated 4 February 1636, in BAV Barb. lat. 6614, n.1). Ironically, as Denzler points out, Magni's bid for the cardinalate coincided with the moment at which the Propaganda Fide had finally decided to dismiss him from his missionary services. See Georg Denzler, *Die Propagandakongregation in Rom und die Kircher in Deutschland im ersten Jahrzehnt nach dem Westfälischen Frieden*, Paderborn: Verlag Bonifacius-Druckerei; 1969 pp. 191–2.

67 'Ho sudato, et affaticato molto: ho sparso con ferite mortali molto sangue per mano de gli heretici et finalmente desidero morire mille volate l'hora per mantenimento di lei. Questa mia V.S. non mostrera ad huomo vivente, che al fratello di Mons. Illmo. Nuntio, quando pero V.S. conosca esser necessità di farlo' ibid.

68 Etienne Noël, *Aphorismi physici seu physicae peripateticae principia breviter ac dilucide proposita*, Flexiae: Apud Georgium Griveau; 1646.

69 Samuel Hartlib, *Ephemerides 1656 Part 3.* (June–September 1656) Sheffield, Hartlib Papers, 29/5/93B–94B.

70 'J'ai lu vos Expériences touchant le vide, que je treuve fort belles et ingénieuses, mais je n'entends pas ce vide apparent qui paraît dans le tube après la descente, soit de l'eau, soit du vif-argent. Je dis que c'est un corps, puisqu'il a les actions d'un corps, qu'il transmet la lumière avec réfractions et réflexions, qu'il apporte du retardement au mouvement d'un autre corps, ainsi qu'on peut remarquer en la descente du vif-argent, quand le tube plein de ce vide par le haut est renversé; c'est donc un corps qui prend la place du vif-argent. Il faut maintenant voir quel est ce corps', Etiènne Noël to Blaise Pascal, Paris; Oct. 1647, in Pascal, *Oeuvres Complètes*, ed. Mesnard, Vol. II pp. 513–518

71 'Présupposons encore une chose vraie, que le verre a grande quantité de pores, que nous colligeons non seulement de la lumière qui pénètre le verre plus que d'autres corps moins solides dont les pores sont moins fréquents, quoique plu grands, mais aussi d'une infinité de petits corps différents du verre que vous remarquerez dans ces tringles qui font paraître les iris, et de ce qu'une bouteille de verre bouchée hermétiquement ne se casse point en un feu lent sur des cendres chaudes', ibid., pp. 514–5.

72 'Si donc on me demande quel corps entre dans le tube et prend la place que le vif-argent quitte en descendant, je dirai que c'est un air épuré qui entre par les petits pores du verre, contraint à cette séparation du grossier par la pesanteur du vif-argent descandant et tirant après soi l'air subtil qui remplissait les pores du verre, et celui-ci,

tiré par violence, traînant après soi le plus subtil qui lui est joint et congéné, jusques à remplir la partie abandonnée par le vif-argent'. ibid., p. 516.

73 'On ne peut vous refuser la gloire d'avoir soutenu la physique péripaticienne aussi bien qu'il est possible de le faire' Pascal, *Oeuvres Complètes*, II p. 527.

74 'Si vous me dites que les espèces du Saint Sacrament ont des parties les unes hors les autres, et néanmoins ne sont pas corps, je répondrai: premièrement, que, par le composé de parties les unes hors les autres, on entend ce que nous appelons ordinairement long, large et profond; secondement, que l'on peut fort bien expliquer la doctrine de l'Église disant que les petits corps qui restent dans les espèces ne sont pas la substance du pain. C'est pourquoi le concile de Trente ne se sert jamais du mot d'accident, parlant du Saint Sacrement, quoiqu'en effet ces petits corps soient vraiment les accidents du pain, selon la définition de l'accident reçue de tout le monde: ce qui ne détruit point le sujet, soit présent, soit absent; troisièmement, que, sans miracle, tout composé de parties les unes hors les autres est corps; et je crois que, pour décider la question du vide, il n'est pas besoin de recourir aux miracles, vu que nous présupposons que toutes vos expériences n'ont rien par-dessus les forces de la nature', Noël to Pascal, Paris; 1 November 1647, in Pascal, *Oeuvres*, cit., Vol. II pp. 528–540, on p. 531.

75 ARSI F.G. 657 p. 171: '2a Tam forma substantialis materialis, quam accidentalis, habens contrarium, habet materiam ex qua constituitur: Secus forma substantialis immaterialis, & accidentalis materialis non habens contrarium'. This censure is discussed in Gorman, *A Matter of Faith?* cit.

76 ibid. p.171: 'Ad 2am dicimus; Itidem improbabilem, & retractandam esse: idque eo magis, quod quaedam inde absurda, contra Mysterium Eucharistiae deduci possint; v.g. Albedinem S. hostiae (tanquam formam materialem accidentalem, habentem contrarium) constitui ex materia; adeoque re ipsa esse totum compositum ex materia & forma. Quae quidem opinio, si ab auctore palam defenderetur, non ad nostram, sed Inquisitionis Censuram, pertineret'.

77 Noël, *Le Plein du Vuide*, 1648, published in *Oeuvres de Pascal*, ed. Abbé Bosset, La Haye, 1779, Vol. IV, pp. 108–146 on pp. 108–109.

78 'Nostre Capucin veut faire mettre en latin le petit traitté de Monsr. Pascal et imprimer car il fait bouclier de tout. Son principal dessein ne s'areste pas au vuide, ne voulant pas seulement qu'Aristote vuide les universitez, mais il veut qu'il soit excomunié', Pierre Desnoyers to Pascal, Warsaw, 4 December 1647, in Mersenne, *Correspondence*, XV, pp. 560–563 on p. 561, emphasis added.

79 Valeriano Magni, *Commentarius . . . De homine infami personato sub titulis M. Jocosi Severi Medii*. Prague: In Seminario Sancti Norberti; 1655. Magni wrote this work to answer an attack on his doctrine of the primacy and infallibility of the Roman pontiff written under the pseudonym of Jocosus Severus Medius, which Magni mistakenly thought to mask a Jesuit. In fact the author of the *Vertrauliches Gespräch zwischen vier päpstischen Scribenten* was a Lutheran.

80 [Blaise Pascal], *Les Provinciales ou les Lettres escrites par Louis de Montalte, à un Provincial de ses Amis, et aux R.R.P.P. Iesuites,* Cologne: Pierre de la Vallée [Amsterdam: D. Elzevier]; 1656/57, *Quinzième lettre.*

81 'Il y avoit adjouté un chapitre qu'il adressoit au Pere Mercenne, qu'il intituloit de l'Ateisme d'Aristote, mais ses amis ausy tost qu'il fut imprimé luy firent suprimer, luy disant qu'il se feroit tant d'ennemis que peut estre cela pouroit empescher sa promotion en rouge qu'il espere a la premiere qui se fera', Desnoyers to Mersenne, 4 December 1647, cit., p. 561.

82 See Pascal to Le Pailleur, February 1648, in Pascal, *Oeuvres Complètes,* cit., Vol. II, pp. 559–576, on p. 572: 'Comme j'écrivais ces dernières lignes, le R.P. Noël m'a fait l'honneur de m'envoyer son livre sur notre sujet, qu'il intitule le Plein du Vide; et a donné charge à celui qui a pris la peine de l'apporter de m'assurer qu'il n'y avait rien contre moi, et que toutes les paroles qui paraissaient aigres ne s'adressaient pas à moi, mais au R.P. Valerianus Magnus, capucin. Et la raison qu'il m'en a donnée est que ce Père soutient affirmativement le vide, au lieu que je fais seulement profession de m'opposer à ceux qui décident sur ce sujet. Mais le R.P. Noël m'en aurait mieux déchargé s'il avait rendu ce témoignage aussi public que le soupçon qu'il en a donné'.

83 On Kojalowicz-Wijuk (Kowno 1609 – Vilnius 1677), see Sommervogel IV, 1166 (*s.v.*) and ARSI Lith. 9, f. 40r, n.21 (*Catalogus Primus et Secundus Collegij Nesuisien Soc. Iesu Anno 1639*): 'P. Albertus Koialowicz. Lituan. Valetudinis infirmae, Annorum 31, Ingressus Sctm. 6 Aug. Anno 1627. In ea Philosophiam 3. Theologiam 4 annis audiuit. Docuit in Grammatica e Presi 1. anno nunc tertium annum finit. Boni est ingenii, iudicii et prudentiae, experientiae parvae. Optimi in his profecti Complexionis sanguiniae. Valet ad docendum Philosophiam et Theologiam et ad concedandum, ad gubernandum et conuersandum', ibid. f.109 v (no.23), and ibid f.237r (no. 3).

84 Albert Kojalowicz Wijuk, *Oculus ratione correctus, seu refutatio demonstrationis vulgaris de vacuo.* Vilnae: typ. Acad. S.J.; 1648.

85 Krüger (Ruthene 1598 – Grodno 1665) published a number of works in optics, geometry, arithmetic and astronomy (see Sommervogel, IV, 1261, 12). On Krüger, see ARSI Lith. 9, f. 25v (Catalogus Primus et Secundus Collegij Vilnensis. Anno 1639) no. 19 'ualet ad docendam Mathematicam, hebraeam, Philosophiam, casis ad concionandum et conversandum', ibid. f.108r, n.10 (Cat. primus et secundus personarum Collegij Vilnensis Ann. 1642), ibid. f.239v: n.23, (Cat. primus et Secundus Collegii Vilnensis Anni 1651): 'Valet eximie ad docendas disciplinas Mathematicas ad gubernandum, ad universandum cum proximis', and ARSI FG 660, f. 318r

86 Oswald Kruger, *Dissertationes de vacuo.* Vilnae: Typ. Acad. S.J.; 1648.

87 'Iudæi duo sacro fonti admoti, et bene collocati. In pagis a frequentia Parochorum subductis clandestina matrimonia viginti circiter ratificata. Multi a bestialitas consuetudine, et Sodomitico flagitio auersi, plures a meretricio quaestu et turpi vivendi licentia ad dignam Christiano hominem vitam reuocatae [...] Mulier

quaedam quo tempore vexata ab incubo continuis attrita molestiis, omnem prope valetudinem amiserat . . .' ARSI Lith. 39, f. 218v.

88 'Quod attinet ad sententiam de quantitate constanti ex indivisibilibus, iam aliquoties scripsi ad Provincias, a me nullo modo probari et per me nulli hactenus licuisse illam proponere ac tueri. Si usquam explicata et defensa fuit, id me inscio factum est. Imo ipsi Cl. Ioanni de Lugo clare demonstravi nolle me ut nostris ea sententia tradatur et propugnetur.' ARSI Gall. 117 p. 144 Resp. Muti. Vitelleschi, Romae, 13 ian. 1633, Dolam.

89 *Elogio Storico scientifico del Padre Paolo Casati Piacentino della Compagnia di Giesu*, ARSI Ven. 121 II, ff. 388r-395v, on f.388v: 'Dal Parmense Collegio de' Nobili, dove era stato Convittore, entrò nella Compagnia; e in essa, dopo compiti con gran lode d'ingegno gli studi di Filosofia e di Teologia, della quale ultima ebbe a Maestro in Roma il Padre Giovanni de Lugo, poi Cardinale di Santa Chiesa, insegnò per un sessennio la Filosofia in Bologna, poi in Roma, per un quadriennio la Mathematica ivi stesso per altrettanto tempo la Polemica Teologia'.

90 'Je vous remercie de ce que vous me mandez du Concile de Constance sur la condamnation de Wiclef; mais je ne voy point que cela fasse rien du tout contre moy. Car il auroit dû estre condamné en mesme façon, si tous ceux du Concile eussent suivy mon opinion; et en niant que la Substance du Pain et du Vin demeure, pour estre le sujet des Accidens, ils n'ont point pour cela determiné que ces Accidens fussent réels, qui est tout ce que j'ay écrit n'avoir point lû dans les Conciles', Descartes to Mersenne, Endegeest, March 1642, in Mersenne, *Correspondance*, Vol. XI, pp. 73–79, on p. 76. See also Durelle to Mersenne, 26 February 1642, in ibid., Vol. XI, p. 46 ff.

91 Roderigo de Arriaga, *Cursus Philosophicus*. Antuerpiae: ex officina Plantiniana Balthazaris Moreti; 1632 (citations are from the second edition, Lugduni, Sumpt. Philip. Borde, Laurent Arnaud, & Petri Borde; 1669)

92 The 15th session of the Council had allegedly condemned the proposition: '*Linea aliqua Mathematica componitur ex duobus aut quatuor punctis immediatis; aut solum ex punctis simpliciter finitis*' as being erroneous in philosophy. In questioning whether the council did in fact condemn the Zenonian proposition, Arriaga reminds his readers that, even if the council of Constance had been sanctioned by papal authority, which it was only retrospectively, it is not the role of the pope or the church fathers to define anything in purely philosophical matters. See Arriaga, *op. cit.*, p. 571 (Liber Quintus et Sextus Physicorum, Disputatio XVI, *De continui compositione* Subsectio VI). On this issue see also Hellyer, Marcus. 'Because the authority of my superiors commands': Censorship, physics and the German Jesuits. Early Modern Science and Medicine. 1996; 1(3): 319-354.

93 'Nihilominus tamen eas ego in hoc Auctario tradere non praesumpsissem, nisi Admodum R.P. Paulus Oliva Societatis nostrae dignissimus Vicarius Generalis, tum ob antiquam, quam dixi, in eis edendis meam bonam fidem, tum quia paucae illae & in hac Pragensi Universitate valde receptae, tum quia in materia pure

Philosophica ad mysteria Fidei & morum nullatemus pertinentia, tum denique, quod mihi gloriosum est, ob singularem benevolentiam, qua[m] me licet omnino indignum a multis iam annis prosequitur, eas recudendi facultatem, humiliter a me rogatus, concessisset gratiosissimè' Roderigo de Arriaga, *Cursus Philosophicus*. Lugduni: Sumpt. Philip. Borde, Laurent. Arnaud, & Petri Borde, 1669 [First edn.: Antuerpiae: ex officina Plantiniana Balthazaris Moreti; 1632], pp. 3-4.

94 Robert Bireley, *Religion and Politics in the Age of the Counterreformation: Emperor Ferdinand II, William Larmormaini, S.J., and the Formation of Imperial Policy*. Chapel Hill, NC: University of North Carolina Press; 1981, p. 32. See also Kroess, Alois, *Geschichte der Böhmischen Provinz der Gesellschaft Jesu*, Vienna, 1910-27 *passim*.

95 Wenzel Wladiwoj Tomek, *Geschichte der Prager Universität*. Prague: Gottlieb Haase Söhne; 1849, pp. 249-270.

96 Cited in Bireley, *op. cit.*, p. 37

97 Magni wrote much later, on 12 March 1658, that 'The first root of this deep resentment [i.e. his unpopularity in Jesuit circles] lies in the fact that I defended the freedom of the Charles University in accordance with the special instructions of Urban VIII' (Archivio della S. Congregazione di Propaganda Fide, *Le Scritture originali riferite nelle Congregazioni Generali*, vol. 324, f. 389v, cited in Denzler, op. cit., p. 211).

98 Arriaga to Vitelleschi, Prague, 1 July 1629, ARSI Austr. 23, ff. 66-67, cited in Cygan, *Valerianus Magni*, p. 238.

99 Vitelleschi to Lamormaini, Rome, 29 July 1629, ARSI Austr. 4, ff. 2v-3r, cited in Cygan, *Valerianus Magni*, p. 238.

100 'Chi vuol esser dal P. Magno aiutato e promosso, basta che si mostri disgustato dei Giesuiti. Tutti quelli che si trovano qui di fresco usciti della Compagnia sono stati dal P. Magno promossi, posso dir concienza sopra i meriti e capacità', Paolo Anastagi to Vitelleschi, Prague; 29 August 1631, ARSI FG 770 3 b, p. [2]

101 Wenzel Wladiwoj Tomek, *Geschichte der Prager Universität*. Prague: Gottlieb Haase Söhne; 1849, esp. pp. 60-61.

102 See Kaspar Schott, *Technica Curiosa sive mirabilia artis*, Würzburg: Jobus Hertz; 1687 [1664], Liber IV, Cap. V §II, on pp. 250-253 (*Vitrum habet poros*).

103 Schott's *Technica curiosa* provides a condensed litany of anti-vacuist readings of the various vacuum experiments (except the new Boylean air-pump, as Boyle himself 'admits that he can never perfectly evacuate his vessel'), see Liber IV, Cap. V, pp. 246-258.

104 'Et quandoquidem Vacuistae tantopere ad oculorum judicium in hoc negotio appellant, cur non hic vel ipso oculorum judicio rem confectam vident? cur non tandem agnoscunt, aërem subtiliorem occulto metu, vel aëri crassiori, vel aquae, vel fumo extracto succedere, & vacuum impedire', Melchior Cornaeus, *De altero*

Experimento, quod per violentam aëris extractionem & exhaustionem sumitur, published in Schott, *Mechanica Hydraulico-Pneumatica,* p. 476. For Cornaeus' discussion of the Magni experiment, intended to precede this discussion, see his *Curriculum Philosophiae Peripateticae,* Herbipoli: Jobis Hertzi, 1675, pp. 384–394, 'De priore experimento quod ab argento vivo per fistulam supra clausam delapso capitur'.

105 'Scribendi occasio haec fuit. Est in supradicti Doctissimi Auctoris Museo sane celeberrimo, frequentatissimoque (quod brevi typis evulgabimus) non exigua Hydraulicarum ac Pneumaticarum Machinarum copia, quas summa animi voluptate spectant atque mirantur ij, quae ex omnibus Urbis & Orbis partibus ad ipsum visendum accurrunt Viri Principes ac LItterati, avideque scire desiderant, & Machinarum constructarum rationes, & machinalium motionum causas. Horum desiderio ut satisfacerem, omnium dicti Musei Machinarum fabricam & quasi anatomiam edocere, aut alicubi iam ab ipso Auctore edoctam enarrarem brevi opusculo aggressus sum' Schott, *Mechanica,* pp. 3-4. On Kaspar Schott's involvement in the discussion of the air-pump and the issue of atmospheric pressure, see Marcus Hellyer, *Catholic Physics,* Chapter 7.

106 Schott, *Mechanica,* pp. 307-8, 'Hinc veluti insolentes & importuni jactatores triumphum ante victoriam canebant, multa effutientes non tantum in Philosophia absurda, sed & in fide Orthodoxa periculosa, ut dum locatum sine loco, accidentia sine subjecto, naturaliter subsistere posse jactitant; nec defuit qui diceret, oculari demonstratione vacuum hoc Experimento comprobari'.

107 Johann Marcus Marci von Kronland carried out an intense correspondence with Kircher on hydraulic and other topics from Prague in the 1640s. On siphons see especially Marcus Marci to Kircher, Prague, 25 January 1642, APUG 557, f. 82r-v. See also John Fletcher, *Johann Marcus Marci writes to Athanasius Kircher.* Janus. 1972; 59: pp. 95–118.

108 J. J. W. Dobrzensky de Nigro Ponte, *Nova, et amaenior de admirando fontium genio (ex abditis naturae claustris, in orbis lucem emanante) philosophia.* Ferrara: Alphonsum, & Io. Baptistam de Marestis; 1657. For a letter from Kinner referring to Marcus Marci's interpretation in terms of an invasion of the tube by 'quintessence' and elucidating Kinner's own interpretation in terms of 'aqua mercurialis' see pp. 27–8, and for a lengthier *Discursus* by Kinner arguing against nature's fear of the vacuum see pp. 34–9. On Dobrzensky see Evans, *The Making of the Habsburg Monarchy,* cit., pp. 337, 339–40, 356, 369–70, 390 and W.R. Weitenweber, *Beiträge zur Literärgeschichte Böhmens,* Sitzungsberichte der kaiserlich Akademie der Wissenschaften, philosophisch-historische Klasse, xix (1856), pp. 144–156.

109 For a perpetual fountain synchronised with the 'motion of the sun 'designed by Moretus, see Dobrzensky, *op. cit.,* pp. 106–8. Stansel is mentioned frequently

throughout the work, and Dobrzensky refers to a hydro-magnetic clock-fountain seen in his *Museum Mathematicum* on p. 46.

110 Cabeo discussed hydraulics at length in his monumental commentary on Aristotle's meteorology, *Nicolai Cabei Ferrariensis Societatis Iesu In quatuor libros Meteorologicorum Aristotelis commentaria*, Rome: Typis haer. Francisci Corbeletti; 1646, in which he criticised Benedetto Castelli's attempt to produce a mathematical model of liquid flow. See Massimo Bucciantini, *Atomi, Geometria e Teologia nella filosofia Galileiana di Benedetto Castelli*. in Bucciantini, Massimo and Maurizio Torrini (eds.), *Geometria e Atomismo nella Scuola Galileiana*. Florence: Olschki; 1993, pp. 171-191.

111 Francesco Lana Terzi, *Prodromo overo saggio di alcuni inventioni nuove* Brescia: Rizzardi; 1670, p. 54.

112 On increased Jesuit discipline during the 1640s see especially Costantini, C. Baliani e i Gesuiti. Florence: Olschki; 1969 esp. pp. 98 ff, Baldini, *Legem*, cit., pp. 75–119, Marcus Hellyer, 'Because the authority of my superiors commands': Censorship, physics and the German Jesuits. Early Modern Science and Medicine. 1996; 1(3): 319-354.

113 'At ego non ob id mixturus [sum] meum amorem erga Societatem illam', Magni to Mersenne, Warsaw; 1648 Apr 14.; BN, Fonds francais, nouvelles acquisitions, 6204, f. 36r-v, publ. in Mersenne, *Correspondance*, XVI, 223–225, on p. 224 (the transcription contains several errors and Albertus Kojalowicz-Wijuk is incorrectly referred to as 'Robertus Koralowicz'). See also Cygan, *op. cit.*, p. 293 (II A no. 48).

114 'Si eis deficiunt rationes, evocant ex theologia non quidem veritates revelatas a Deo, sed propria ac privata commenta super his quae communiter credimus'. ibid. p. 233.

115 'Si in parte tubi vitrei vacua ponatur panis, & lumen producatur in illo, Lux inhaerens pani afficitur mol, figura & colore ipsius panis ita ut ad oculum discernatur illa lux a reliqua quae lucet ex parte tubi, vacua ab omni corpore. Verum, si Deus annihilaret substantiam panis, non annihilata luce, quae illi inerat, ea, in praesentia lucentis, maneret sine miraculo, sed defecata a vitiis contractis in corpore panis, scilicet lux illa non referret molem, figuram, & colorem panis: neque sine miraculo, posset lux illa conservare, sine subjecto pane, speciem molis; figurae & coloris ejusdem panis. Hisce verò obiter explicaverim, quod mea sententia de lumine in vacuo non detrahat Deo creationem, nec SS Sacramento miracula. Omnia vero sunt subjecta censurae Ecclesiastice' Valeriano Magni, *Principia et specimen Philosophiae axiomata*. Coloniae: apud Jodocum Kalcovium; 1661, pp. 71–72.

116 Jerzy Cygan, *Valerianus Magni (1586–1661). 'Vita Prima'*, cit., p. 281.

117 The reports of Gabriel Beatus (1st January 1655) and Paolo Casati (14 January 1654) on Schott's *Mechanica Hydraulico-Pneumatica* are in ARSI FG 661 ff. 482r-484r.

118 ARSI Rom. 81, ff.64v, 88v, 114v: 1652–54: 'P. Gaspar Sciot, socius P. Athanasii', 'P. Athanasius Chircher, scribit imprimenda'.
119 ARSI Rh. Sup. 7, f.68, n.25
120 Schott to Kircher, Mainz; 15 July 1655; 'hanno un instrumento nuovo per mostrare quod possit dari vacuum, o che voleva mostrarmelo', APUG 567 f.47r.
121 'Reverentia Vestra opportunissimè id praestare poterit in Hydraulica, ubi de vacuo tractat', Kircher to Schott, Rome, 26 February 1656, in Schott, *Mechanica*, p. 453.
122 ibid., title page.
123 Guericke, *Experimenta Nova*, p. 114, II, ch. 2.
124 On the political career of Johann Philipp von Schönborn (elector of Mainz from 1647 to 1673) see G. Mentz, *Johann Philipp von Schönborn*, i–ii (Jena, 1896–99), Evans, *The Making of the Habsburg Monarchy*, pp. 278–79 and, especially, F. Jürgensmeier, *Johann Philipp von Schönborn (1605–1673) und die Römische Kurie: Ein Beitrag zur Kirchengeschichte des 17. Jahrhunderts*, Mainz: Selbstverlag der Gesellschaft für Mittelrheinische Kirchengeschichte; 1977. Schönborn was a keen supporter of Magni in the early 1650s and expressed sharp disapproval of the Jesuit anti-ecumenist stance at the close of the Thirty Years' War, so Schott may have had important reasons for attempting to curry favour with him as he acquired power in the years after the Regensburg *Reichstag*. See Jürgensmeier, *op. cit.*, pp. 200–203.
125 Schott, op. cit., p. 444. 'Sunt qui huiusmodi Machinamento vacuum (quod hactenus phantasma fuit sive tentasse, sive sperasse, obsistente invulnerabili, vel ab Angelo, plenitudine Naturae) modis omnibus evincere tentant; alij vero non alio efficacius quam hoc ipso Experimento eliminari id posse autumant'.
126 'Machinam ipsam quà verbis, quà pictura subjicio, pugnantium utrimque argumenta affero, judicem Lectorem meum constituo, si modo nulla praeoccupatus opinione accesserit.' ibid.
127 The language used by Schott – '*machinamentum*', '*machina*' – to describe the *antlia*, and even his description of Guericke as '*ingeniosus*', emphasise the artificiality of the experiment and its pertinence to the domain of engineering rather than natural philosophy.
128 Schott, op. cit., p. 452
129 'Ex quibus comprobatur virtus, quam in litteris impressis ad Patrem Grandamy probavi inesse corporibus, pro servanda [sic, for preservanda?] unitate contiguitatis in universo, ad sistendum corpus proximum, quoties ad remotionem illius non potest succedere aliud', ibid. p. 464
130 'Vidi ego Paderbornae ante annos 37. talem Naturae victoriam. Paraverant Nostri in Collegio puteum ingentis prorsus profunditatis, ex quo per machinas multasque rotas arduo molimine aqua hauriebatur. Ergo compendii studio, constructa est antlia, eximo fundo erecta, arboribus integris in tubos excavatis, & invicem innexis.

Quia vero profunditas aquae erat major, quam pro consueto Naturae modulo, & quam ut aër extrahi, per pistilla & assaria posset; nulla vi moveri illa potuere. Comque demum & lateris & machinis homines plurimi & robustissimi extreme contenderent, hoc effecerunt, ut tubi, licet densissimi, ex truncis arborum confecti, cum terribili fragore crepuerint, aërique viam patefecerint', ibid. p. 474

131 'Caeterum quoniam varij rerum curiosarum Amatores mecum de dicto Experimento per Litteras egerunt, variaque responsa extorserunt; existimo ea, praesertim quae praestanti Philosopho cuidam non ita pridem communicavi, non contraria fore ijs quae RV. imprimenda curat', ibid. p. 455.

132 Ibid., p. 443

133 'Resistentia itaque & reluctantis Naturae impetus potius in rarefactionem aëris, quam in aliam rem conferendus est, ut centies ego in similibus Machinis expertus sum', ibid., p. 452.

134 'Sed doleo non mihi tempus superesse, ad totam machinationem ex fundamentis confutandam', ibid., p. 453

135 'Caeterum salutat Reverentiam Vestram P. Valentinus Stansel Socius meus, ac is quoque in Hydrostatica versatissimus', ibid. loc. cit.

136 'Reverende in Christo Pater, Pax Christi. Cum inaudissem, scriptas Reverentia Vestra litteras ad Patrem Assistentem, quibus tum a peritioribus, tum a me quoque judicium requirebat de novis Experimentis, ad exhibendum vacuum istic propositis', Schott, *Mechanica*, p. 463.

137 BNR F.G. 1323, f. 127r

138 'Cuperem ut Vestra Reverentia curaret experimentum postremum retentari, sed in tenebris revolvi clavem colli in vase aereo, & in eo aquam e vitreo descendere, ut notari posset, an in illis radiis candicantibus fulgor aliquis appareret', Schott, *Mechanica*, p. 465.

139 Orazio Grassi, *Libra astronomica ac philosophica....*, Perugia (1619), OG VI: pp.109–179.'But why was it so readily believed that this Gregoriana of ours, renowned for the many interests of its academicians should be considered as, among other things, the eyes of all, and that it ought especially to be consulted and its answers awaited?', transl. Drake and O' Malley, *The Controversy on the Comets of 1618*, Philadelphia: University of Pennsylvania Press;1960, p. 69.

140 Chigi to Kircher, Münster; 29 May 1648, cit. above note 50.

141 On alchemy and the Jesuits see Martha Baldwin, *Alchemy in the Society of Jesus.* in Z.R.W.M. von Martels, ed. *Alchemy revisited: Proceedings of the International conference on the history of alchemy at the University of Groningen; 1989 Apr 17*; Groningen. Leiden: Brill; 1990: pp. 182–187. For a Jesuit response to van Helmont's teachings on the weapon-salve see J. Roberti, *Curationis Magneticae, & Unguenti Armarii Magica Impostura ... Modesta Responsio Ad perniciosam Disputationem Io. Baptistae ab Helmont Bruxellensis Medici Pyrotechnici* Luxemburgi: Excudebat

Hubertus Reuland; 1621, on the Rosicrucians see Kircher, *Mundus subterraneus*, Amsterdam: Joannem Janssonius & Eliseum Weyerstraten; 1665, Tom. 2, p. 280, on judicial astrologers see Benito Pereira, *Adversus fallaces & superstitiosas artes*, Venice: Apud Ioan. Baptistam Ciottum, Senensem.; 1592 pp. 164 ff.

142 'Il monopolio di questa preziosa merce non è conceduto ad alcuni, che, a guisa a punto de' negromanti, con certi vocaboli orrendi ed oscuri si rendono venerabili al volgo per singolarità di sapienza'Pallavicino, Sforza. *Del Bene. Libri Quattro*. Venice: Lorenzo Basegio; 1648.

143 On disinterestedness, see Peter Dear, *From Truth to Disinterestedness in the Seventeenth century.* Social studies of science. 1992; 22: pp.619–631.

144 Boyle, *Defence ... against the Objections of Franciscus Linus*, in Boyle, *Works*, I, pp. 118 ff. 'this opinion [i.e. the presence of an extremely rarefied substance in the Torricellian space] being approved by many eminent scholars, especially of that most learned order of the Jesuits (to whom perhaps its congruity to some articles of their religion chiefly recommends it)'.

5

The Angel and the Compass: Athansius Kircher's Geographical Project[1]

Lost at sea – magnetic declination and navigation

In October 1639, the Jesuit missionary Martino Martini found himself adrift in the Atlantic. On route to Goa, the Portuguese vessel that contained nine Jesuits destined eventually for the Chinese mission had met with catastrophic conditions. The boat, along with its companion vessel, was forced to make an unplanned forty-six day stop on the Guinean coast that drained its supplies, infected its passengers with horrific maladies and forced a return to Lisbon. 'To tell the truth to Your Reverence', Martini wrote to his erstwhile mathematical mentor at the *Collegio Romano*, Athanasius Kircher, 'the land and sea along that coast generally called Guinea appear to have been damned from all eternity, such are the heat, the rain, the pestilence, things that you would never believe'.

Dejected at their aborted mission, the Jesuits and their companions turned back towards Portugal, passing close to the Azores, where Martini noticed the abundant Sargassum grass floating in the water. In addition to indicating to the mariners their position with respect to the islands, Martini noted, the round berries of the flax-like sea-grass were reputed to be an indispensable remedy for gallstones.

On October 1, the vessel was hit by a violent storm. 'The water was higher than mountains.' With all sails taken down except for one the size of a sheet, the boat was driven along by the wind for 'almost seventy leagues'. After the winds had finally subsided, the nobles and sailors on board entertained themselves by making bets as to their distance from the Portuguese coast.

Martini, armed only with a chart on which he had been tracking every step of the ship's voyage, and a compass specially adapted to allow him to calculate the declination of the magnetic needle from true North, defeated both noblemen and mariners in his calculations. 'I said that we were to the East of the island of Terceira and only one hundred leagues from the mainland'. In exact accordance with Martini's predictions, the ship arrived in Portugal early in the morning of 14 October.

How did a Jesuit priest with almost no seafaring experience defeat the estimates of seasoned navigators with expert knowledge of sea currents, winds and marine phenomena? Martini's reasoning went thus:

> if we had been to the West of the Azores, the magnet should have declined to the West, but as it declined to the East, we could not have been to the West. Some said that we were in the midst of the Islands, but I demonstrated that this could not be true, as, even though we were at their latitude we did not see them, and that was impossible.

Sceptics challenged Martini, wondering why, given that the islands extended for 120 leagues from East to West, they had never been seen during the course of the storm. 'Precisely because even before the storm we were to their East', rebutted Martini, supporting his claim with a detailed mathematical analysis of the ship's meandering route.

'I write this', Martini flattered Kircher, 'not so as to praise myself, but so that Your Reverence may see all that I have learned from you, especially in the field of magnetic declination'.[2]

Martini had spent a mere two months as Athanasius Kircher's 'private disciple in mathematics' in the Jesuit *Collegio Romano*, but this brief apprenticeship, occurring shortly after Kircher had taken up the post of mathematics professor, apparently had a transformative effect on him. At the end of the sixteenth century, Christoph Clavius had created a private mathematical academy in the *Collegio Romano*, with the express goal of providing advanced training to those destined to teach mathematics in Jesuit colleges in the different provinces, and to those destined for the Chinese mission, for which mathematical skills were regarded as particularly relevant. Matteo Ricci, the most famous representative of the first generation of Jesuit missionaries to China, was himself an alumnus of Clavius's original academy. The 'academy' really consisted in informal advanced training that took place in the bedroom of the senior mathematician of the College, also known as the 'mathematical museum' (*musaeum mathematicum*), where valuable instruments and mathematical manuscripts were kept under lock and key.[3]

Martini's floating microcosm – his cabin aboard ship, filled with charts, astrolabes, quadrants, compasses and the astronomical works of Clavius, Peter Apian and Tycho Brahe – is a fascinating mirror of Kircher's *cubiculum* in the *Collegio Romano*, which served as an advanced training ground for Jesuit mathematicians. While Kircher would draw heavily on Martini's reports in compiling his *Magnet, or on the Magnetic Art*, Martini used the mathematical techniques he had learned from Kircher during his two-month apprenticeship in Rome to demonstrate his navigational superiority over the ship's pilots. Pitting his own book-knowledge, charts and instrumental abilities acquired from Kircher against the accumulated experience, dead-reckoning and reliance on natural signs of the Portuguese mariners, Martini claimed multiple victories. On his subsequent voyage to Goa, his judicious use of the magnetic needle saved his ship, carrying the Viceroy of the Indies, from certain destruction on a shoal of treacherously sharp rocks.[4]

Kircher's geographical plan

In his aspirations to universal knowledge, Kircher relied crucially on Martini and his ilk, Jesuit missionaries inflamed by their Ignatian training to endure every sacrifice to advance the glorious achievements of their Order. Conversely, the mathematical skills of Jesuit missionaries, in addition to their willingness to nurse the sick, hear confessions and even parade as flagellants during Easter week, helped to ensure them a welcome place aboard the heavily charged Portuguese ships destined for the Indies.[5]

In particular, Kircher's audacious attempt in the late 1630s and early 1640s to carry out a great 'Geographical Plan' (*Consilium Geographicum*), aimed at harnessing the global network of Jesuit missionaries in order to reform geographical knowledge and to resolve the problem of calculating longitude at sea constitutes a vivid demonstration of the nature of the organic connections between Kircher's Roman cell, on the one hand, and the missionary space inhabited by Martini, on the other. The global distribution of Jesuit missionaries was absolutely essential to Kircher's attempt to reshape terrestrial geography – by fixing the longitudes and latitudes of Jesuit missions and colleges – and to reform navigation – by devising a method for calculating longitude at sea.

The primary 'enabling technology' for Kircher's project was correspondence – frequent epistolary contact with mathematically trained Jesuits. Kircher's deployment of correspondence as an instrument with which to gather scientific data will be a primary concern of this chapter. Noel Malcolm argues convincingly that Kircher's 'oracular' correspondence was atypical of the fluid, multi-directional model of correspondence endorsed by the seventeenth century Republic of Letters.[6] Kircher's Geographical Plan constitutes a particularly striking example of his conception of the role of the centralised accumulation of correspondence in the reform of natural knowledge, and makes explicit the monarchical power-structure that characterised his epistolary community. The ultimate failure of his geographical project, which quite literally vanished, as we will see, and his dispute with Jesuit astronomer Giambattista Riccioli over the relative merits of global correspondence and expensive local instrumentation, illustrate a clash between two contrary social models for the prosecution of research in astronomy and geography.

Correspondence in Jesuit culture

Letter writing had a special place in early modern Jesuit culture. Indicatively, Ignatius Loyola's correspondence constitutes the largest extant correspondence of any sixteenth-century figure. Loyola's prolific secretary Juan de Polanco emphasised the enormous importance of letter writing to maintaining the spiritual unity of the globally distributed members of the Society of Jesus, exhorting Jesuits in the different geographical Provinces of the order to maintain frequent epistolary contact with the Roman centre and providing them with complex guidelines concerning all aspects of letter writing.[7]

On 25 July 1547, Polanco, who had just assumed the post of secretary to Ignatius Loyola,[8] wrote a circular letter addressed to each of the members of the Society of Jesus in which he outlined the various functions of frequent correspondence amongst the members of the order.

> Although we do not know one another by sight, for a long time our Redeemer and Lord Jesus Christ, who reinforces the link of common charity that unites us together like members of his body, has joined me closely to Your Reverence [...] There is thus no reason for me to find anything strange about writing to Your Reverence[9]

Polanco continued his letter to outline the immense importance of frequent letter writing between Jesuits. Merchants, well-practised in the epistolary art for their own 'miserable interests' were putting the early Jesuits, driven by far more pious goals, to shame.[10] Polanco provides a list of no less than twenty benefits achieved through the frequent exchange of letters between Rome and the periphery. As well enhancing the unity of the Society, nurturing mutual love between its members and increasing their humility by keeping them punctually informed of the worthy deeds of their confrères, letter writing would aid the growth of the good reputation of the Society and increase the efficiency of its government.[11] Polanco noted that Jesuits in the provinces were being asked to perform a far less onerous task than the bureaucrats of the Roman centre of the order, who 'occupy ourselves willingly with this task of writing, which is our principal, and almost exclusive activity'.[12] While those on the periphery need only write to give account of themselves to those in Rome, the latter were obliged to satisfy the needs of all of the far-flung places where the Jesuits were stationed. A brief set of rules for correspondence between Rome and the periphery was attached to Polanco's letter, and instructed Jesuits on the practicalities of sending and receiving letters, by means of travelling clerics or merchants, and keeping them from falling into the wrong hands.[13]

Polanco's letter illustrates the immense importance invested in letter writing in the early years of the Society of Jesus. Polanco's model of epistolary exchange, moreover, was emphatically centralised in Rome. During the years that followed, practices of letter writing were gradually refined to deal with the immense expansion of the Society, but centralised correspondence remained at the core of the Jesuit apostolate. One might even say that in the Jesuit order, correspondence substituted the collective prayer of the cloistered medieval orders as the primary expression of the *opus divinum* and the ultimate bond between the globally distributed members of the *corpus christi*, as the order frequently described itself. All aspects of the Jesuit ministry, from the construction of new churches and colleges to the elaboration of the educational structure published in the *Ratio Studiorum* were carried out through the accumulation and evaluation of letters, reports and projects in Rome.[14]

Kircher's magnetic geography

In his 1641 work *The Magnet, or on the Magnetic Art*[15] Kircher outlined his proposal for a *Magnetic Geography* that would be magnetic in two respects – both in seeking magnetic solutions to geographical and navigational problems and in drawing the observations performed by mathematicians, navigators and missionaries throughout the world together in Rome, as if by some occult force of attraction.[16] Kircher likened his project to the reform of the calendar reform carried out under Pope Gregory XIII, suggesting that a similar initiative might allow geographical knowledge, clearly in disarray, to be reformed, just as the convergence of the authorities of Pope, princes and universities had reformed the temporal order governing religious and civil affairs. [**See Figure 5.1**]

Figure 5.1 A magnetic Habsburg eagle, from Kircher, *Magnes, sive de arte magnetica* (1643 edn.), frontispiece

Like the Gregorian reform of the calendar, Kircher argued, geographical reform could not be carried out by a single individual. Instead, it was seen to require a 'unanimous conspiracy of mathematicians'. The religious orders were particularly suited to such a task, but most appropriate of all was the Society of Jesus, 'distributed throughout the whole globe, provided with men skilled in mathematics and, above all, enjoying a unanimous harmony of minds'.

Kircher was urged to embark upon the reform of geographical knowledge through the use of Jesuit informants by a number of sources, and especially by the General Muzio Vitelleschi, who ordered him to compose a Geographical Plan' (*Consilium Geographicum*), 'a treatise in which I would display the methods and procedures for restoring Geography, and would explain by what means, with which instruments, and in which place, state and time observations might be carried out fruitfully. I would try to show briefly and clearly that this business would not be difficult work for the religious orders'. Kircher's plan for a Jesuit-led global observational imperative would go far beyond mere cartography: 'I would also provide instructions for what they should observe about the flux and reflux of the tides, the constitution of lands and promontories, the natures and properties of winds, bodies of water, rivers, animals, plants and minerals, and, finally, about the customs, laws, languages and religious rites of men'.[17]

Although Jesuit missionaries, from Matteo Ricci to José de Acosta, had been enormously active in accumulating observations of just this kind in the first century of the Society's existence,[18] at the beginning of the second century Kircher wished to discipline and coordinate such reports. By doing so he would avail of the mobility, mathematical expertise and self-effacing obedience of his Jesuit colleagues, a human resource generated largely through the political and pedagogical adroitness of Christoph Clavius during the last decades of the sixteenth century. Inscribed into Kircher's larger geographical project was an attempt to resolve the recalcitrant navigational problem of calculating longitude at sea.

The problem of determining longitude while at sea was of the utmost importance for navigation in the seventeenth century, given the absence of a mechanical clock that could remain reliable during a sea-voyage.[19] A huge number of solutions were proposed after Philip III offered a perpetual pension of 6,000 ducats to anyone who could find a workable method of maritime longitude determination in 1598. Galileo had proposed using the eclipses of the newly discovered satellites of Jupiter as a 'celestial clock' which sailors might consult to determine their position, a project frustrated by the difficulty of making accurate telescopic observations of the Jovian moons aboard a moving ship.[20] Giuseppe Biancani wrote to Clavius to suggest a mechanical solution involving synchronizing a large number of on-board water clocks.[21] Oronce Finé, followed by Jean-Baptiste Morin, proposed an immensely complicated method involving the movement of the moon against the background of the fixed stars, of which Kircher later complained that its use required the mathematical ability of a Euclid or a Ptolemy.[22] Michael Florent van Langren attempted to use the motion of the terminator shadow across the lunar disc as a painfully slow celestial sundial.[23] Kircher approached the problem in a different way, through magnetic variation.

The idea of determining longitude by using the deviation of a compass needle from North as determined by the pole star, or by observing the sun at equal intervals before and after noon and taking an average was not original to Kircher, having been suggested by the Neapolitan magus Giambattista della Porta in the late sixteenth century and by mathematicians and navigators in England.[24] The famous series of engravings carried out in the late sixteenth-century by the Flemish artist Jan van der Straet, or Stradanus, and printed by Jean Galle included, along with such celebrated inventions as gunpowder, eye-glasses and the printing-press, an illustration of 'the longitudes of the globe discovered by the declination of the magnet from the pole'. In the illustration, a sailor aboard a ship in stormy seas, calculates the position of the meridian by observing the position of the sun, and compares it to the direction of the magnetic needle to calculate the declination [**See Figure 5.2**].

Despite the optimism of Jan van der Straet, however, it was by no means obvious to most navigators in the early seventeenth century just how the

Figure 5.2 Jan van der Straet (Stradanus), *The longitudes of the globe discovered by the declination of the magnet from the pole*, from the series *Nova Reperta*, circa 1600.

measurement of magnetic declination could allow longitude to be calculated at sea. The Jesuit missionary Cristoforo Borri, who travelled to Macao and Indochina between 1615 and 1622, was reputed to have discovered a method. He was nominated to the position of technical adviser to the Royal Council in Lisbon, and later also in Madrid. His negotiations to sell his technique to the Spanish crown and to the papacy met with little success, however, and may have led to his sudden departure from the Jesuit order in 1632. Kircher clearly knew about Borri's efforts, and endeavoured to use Martini to discover further details of his method. The technique, at least according to Martini, seems to have involved the construction of a chart mapping points of equal magnetic declination, an azimuthal compass (i.e. a magnetic compass equipped with a sighting device or shadow-casting device to allow the astronomical meridian to be determined) and a technique for measuring the declination at any time of day.[25]

Acknowledging Kircher's privileged position in Rome, Marin Mersenne wrote to Gabriel Naudé in Rome in 1639 to suggest that Kircher should 'order some Reverend of the Society in each college, by whatever means possible, to note the variation of the magnet and the height of the pole star accurately. Let him order that one or another lunar eclipse be observed in these same houses and colleges'. 'If this task were completed', Mersenne continued, 'and if the authority of the supreme pontiff would lend itself to this task, the result would be that some time under the happy auspices of Urban VIII we would know the magnetic variation of the whole world, the altitudes of the pole star, and the longitudes so long sought after'.[26] Kircher responded to Mersenne to inform him that he had already embarked on just such a project.[27]

Mersenne's suggestion was similar in tone to one made some years before by Gassendi, who proposed to Kircher's patron Nicholas Claude Fabri de Peiresc that either Urban VIII or his nephew Cardinal Francesco Barberini should incite missionaries to make accurate eclipse observations to reform the geographical art.[28] Interestingly, Gassendi did not restrict his suggestion to the Jesuits, having made previous use of the observational powers and mathematical expertise of other peripatetic counter-reformation orders such as the Capuchins and the discalced Carmelites in collecting reports of eclipses.[29]

Kircher's own appreciation of the power of a distributed collective of observers seems to owe much to his stay in Provence from 1632–1633, and in particular his contact with the Peiresc circle. As Nick Wilding has shown, Peiresc's attempts to train Capuchin missionaries bound for the Orient to perform coordinated eclipse observations involved Kircher, not as coordinator, but as obedient, if not always accurate, observer.[30] The observations of solar and lunar eclipses, while they were of little use to a sailor lost at sea, could nonetheless establish precise differences in local time between two land locations, and thereby allow their difference in longitude to be calculated.

In 1635, after Kircher's departure from Provence, by coordinating the eclipse observations of French Capuchins in Aleppo and Cairo, and Gassendi in France, Peiresc 'changed the size and shape of the Mediterranean and mercantile practices', as Wilding argues. As Peiresc put it, in a letter to the Capuchin P. Césaire de Roscoff:

> Anything you might say by calculation can be destroyed by the smallest of real observations – as happened with the distance between Marseilles and Palestine, which maritime charts reckoned to be 2,700 miles, or 900 leagues, but which, from the last eclipse of 28 August 1635, observed by your Reverend Fathers in Aleppo and Cairo, and at the same time by Gassendi in this country, and by other friends of ours in Rome and Naples, is found to be at least a third shorter. To remedy this, sailors had the habit of taking the wrong route, and tending left, both on the way out, and on the way back, which got them, without thinking, to the place they were looking for in its actual location, much better than maritime charts did[31]

Peiresc also attempted to involve Kircher in this crucial observation, but the Jesuit's data turned out to be unusable, due to his use of inappropriate instruments of time measurement. As the legally-trained Peiresc wrote stiffly to Kircher,

> I thank you most humbly for remembering to observe the eclipse for the love of M. Gassendi, who is in Digne and will not neglect to thank you in time. Mainly for your demonstration of good will. For I do not think that he will be able to trust your observation completely, given the small degree of precision of sand-glasses and mechanical clocks.[32]

Instead, Kircher should have used his astrolabe to measure the time of the lunar eclipse, by taking a sighting on a fixed star.

While Peiresc and Gassendi could use Capuchins and discalced Carmelites to transfigure the Mediterranean, however, the Atlantic space remained far less accessible to their network of informants. Additionally, while eclipse observations might allow longitude to be established at a terrestrial location, they were of little use to a lost ship's captain unless his predicament happened to coincide with a lunar eclipse.[33]

Kircher put Mersenne's proposal into action as part of his geographical project. Having performed numerous observations of the magnetic declination during his own peregrinations through Europe, and armed with the observations collected by his predecessors in the *Collegio Romano*, he wrote to distinguished mathematicians throughout Europe to solicit their measurements of the magnetic variation of their place of residence. He hoped that in this way they 'would all be inspired to perform careful observations to determine this variation and other matters with which our Geographical Plan is concerned'.[34] The outcome of this first attempt was disappointing. Kircher had 'almost no news at all from the more famous mathematicians',[35] despite his entreaties. This required a change of plan. Taking advantage of a meeting of the Procurators (responsible for the financial affairs of each Province of the Jesuit order) in Rome in November 1639, Kircher asked each Procurator to solicit observations of local magnetic declination from the Jesuit mathematicians resident in the different cities of his Province.[36] In addition to sending observations, each mathematicians was to explain in detail exactly what precautions had been taken, and what type of equipment had been used. Unlike the more famous mathematicians, a great number of their Jesuit contemporaries responded immediately.[37]

Kircher published their observations along with those made by others in his *Magnes, sive de arte Magnetica*. In recognition of the labours of his Jesuit helpers, performing observations of the magnetic variation in places as far apart as Goa, Paris, Macao, Alexandria, Constantinople, and Vilnius, Kircher published their names in a large table [**See Figure 5.3**] reporting the magnetic declination and the latitude of the place at which the observation was made. Behind this table lies an enormous amount of labour, in the performance of observations in different urban centres, their transmission to Kircher and their tabulation.

Politically, it has often been observed that the Jesuit order has a monarchical organizational structure, with great emphasis on obedience to commands issued to the periphery from the Roman centre.[38] Such a structure, to be contrasted with the capitular structure of the older monastic and mendicant orders, clearly lends itself extremely well to projects like the measurement of global magnetic variation.[39] One of Kircher's more expert correspondents on magnetic matters, the French Jesuit Jacques Grandamy, made the congruence of absolute power and global observation very explicit when he suggested in a book published four years after Kircher's *Magnes* that kings and princes should order their subjects to measure magnetic variation diligently in the cities of under their rule, and that the General of the Society of Jesus should order his subordinates – Jesuit priests and lay brothers in different parts of the world – to do the same.[40] Although Kircher makes frequent reference to a 'Republic of Letters' in his works, both he and Grandamy are clearly conscious that in the world in which they live, the

GEOGRAPHIA MAGNETICA. 401

Tabula III. Declinationum Magneticarum à Mathematicis per Europam ad inſtantiam Authoris obſeruatarum.

Nomina obſeruatorum	Locus Obſeruat.	Declin. G.	M.	Latit. loci G.	M.
P. Chryſoſtomus Gallus S.I. P. Chriſtophorus Burrus	Vlyſsipone relatione aliorum	7	30	39	38
P. Martinus Martinius S.I.	Eboræ in Luſitania	6	12	39	0
	Conimbricæ in Luſ.	6	3	40	30
P. Chriſtophorus Burrus S.I.	Madriti	5.circit.		40	45
R.P.F. Marinus Merſennus, & P. Petrus Bourdinus S.I.	Pariſijs	31	0	48	40
P. Iacobus Grandamicus S.I. P. Vincentius Leotaudus S.I.	Turonibus in Francia	4	50	43	43
	Dolæ in Burgundia	5	14	46	48
Author	Veſontione	5	0	47	15
	Lugduni	4	30	45	10
P. Antonius Lalouure S.I.	Turnoni	3.51.Or.		45	16
Clar. D. Ant. Franc. de Payen.		3	10		
Prænob. D. Petrus de S. Eligio Author	Auenione	4	0	43	42
		4	30		
P. Guglielmus Degner S.I.	In Monte Peſſulano	3	30	43	30
Author	Arelati	3	30	43	38
	Maſſiliæ	2	40	43	20
R.D. Petrus Gaſſendus	Diniæ	2	40	43	0
	Aquis Sextijs	2	36	43	25
Author	Villæ Francæ prope Niſſam	226.Oc.		43	30
P. Fr. Franciſcus Niceron.	Liburni, vel Ligurni	5	0	42	30
P. Martinus Martinius S.I. Clar. D. Hieronymus Bardius	Genuæ	5	58	43	50
		5	30	43	50
P. Carolus Moneta S.I.	Mediolani	2	30	45	6
D. Franciſcus Iardinus	Mantuæ	0	30	45	12
Clar. D. Io. Baptiſta Manzinus	Bononiæ	3	0	44	16
P. Nicolaus Cabæus S.I.	Ferrariæ	5	50	44	10
P. Ioſephus Blancanus S.I.	Parmæ	6	0	44	30

Eee   Re-

Figure 5.3 Table of magnetic declinations, from Kircher, *Magnes, sive de arte magnetica*, 1643, p. 401

command of an absolute authority, whether secular or clerical, was the most effective way of galvanising observers into action.

The letters sent to Kircher by his Jesuit informants reveal the difficulties of constructing a collective experimental enterprise [**See Figure 5.4**]. Joannes Ciermans, writing to Kircher from Louvain, writes in highly charged language:

Although the sky here is cold and cloudy, this is not true of my breast, under which something is warm and lives in ready obedience to Your Reverence. To accumulate together in the Father that which you estimate to bring splendour to his name and to that of our Mother, the Society, you will have a strong helper in me if you wish. For we know that it is not for one man to repair [*instaurare*] astronomy and geography, but requires the works of many mathematicians to be gathered together in one.[41]

In Lithuania, on the request of the Provincial, Oswald Krüger took time away from his cooking-duties to observe the magnetic declination of Vilnius and two

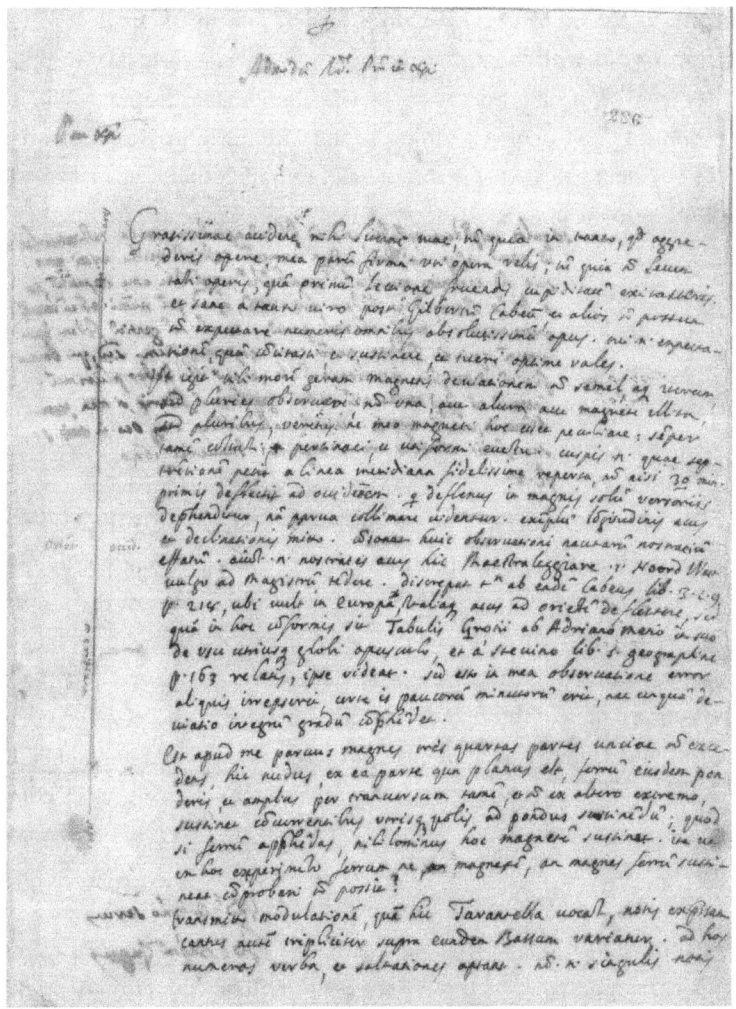

Figure 5.4 Letter from Giovanni Battista Zupi to Kircher, Naples, 21 January 1640, APUG 567, f. 286r.

neighbouring towns and wrote to the Polish Provincial to encourage Jesuit mathematicians in the Polish province to do likewise.[42]

The letters sent by Kircher's magnetic correspondents give a picture of an increasingly fervent exchange of magnetic needles, books, and observations in the early 1640s, continuing well after the appearance of the first edition of Kircher's Magnes. A correspondent in Mainz, a city where Kircher had previously taught for several years, though keen to send Kircher his measurements, was unable to be of any use because the marauding Swedish armies had taken every mathematical instrument in the Jesuit college, down to the last pair of compasses.[43] At the other end of the scale, Jacques Grandamy boasted of a new instrument which he had designed to measure both magnetic declination and inclination, or dip, with the utmost accuracy.[44] Others clearly didn't understand what they were supposed to do, and asked Kircher for clarification, while sending observations of questionable meaning. Along with the numerical measurements, Kircher's obedient observers often sent diagrams and other information to make their observational practices as transparent as possible to the 'mathematical prince of our Society' in Rome.[45]

Occasionally the task of observation was delegated by Kircher's correspondents to their subordinates: 'The declination of the magnet from the Meridian, required by Your Reverence, has been investigated by Master Gaspar Schiess, the private mathematical disciple of Fr. Cysat', Jacobus Imhofer wrote to Kircher from Innsbruck on 15 January 1640. 'He has used various needles, all of which disagree with each other, some indicating 4, some 6 and some 10 degrees [of declination]. He says that he is waiting for the arrival of Fr. Cysat, who has the best magnets locked-up, and that he will then make observations most diligently and send them to Your Reverence.'[46] Jesuits world-wide begged Kircher to turn them into more efficient measurers. 'If Your Reverence has some information about this practice', wrote Jacques Durand, 'I would be most grateful if you could send it to me.'[47] Some sent reflections of a philosophical nature, querying the source of terrestrial magnetism and Gilbert's suggestion that the earth was a large magnet. Others reported on magnetic magic, particularly Francis Line's magnetic clock composed of globe suspended in water that rotated to indicate the hours of day and night.[48]

Martino Martini himself provided Kircher with a vast number of measurements made during his voyages, from Portugal to Cape Verde and the Azores, from Goa to Macao. Martini was also perhaps most optimistic amongst Kircher's correspondents of the possibility of solving the famous problem of longitude. He wrote from Goa to Kircher in a letter that Kircher published proudly in the *Magnes*, claiming that 'the discovery of longitudes by the magnet is no longer held by me to be impossible, indeed, I believe it has already been discovered', and following his claim with a description of a technique for using a chart marked with magnetic meridians to calculate longitude.[49]

However, a number of correspondents wrote independently to advise Kircher of some anomalous observations recently performed in England. The measurements of magnetic declination performed in Limehouse by William Borough, Edmund Gunter and Henry Gellibrand appeared to show a decrease in magnetic declination between 1580 and 1634.[50] Mersenne, Gassendi, Pierre Bourdin and Jacques Grandamy all reported the same phenomenon to Kircher in their letters and speculated on its possible causes.[51] Similar changes had been observed by Jesuit mathematicians in Rome and Bologna. Although Kircher recognised the difficulty which such observations posed to his project of using tables of declination to calculate longitude, he was hesitant to pronounce on the cause of this phenomenon, and effaced many of the cosmological speculations of his informants from the published work.

Angelic observation

There is a fine balance, in this whole episode, between acknowledging the fallibility of the single observer or instrument and emphasising the immense power of a Jesuit experimental collectivity. Kircher's reaction to Gellibrand's observations, which were eventually to quash hopes for a geomagnetic method of marine longitude determination, is indicative of this tension. Every observer was born with original sin in Kircher's world. 'A perfect observation, free of all error and falsehood could only be carried out by an angel', he claims in *Magnes*, so mere mortals must acknowledge their fallibility before jumping to conclusions of the nature of terrestrial magnetism or other questions of cosmological import. 'While I assert this', Kircher continues, 'nobody should think that I wish to detract from the most useful and absolutely necessary study of observations. I only wish to show how much caution, circumspection, industry and indefatigable labour is required in making observations, for them to be reliable.'[52]

Kircher's 1646 *Ars Magna Lucis et Umbrae* renewed Kircher's promise to publish his *Consilium Geographicum* for the collective restoration of all terrestrial knowledge. In the meantime, he provides his readers with a *Horoscopium Catholicum* – a composite sundial in the form of an olive tree representing the different provinces of the Jesuit order that Kircher displayed to visitors to his museum in the *Collegio Romano*.[53] When a stylus was placed in each Province, and the device was positioned vertically so that the Roman time was given correctly, the clock allowed the time in all the different Jesuit provinces to be read correctly. In this way, the viewer could perceive that the Society of Jesus was performing its religious duties – masses, confessions, sermons and catechesis – throughout the world, day and night, with no interruption and in all known languages.[54]

Following emblematic themes developed in the 1640 *Imago Primi Saeculi Societatis Iesu, [Image of the First Century of the Society of Jesus]* celebrating the first centenary of the Jesuit order, Kircher's universal horoscope is the apotheosis of Jesuit globalism and pious synchronicity. Initially a cruciform version of the paper instrument was displayed, and dedicated to the new General Vincenzo Carafa on the day of his election.[55] Surmounted by a Habsburg eagle, carrying an Austrian [*Austri-acus*] compass needle, a feature removed from the Amsterdam edition of the *Ars Magna* for the peace of mind of a Protestant readership, the olive tree sundial was designed so that the shadows of the small gnomons, when aligned, spell the abbreviated name of Jesus, IHS, which appears to 'walk over the world' with the passing of time, like the synchronized, uniformly trained members of the Jesuit order who used the abbreviation as their symbol. Kircher's idealised Jesuit geography, placed on display to visitors in the Roman centre, situated the prime meridian emphatically in Rome [**see Figure 5.5**]

But what of the great Geographical Plan? Giambattista Riccioli wrote to Kircher in 1642 to ask about when the *Consilium Geographicum* might at last appear in print. Riccioli had collected a vast number of observations himself, and conducted a lengthy series of experiments on precision time-measurements using pendulums which he applied to making eclipse observations. In some ways providing a competing model to Kircher's distributed information-community, Riccioli surrounded himself with local disciples willing to observe pendulum oscillations for consecutive periods of up to twenty-four hours at a time, and extremely precise observational instruments.[56] Riccioli's impatience to see Kircher's *Consilium Geographicum* in print was in vain. In the 1654 edition of the *Magnes*, edited and amplified by Kircher's disciple Kaspar Schott, it became clear that the Great Geographical Plan would never be revealed. 'When I was keeping the work, composed with no small effort, amongst other things, in my Museum, and waiting for the right moment to publish it for the good of the Republic of Letters', Kircher wrote, 'it was secretly removed by one of those people who come to me almost every day from all over the world to see my Museum'.[57] Kircher's project for a universal reform of terrestrial knowledge through the concerted agency of the Jesuit order was stolen!

Figure 5.5 The Catholic Horoscope of the Society of Jesus, from Athanasius Kircher, *Ars Magna lucis et umbrae*. Romae: Ludovico Grignani; 1646, facing p. 553

The mysterious theft of the *Consilium* from Kircher's museum conveniently relieved him from the need to produce a method for determining longitude by magnetic declination, an obligation that had become increasingly complicated with further observations of the temporal instability of declination, despite the optimism of Kircher's Jesuit disciples for the magnetic reform of geography and hydrography. Even before the disappearance of the *Consilium*, Kircher's longitudinal concerns had swung decisively landwards. He wrote to Gassendi in 1642 to say that Cardinal Francesco Barberini was urging him to coordinate eclipse observations, in the same way that he had coordinated measurements of magnetic declination two years previously.[58] As with the declination observations, Kircher demanded that his informants on eclipses provide him with all of the details of the circumstances under which the observations were carried out, the names of those who were present as 'indicators [*indices*] and witnesses of the said eclipses'.[59]

Riccioli's exquisite instruments

Riccioli probably received a similar request at this time. In any case, he wrote to Kircher shortly afterwards to say:

> I have exquisite instruments [*organa*] in which, for reasons explained in an astronomical work that I have in my hands, I place my trust more than in those of Tycho himself, even if that great man got very close to the truth. I also have four of ours [i.e. Jesuits] who are extremely well trained and are both my witnesses and my assistants in conducting observations.[60]

In the end, it was Riccioli, not Kircher, who published a *Reformed Geography*, incorporating many of the observations previously published by Kircher into his tables and adding observations performed by himself and supported by the financial resources of the extremely wealthy Grimaldi family of silk-merchants.[61] Well before he did so, however, he was subjected to a process of censorship that reveals something of the tension between local and non-local modes of natural investigation in the Jesuit order.

On 24 November 1646, Riccioli was forwarded a copy of an anonymous censure from Rome. The letter requested him to 'send to Rome that part of his work which is entitled "On my own Discoveries", so that it can be known what he will put forward that is new with respect to the most excellent artificers Tycho, Kepler and Lansberg whose expenses in this matter of such great importance were supported for all their lives by Emperors and Kings'. The anonymous Censor also asked 'What methods and instruments were used to observe the motions of the stars', and insisted that Riccioli 'should also send that part of the work which he calls Instrumental Geography, so that it can be known from this what method he will use in emending and assigning the true longitudes of regions. For this is a task not for a single man, but such as deserves *the unanimous collaboration of all the mathematicians of the Society*.⁶²

The tone of the censure clearly recalls Kircher's geographical project, and, indeed the handwriting of the anonymous text is a convincing match with Kircher's letters from the period, providing further confirmation of his authorship.⁶³ Riccioli sent a chastened official response to the Roman Censor, but Kircher sent a further, private letter to him at this time that included a number of more damning criticisms voiced by other people both inside and outside the Jesuit order.⁶⁴ To this second letter, Riccioli responded at some length.⁶⁵

Dismissing as absurd the criticism that Riccioli, a theologian, should not engage in mathematics because it was 'unbecoming for a single person to profess two different faculties', Riccioli invoked a number of illustrious polymaths, ranging from Thales to Tycho Brahe and Kircher himself. 'To speak freely to you', he continued to Kircher,

> it was worthwhile procuring a vacation from theology, and refusing the administrative offices that I was offered more than once, acquiring from whatever source the money necessary for the construction of instruments and observational glasses, and wearing away my health by so many long night vigils, that all of whatever mind I had, nay, not mind, but back and upper-arms, has been expended as if from rolling a great weight ahead of me.⁶⁶

Riccioli also defended himself strenuously against the accusation that he relied solely 'on the judgements of [his] pupils', inverting the traditional Jesuit hierarchy of authority.⁶⁷ The following objection, however, was that Riccioli was a 'private man' – 'that is, as I interpret it, that I do not supply the expenses necessary for this business, but that they are supplied by my disciples from most noble families, Fr. Alfonsus Gianoti rector of this College, Marquis Cornelius Malvasia and, in the first place, by the Grimaldi, a most opulent family of this city'.⁶⁸ Riccioli did not deny the charge – 'Certainly our metal instruments are present in the college, and I did not create them out of nothing'.⁶⁹ However, the expenses incurred in instrument-building were justified by their capacity to enhance the reputation of the Society for mathematics and to bring direct returns:

> To inspect and to be witnesses on one occasion or another, were not only ours [i.e. Jesuits], but also other men of this city, and they were astonished by the agreement of the different instruments, directed towards the same star, to the minutes. And, among others the same Rocca [i.e. Giannantonio Rocca] remarked

that he would trust (hold back your envy of the word) my observations no less than those of Tycho himself. Dr. Antonio Roffini was so captivated by [the instruments], that although he was previously hostile to ours [i.e. the Jesuits], he will bequeath his library, most richly provided with mathematical books, to our College.[70]

Perhaps most revealingly, Riccioli politely refused Kircher's request that he should move to Rome:

I say sincerely that there are reasons why I cannot do so without great damage to my work. Where you are, I cannot hope for the instruments and the books which, in addition to the library I already mentioned, I am given freely by the Marquis Malvasia, P. Cavalieri, P. Ricci, Dr. Manzini and others who are extremely well provided with them, far less the enormous gnomon which I use in the church of S. Petronius. Two Coriolians, engravers of figures in wood which are so fine that they seem to be in copper, and who are now obliged to me, as is the caster of new print-characters; the said D. Cornelius Malvasia Vexillifero, now a Senator, who encourages me and helps to cover my expenses together with the Most Eminent Cardinal [Girolamo Grimaldi], who also expects the book to be dedicated to him – all of these, I say, I cannot hope to find elsewhere.[71]

Where Kircher saw the acquisition of natural knowledge as operating through a centralised global epistolary network of Jesuits, Riccioli's project was irretrievably local. Apart from his own body, he could not even send the parts of his book that Kircher requested from Bologna to Rome because 'the affectations of my health and my stomach pains' rendered copying out the different parts of the book an impossibly arduous task.[72]

Local patronage, books, instruments, artisans, and Ignazio Danti's utterly immobile meridian line in S. Petronio – a fitting foil, perhaps, to Kircher's universal Jesuit horoscope – conspired to prevent his removal to Rome.[73] Where Kircher concentrated his energies on marshalling a distant community of observers, Riccioli cultivated close local friends and disciples. Too close, occasionally – his celebrated relationship with Francesco Maria Grimaldi extended to allowing the latter to shave him and cut his hair, and the tendency for the older Jesuit to entertain his younger disciple in his bedroom late at night, after the other members of the community had gone to bed, led to rumours reaching the ears of the General, who obliged Riccioli, against his protestations of health problems, to move from Parma to Bologna,[74] where Grimaldi would eventually join him.

Figure 5.6 Frontispiece showing Astrea, goddess of justice, as a winged angel, from Giambattista Riccioli, *Almagestum Novum*, 1651

When Riccioli published his extremely influential *New Almagest (Almagestum Novum)*, stripped of the part containing the descriptions and illustrations of his expensive instruments that had so worried the Roman censors, he acknowledged his human fallibility in the frontispiece, by giving wings to the figure of the goddess Astrea, in explicit acknowledgement of the truth of Kircher's claim that perfect observations were only possible for an angel. [**See Figure 5.6**][75]

How should nature be investigated?

We have seen that for Athanasius Kircher, the Jesuit role in reforming natural knowledge was utterly continuous with the other parts of Jesuit ministry, a continuity that is emblematically represented by the marriage of longitude and prayer in Kircher's *Horoscopium Catholicum Societatis Iesu*. The self-effacing, centralised bureaucracy of natural knowedge encapsulated in this idea implies a certain reworking of the relationship between human expertise, observational instruments, techniques for reporting and transmitting observations and local factors. What makes Kircher's project specifically Jesuit? One might approach this question on a number of levels: Kircher's harnessing of an existing Jesuit infrastructure – the three-yearly congregations of provincial procurators in Rome – to disseminate his observational imperative and his apparent reliance on Jesuit couriers to receive his responses. While central Europe was undergoing a revolution in postal service from the late sixteenth century on, with the rise of the Habsburg Taxis,[76] the Jesuit 'monarchy' generally functioned as its own postal service, through the extreme mobility of its subjects.[77] Apart from pointing to the importance of the Jesuit infrastructure for Kircher's project, one might point to the ethical codes enforced by Jesuit spiritual training, and the Ignatian *Constitutiones*, particularly as they relate to modesty, obedience and the union of souls in the Society through the frequent exchange of letters. The type of humility and self-abnegation involved in surrendering the results of one's labours to be published by someone like Kircher was particularly encouraged in the Jesuit order, as we have seen for the case of Christoph Grienberger. One might point to the vantage point enjoyed by Kircher – his room in the Collegio Romano, his ready supply of willing collaborators, his 'mathematical archive' inherited from Clavius and Grienberger, his access to a vast stock of printed and manuscript works of a mathematical nature. One might look to the common culture, in terms of mathematical training, instrument-making and literary ability that gave Kircher and his distributed correspondents a shared basis for communication, a culture fostered by the Jesuit collegiate structure. Finally, one might look to a Jesuit tradition of precision measurement – ranging from Juan Bautista Villalpando's recovery of ancient metrology in his *Apparatus Urbis ac Templie Hierosolymotani*[78] to Grienberger's trigonometric tables and star-charts, and the measurements performed by the 'human-metronomes' who aided Riccioli in the preparation of his *Almagestum Novum*.[79]

Kircher's angelic observer was not an expert individual but a distributed collectivity of disciplined Jesuits. Equipped with mathematical skills, azimuth

Figure 5.7 Frontispiece of Athanasius Kircher, *Ars Magna Lucis et umbrae*, 1646

compasses and an efficient postal system, the humble, mathematically trained individuals that constituted the limbs and senses of the Jesuit order could reform astronomy and geography just as they could use prayer, confession and instruction to heal the spiritually ill and spread the Gospel [**See Figure 5.7**].

Notes

1. A version of this chapter previously appeared as 'The Angel and the Compass: Athanasius Kircher's Magnetic Geography' in *Athanasius Kircher, The Last Man Who Knew Everything*, edited by Paula Findlen, New York and London: Routledge, 2004, 239–259. I am grateful to Paula Findlen for permission to include here.
2. Martino Martini to Kircher, Evora, 6 February 1639 APUG 567, ff. 74r–75v, published in Martino Martini, *Opera Omnia*, ed. Franco Demarchi, Trento: Università degli Studi di Trento, 1998. 61–69. On Martini, see also Franco Demarchi and Riccardo Scartezzini, eds., *Martino Martini: A Humanist and Scientist in Seventeenth Century China*, Trent: Università degli Studi di Trento, 1996.
3. On Clavius's mathematical academy, see CC I.1, 59–89.
4. Martino Martini to Muzio Vitelleschi, Goa, 8 November 1640, ARSI, Goa 34 1, ff.81r–86v, in Martini, 1998, pp. 97–140.
5. On Jesuit involvement in Portuguese trade networks, see Dauril Alden, *The Making of an Enterprise: The Society of Jesus in Portugal, Its Empire, and Beyond: 1540–1750*, Stanford, CA: Stanford University Press, 1996.
6. Noel Malcolm, 'Private and Public Knowledge: Kircher, Esotericism and the Republic of Letters', in *Athanasius Kircher: The Last Man Who Knew Everything*, edited by Paula Findlen (New York and London: Routledge, 2004), 297–308
7. On the development of Jesuit correspondence, see John Correia-Afonso, *Jesuit letters and Indian history a study of the nature and development of the Jesuit letters from India (1542–1773) and of their value for Indian historiography*, Bombay: Indian Historical Research Institute, 1955. Chapter 1, Steven J. Harris, "Confession-Building, Long-Distance Networks, and the Organization of Jesuit Science", *Early Science and Medicine* 1 (1996): 287–313. 1996, and Nick Wilding, *Writing the Book of Nature: Natural Philosophy and Communication in Early Modern Europe*, PhD dissertation. Fiesole: European University Institute, 2000, 96–123. For a recent discussion on correspondence networks in Jesuit culture see Paula Findlen, "How Information Travels: Jesuit Networks, Scientific Knowledge, and the Early Modern Republic of Letters, 1540–1640," in Findlen, ed., *Empires of Knowledge: Scientific Networks in the Early Modern World*, London: Routledge, 2018.
8. On Polanco and his role as secretary see O' Malley, *The First Jesuits* 9–11 and André Ravier, *La Compagnie de Jésus sous le gouvernement d'Ignace de Loyola (1541-1556): D'après ;es Chroniques de J.-A. de Polanco*, Paris: Desclée de Brouwer, 1990.
9. Juan de Polanco to the whole Society of Jesus, Rome 27 July 1547, in Monumenta Ignatiana, Series Prima, Sancti Ignatii de Loyola Epistolae et Instructiones 1903, Vol. 1, pp. 536–541
10. ibid.
11. ibid.

12 ibid., on p. 540.
13 Polanco, Reglas que deven observar acerca del escribir los de la Compañia que están esparzidos fuera de Roma, in Monumenta Ignatiana, cit, part 1, vol. 1, pp. 542-549.
14 For architectural projects see Jean Vallery-Radot, *Le recueil de plans d'édifices de la Campagnie de Jésus conservé à la Bibliothèque nationale de Paris*, Rome : Institutum Historicum S.I., 1960, especially pp. 6*-18*. For the development of the *Ratio studiorum*, see the new French translation and critical edition, Adrien Démoustier and Dominique Julia, eds., *Ratio studiorum: [Version de 1599], édition bilingue latin-français*, Paris: Belin, 1997.
15 On Kircher's *Magnes*, and his magnetic philosophy in general, the most comprehensive study remains Martha Baldwin, *Athanasius Kircher and the magnetic philosophy* [Ph.D. thesis]. Chicago: University of Chicago, Department of History; 1987. See also William Hine, *Athanasius Kircher and magnetism*. in John Fletcher (ed.), *Athanasius Kircher und seine Beziehungen zum gelehrten Europa seiner Zeit*. Wiesbaden: Harrassowitz; 1988; pp. 79-99. For a different interpretation of Kircher's collection of data on magnetic declination see Martha Baldwin, "Kircher's Magnetic Investigations", in Stolzenberg, ed., *The Great Art of Knowing*, cit., pp 27-36. (on p. 33).
16 Kircher, Magnes, 1641, Lib. 2, Pars Quinta. *Geographia Magnetica*.
17 Kircher, Magnes, 1654, p. 293. The first reference to Kircher's Geographical Plan that I have been able to find is in a letter from Martino Martini to Kircher, sent from Evora in Portugal on February 6, 1639. In this letter, Martini writes: 'I am awaiting the Magnetic Philosophy [i.e. the Magnes] and the Mathematical Plan [Concilium Mathematicum]' (Martini, 1998, 57-70). The 'mathematical plan' to which Martini alludes is almost certainly Kircher's geographical plan, suggesting that Kircher may have conceived it in as early as 1637, when Martini was studying magnetic declination with him in Rome.
18 For an analysis of the relationship between travel and data-gathering in Jesuit culture see Harris, "Confession Building", cit., Harris, "Mapping Jesuit Science: The Role of Travel in the Geography of Knowledge", in O'Malley et al, *The Jesuits: Cultures, Sciences and the Arts*, cit., pp. 212-240. and Florence Hsia, "Jesuits, Jupiter's Satellites, and the Académie Royale des Sciences", in O'Malley et al., *The Jesuits*, cit., pp. 241-257.
19 There is an enormous literature on the longitude problem, but see especially, William J.H. Andrewes, ed., The Quest for Longitude: The Proceedings of the Longitude Symposium. Cambridge, MA: The Collection of Historical Scientific Instruments, Harvard University, 1996 and Silvio A. Bedini, *The Pulse of Time: Galileo Galilei, the Determination of Longitude, and the Pendulum Clock*, Florence: Olschki, 1991.
20 Albert van Helden, "Longitude and the Satellites of Jupiter", in Andrewes, ed., The Quest for Longitude, cit., pp. 81-100.
21 Giuseppe Biancani to Christoph Clavius, Padova; 28 February 1598, in CC 1992, IV.I: 34-37.

22 Kircher, *Ars Magna Lucis et Umbrae*, 1646, p. 552.
23 See Omer van de Vyver, "Lettres de J.-Ch. della Faille S.I., cosmographe du roi à Madrid, à M.-F. Van Langren, cosmographe du roi à Bruxelles, 1634-1645", Archivum Historicum Societatis Iesu 46 (1977): 73-183.
24 See Bennett, 1987, pp. 53-55.
25 Martini to Kircher, Lisbon, 16 March 1640, in Martini, 1998, 87-92. On Borri, see L Petech, "Borri, Cristoforo", in Dizionario biografico degli italiani., Vol 13: 3-4 Rome: Istituto della Enciclopedia Italiana, 1971 and Angelo Mercati, "Notizia sul gesuita Cristoforo Borri e sue 'inventioni' da carte finora sconosciute di Pietro della Valle, il pellegrino", *Pontificia Academia Scientiarum, Acta* 15 (1951): 25-45.
26 Marin Mersenne, [Treatise on the magnet, 1639?], BL Add. ms. 4279, ff. 145r-146v, in Mersenne, *Correspondance*, 1932-, VIII, 754-762, on p. 761.
27 Kircher to Mersenne, Rome, 23 December 1639, Houghton Library, Harvard University, Fms. Lat. 306. 1 (3) [A copy, apparently in the hand of Gabriel Naudé]. This letter is not published in the Mersenne correspondence
28 Gassendi to Peiresc, n.d., n.p., published in Gassendi, *Opera Omnia*, 1658, Tom. VI, p. 90.
29 See Gassendi to Diodati, Aix, 23 April 1636, in Gassendi, *Opera Omnia*, 1658, Tom. VI, pp. 85-90, on p. 88. On Peiresc and Gassendi's attempts to coordinate eclipse observations made by missionaries, see especially Wilding, *Writing the Book of Nature*, cit., 132-139.
30 Wilding, *Writing the Book of Nature*, cit.
31 Peiresc to P. Césaire de Roscoff, September 1 1636, translated in Wilding, *Writing the Book of Nature*, cit., p. 136.
32 Peiresc to Kircher, Aix, October 8, 1635, APUG 568, ff. 368r-369v, transcribed in Wilding, *Writing the Book of Nature*, cit., 377-383.
33 Before changing to magnetic variation, Kircher also attempted to use Jesuit missionaries to gather measurements of lunar eclipses with the help of a paper *Rota Geographica* which he distributed to correspondents. See Kircher to anonymous Jesuit priest, Rome, 14 October 1636, APUG 561 ff. 83r-84v.
34 Kircher, 1641, p. 430
35 Kircher, 1641, p. 430
36 ARSI Congr. 7 ff. 46r-48v: *Acta Congregationis Procuratorum anni 1639*. Two of the Procurators present at this congregation, P. Pierre Cazré and P. Nithard Biber subsequently corresponded with Kircher directly. See APUG 567 f. 192r (Cazré) and APUG 567 ff. 128r, 172r (Biber)
37 Kircher, *Magnes*, 1641, p. 430
38 Adrien Démoustier, "La distinction des fonctions et l'exercice du pouvoir selon les règles de la Compagnie de Jésus", in Luce Giard, ed., *Les Jésuites à la Renaissance*, cit., pp. 3-33.

39 On this point see O' Malley, *The First Jesuits*, p. 354 and the revealing comments made by Jeronimo Nadal in his Dialogus II (1562-1565), in Michael Nicholau, S.J., ed., *P. Hieronymi Nadal Commentarii de Instituto Societatis Iesu (Epistolae et Monumenta P. Hieronymi Nadal, Tomus V)*, Rome: Monumenta Historica Societatis Iesu, 1962, pp. 601-774, on pp. 764-770 (*De ratione gubernationis*), especially p. 767

40 Jacques Grandamy, *Nova demonstratio immobilitatis terrae*, La Flèche, 1645, p. 83

41 Ciermans to Kircher, Lovanij 7. Martij 1640, APUG 567 f. 90r

42 Oswald Krüger to Kircher, Vilnius, 21 July 1639, APUG 567 f. 53r

43 Henricus Marcellus to Kircher, Mainz, 1 May 1640, APUG 567 f. 213r

44 Grandamy to Kircher, Touron, 9 May 1640, 557 ff. 400r-401v, on f. 400r

45 Henricus Marcellus to Kircher, Mainz, 1 May 1640, cit., my emphasis.

46 Jacobus Imhofer to Kircher, Innsbruck, 15 January 1640, APUG 567, f. 177r

47 Jacques Honoré Durand to Kircher, 12 March 1640, APUG 567 f. 202r

48 Lorenz Mattenkloth to Kircher, 8 March 1640, APUG 567 f. 159r, P. Grégoire a St Vincent to Kircher, 8 March 1640, APUG 567 f. 24r-v

49 Martino Martini to Kircher, Goa, 8 November 1640, in Martini, 1998, 71-86.

50 Henry Gellibrand, *A Discourse Mathematical On the Variation of the Magneticall Needle*, London, 1635. On this episode, see Stephen Pumfrey, "*O tempora, O magnes!*" *A Sociological Analysis of the Discovery of Secular Magnetic Variation in 1634*. British Journal for the History of Science. 1989; 22: 181-214.

51 See APUG 557 ff. 41r-56v, and Kircher, 1654, Lib. II. Pars V, Caput VI, p. 340.

52 Kircher, 1641, p. 483

53 See Kircher, 1646, p. 553

54 Ibid.

55 Kircher, 1646, facing p. 554.

56 On Riccioli's time measurements see Alexandre Koyré, *An experiment in measurement*. Proceedings of the American Philosophical society. 1953 Apr; 97(2): pp. 222-237 and Paolo Galluzzi, Galileo contro Copernico. Annali dell' Istituto e Museo di Storia della Scienza di Firenze. 1977; 2: pp. 87-148. On Riccioli's early training, see Ugo Baldini, *La formazione scientifica di Giovanni Battista Riccioli. in Copernico e la questione copernicana in Italia*, a cura di Luigi Pepe, Firenze: Olschki; 1996: 123-182. On Riccioli's cosmology see Alfredo de Oliveira Dinis, *The cosmology of Giovanni Battista Riccioli (1598-1671)*. PhD thesis, Cambridge: Cambridge University; 1989, which includes an extremely useful intellectual biography (Chapter 1). For a more recent discussion of Riccioli's position on Copernicanism, arguing that Riccioli's criticisms of heliocentrism in favour of a modified Tychonic system were grounded in astronomical arguments rather than doctrinal authority, see especially Christopher M. Graney, *Setting Aside All Authority: Giovanni Battista Riccioli and the Science against Copernicus in the Age of Galileo*, Notre Dame, Indiana: University of Notre Dame Press, 2015.

57 Kircher, *Magnes*, 1654 p. 294: 'Dissimulare hic non possum animi mei iustum dolorem, quem ex iactua praefati consilii Geographici precepi: cum enim opus non sine vigilijs elaboratum, inter alia in Musaeo meo conservarem, tempusque opportunum in lucem publicam litterariae Reipublicae bono emittendi praestolarer; ab uno illorum, qui quotidie paene Musaei inspiciendi causa ad me undique confluebant, clam subductum est'.

58 Kircher to Gassendi, Rome, 13 February 1642, published in Gassendi, *Opera Omnia*, 1658, Vol. VI, p. 446.

59 ibid.

60 Riccioli to Kircher, Bologna, 5 July 1642, APUG 561 ff. 177r–178v, published in Gambaro, 1989, pp. 44–52, on p. 44.

61 Riccioli, 1661, For Riccioli's consideration of the longitude problem and magnetic declination see Lib. VIII, *Geomecographus*, Cap. 12–16.

62 ARSI FG 662, f. 477 r, published in Gambaro, *Astronomia e Tecniche di Ricerca*, cit., p. 40, and re-transcribed (with amendments) in Baldini, *La formazione scientifica di Giovanni Battista Riccioli*, cit., p. 176, note 55, emphasis added.

63 Gambaro (1989) adduces no hypothesis concerning the authorship of the censura, whereas Baldini (1996) explicitly dismisses the possibility of Kircher's authorship on the basis of a later letter from Riccioli to Kircher. However, the letter in question, discussed below, refers not directly to the anonymous censura, but to a letter, now lost, from Kircher to Riccioli reiterating some of the points in the original censure and adding a number of other points of contention concerning Riccioli's way of life. It is from these other points (particularly the inability of a single person to be proficient in two different faculties simultaneously) that Riccioli dissociates Kircher. Taken in its entirety, the existing evidence is entirely compatible with Kircher's authorship of the original anonymous censura of ARSI FG 662, f. 477 r.

64 Riccioli to the Roman Censor, n.p., n.d. [Bologna, between 24 November and 22 December 1646?], published in Gambaro, 1989, pp. 70–76.

65 Riccioli to Kircher, Bologna, 22 December 1646, published in Gambaro, 1989, pp. 77–81.

66 ibid.

67 ibid., on p. 78.

68 ibid.

69 Ibid.

70 ibid.

71 ibid., p. 81. See Riccioli, 1651, Sig. *Ar – A2r, letter of dedication to Prince Cardinal Girolamo Grimaldi. For the involvement of Francesco Maria Grimaldi in the work see Sig. A2r

72 ibid. Riccioli's decline in bodily powers during the period prior to the publication of the *Almagestum Novum* is corroborated by the *Catalogi Triennales* for the period: on

15 May 1645 his '*vires*' are reported to be '*mediocres*', by 15 September 1649 they have become 'imbiscilles', and by 1 October 1651 they are reduced to '*debiles*'. See ARSI Ven. 40 ff.18v, 48v: #11 (for 1645), ibid., ff. 94v, 125v: #16 (for 1649), ibid., ff.178r, 204r: #14 (for 1651).

73 On the use of meridian-lines in churches, including S. Petronio, to perform astronomical observations see John L. Heilbron, *The Sun in the Church*, Cambridge, MA: Harvard University Press, 1999, pp. 82–119.

74 See Muzio Vitelleschi to the Provincial for the Veneto, 13 September 1636, ARSI Ven. 1, f. 318v, cited in Baldini, 1996, p. 174, note 40.

75 Riccioli, 1651, Pars Prior, XVII.

76 See Gerhard Dohrn – van Rossum, *History of the Hour: Clocks and Modern Temporal Orders*, cit., pp. 335–340, E. John B. Allen, *Post and Courier Service in the Diplomacy of Early Modern Europe*. The Hague: Martinus Nijhoff; 1972.

77 Mobility was central to the Jesuit apostolate from the very beginning, as Ignatius explained to Ferdinand I, King of the Romans in 1546: 'Thus the spirit of the Society is to move on from one city to another in complete simplicity and modesty, and from one district to another, not to settle ourselves in one specific place' MHSI, I, 450–53, translated in *Saint Ignatius of Loyola: Personal Writings*, transl. Joseph A. Munitz and Philip Endean, London: Penguin; 1996, pp. 168–170.

78 Villalpando, *Apparatus Urbis ac Templie Hierosolymotani* Rome, 1604, pp. 433–550. For the metrological/experimental aspects of Villalpando's work see the discussion in P. D. Napolitani, *La Geometrizzazione della realtà fisica: il peso specifico in Ghetaldi e in Galileo*. Bolletino di Storia delle Scienze Mathematiche. 1988; VIII: pp.139–237.

79 See Alexandre Koyré, *An experiment in measurement*, cit., and Galluzzi, *Galileo contro Copernico*, cit.

6

Between the Demonic and the Miraculous: Athanasius Kircher and the Baroque Culture of Machines[1]

Introduction: serious jokes

From the magnetic Jesus walking on water described in his very first published book, the 1631 *Ars Magnesia*, to the unfortunate cat imprisoned in a catoptric chest and terrified by its myriad reflections shown to visitors to his famous museum, the peculiar mechanical, optical, magnetic, hydraulic and pneumatic devices constructed by Athanasius Kircher continue to defy the analytical categories used in both traditional museum history and history of science.[2] Although Filippo Buonanni (1638–1725) later attempted to reduce the machines of the Kircherian museum to the status of mechanical demonstrations, even adding some of his own,[3] it is clear that for Kircher and his immediate entourage, these machines were, in some real sense, magical. Far from being trivial addenda to a collection of antiquities and *naturalia*, the documents suggest that Kircher's machines were utterly central to any seventeenth century visit to the *Musaeum Kircherianum* [**See Figure 6.1**]. But, from the point of view of traditional histories of science, Kircher's machines remain defiantly perplexing. Their emblematic, ludic, and deceptive connotations sit ill with any attempt to place them within grand histories of 'experimental science' emphasizing the demise of Aristotelianism through the triumph of an 'experimental method' during precisely the period in which the Kircherian museum enjoyed its exuberant heyday. From the point of view of the history of collections, the machines accumulated by Kircher and his disciples in Rome cannot merely be treated as objects removed from circulation, or from their original context of usage, as these machines had no original context of usage, and did not circulate prior to their display in the museum.[4] Rather, we are dealing with purpose-built installations, constructed ad hoc by Kircher and his changing body of assistants, technicians and disciples in the *Collegio Romano*.

Figure 6.1 Frontispiece depicting the *Musaeum Kircherianum*, from G. de Sepibus, *Romanii Collegii Musaeum Celeberrimum cuius magnae antiquariae rei* ... Amsterdam: Ex Officina Janssonio-Waesbergiana, 1678

So what are we to make of these magical machines? This article attempts to situate Kircher's machines in a Baroque culture of artificial magic. Using contemporary accounts of visits to Kircher's museum and other documents, it aims to recover the purpose of these devices, to understand how they worked, not only by peering inside them to examine their secret workings, but also by looking outside them at how people responded to them, and at how Kircher and his Jesuit companions placed this part of their output in a rich tradition of artificial magic that has commonly been overlooked or trivialised by historians of science. We will argue that Kircher's machines found their meaning in a flourishing Baroque culture of special effects. In the same way that 'inside jokes' confirm the identity of a particular social group, while excluding the majority of people who are not privy to the assumptions on which the joke is based, the machines of Kircher and his disciples provided an elite social group with self-defining puzzles and enigmas.

The game of deducing the natural causes behind the strange effects produced by Kircher's magical machines, such as a clepsydra apparently pouring water upwards into a 'watery heaven', really caused by a hidden mirror, was somewhat akin to fox-hunting or golf in our society: if you could play the game, your identity as part of a particular social elite was confirmed. If you could not play the game, and had to assume that demonic forces were responsible for the strange effects that you were witnessing, you were doomed to the ranks of the vulgar masses. In this respect, Kircher's machines had much in common with courtly emblems and enigmas, and the culture of '*sprezzatura*' which countless behaviour-manuals vainly attempted to divulge in the sixteenth and seventeenth centuries.[5] Like many types of joke, Kircher's machines are, we argue, inherently conservative. They rest on a shared mystery – the hidden causes behind the visible effects. To challenge the received picture of the causes operating in the natural world in response to such a machine would thus amount in a strong sense to spoiling the joke for everybody else.[6]

At the core of Kircher's marvellous machines, then, lies a robust epistemological conservatism. Kircher's machines thus offer us an alternative to conventional stories of the inevitable collapse of Aristotelian natural philosophy through direct experimentation, and require us to refine our understanding of the roles played by machines, experiments and instruments in seventeenth century natural philosophy. The culture of the elite audience for which Kircher's machines were designed is inscribed graphically on the machines themselves – one need only consider such items as the water-vomiting two-headed Imperial Eagle [**See Figure 5.1**] or the perspectival trick unjumbling an image of Pope Alexander

VII. Indeed, one could arguably take this further and view the *Musaeum Kircherianum* as a whole as something of a self-portrait of an elite, primarily a Roman Catholic elite centred around the twin poles of the courts of Rome and Vienna. This elite was not a 'given' quantity when Kircher's museum came into existence – rather the museum helped to construct and consolidate the elite while the elite helped to construct the museum by corresponding with Kircher and providing him with portrait medals, natural curiosities and other objects for his collection.

At the centre of a vast correspondence network, and increasingly famous through his lavishly illustrated encyclopedic publications, Kircher wielded considerable power to shape the social group represented in his museum. Limited only by his religious poverty, Kircher extended his network at will to include powerful Protestants such as Duke August of Brunswick-Lüneburg or Queen Christina of Sweden, prior to her conversion. In a revealing letter to Duke August's librarian Johann Georg Anckel, Kircher wrote that he had immediately had Duke August's portrait 'framed in gold and put up in my Gallery as a Mirror of the magnanimity, wisdom and generosity of the high-born prince', adding that 'my Gallery or museum is visited by all the nations of the world and a prince cannot become better known *in hoc Mundi theatro* than have his likeness here. And if the expense were not so great I would do this for all Germans, but I must cut my coat according to my cloth'.[7]

As well as holding up a trick-mirror to an elite audience, Kircher's museum also emblematised the Jesuit order itself. Many of the curious natural objects and artefacts of remote cultures present in the museum were sent to Kircher by Jesuit missionaries, who constitute the single most numerous group of his correspondents. Some of Kircher's machines provide striking emblematic depictions of his order – his universal catholic horoscope of the Society of Jesus was a large sundial representing the Jesuit order as an olive tree, with the different Assistancies or administrative divisions of the order represented as branches, and the different colleges represented as leaves. Tiny sundials placed in each province give the local time, and the shadows of the gnomons of the sundials, when aligned, spelled 'IHS', the abbreviated name of Jesus and symbol of the Jesuit order, which appears to 'walk over the world' with the passing of time [**See Figure 5.5**].

One of the aims of this chapter is to understand the relationship between such artefacts and Kircher's position in the Jesuit *Collegio Romano*. The moment of the creation of the *Musaeum Kircherianum* coincided with a disciplinary crisis in Jesuit education that led the superiors of the order to condemn departures from Aristotle in philosophy, including natural philosophy or physics, and from Thomas Aquinas in theology. The works of Jesuit authors on natural philosophy during this period were closely scrutinised for anti-Aristotelian views.[8] The exotic publications of Kircher and his disciples seem to contradict this doctrinal fundamentalism, but we will suggest that the contradiction is only apparent. The treatment of machines and instruments, even those associated with criticisms of Aristotle, in the works of Kircher and his Jesuit apprentices in magic was designed to avoid conflict with fundamental Aristotelian principles.

The machines

Before taking a look at the the magical and mathematical traditions from which Kircher's machines emerged and the functions, mechanical and social, that they performed, it might be opportune to have a first glance at the machines themselves. In 1678 Giorgio de Sepibus (fl. 1678), Kircher's 'assistant in making machines' published the first catalogue of the *Musaeum Kircherianum*.[9] Little is known about De Sepibus, from the Wallis (Valesia) canton in Switzerland, who seems to have been an intermittent companion of Kircher, and is first mentioned ten years earlier in a letter from the Oratorian priest Francesco Gizzio to Kircher. In 1670 Kircher sent De Sepibus to Naples, where he brought a number of machines to perfection, with the exception of a 'versatile pulpit' that was left incomplete. It is not clear when De Sepibus left Kircher's service, but by 1674 Kircher seems to have feared him dead, so with all likelihood the catalogue was completed well before its publication.[10] De Sepibus provides us with a summary list of the machines present in Kircher's museum, which may serve as our starting point:

1. Two helical spirals most skilfully measuring cycles with the twisted coils of snakes. An organ, driven by an automatic drum, playing a concert of every kind of birdsong, and sustaining in mid-air a spherical globe, continually buffeted by the force of the wind.
2. A hydrostatic-magnetic machine, representing the hours, zodiac, planets and the whole fabric of the heavens. The hours are described by means of a very simple motion, in which images of the Sun and Moon alternately ascend and descend vertically. The divisions of the hour are marked by the sympathetic motion of the flight of small birds.

3. A magnetic-hydraulic machine displaying the time all over the world, as well as the astronomical, Italian, Babylonian and ancient hours.
4. A little fountain moving the globe weighing down on the head of Atlas in a circle by hidden movements.
5. A fountain lifts a genie fixed in the water up and down, with a perpetual motion of tossing about and turning.
6. A fountain in which the Goddess Isis, contained in a crystalline sphere, is sustained, and greets guests by spraying water everywhere.
7. A hydraulic machine that apes perpetual motion, recently invented by the author, consisting of a clepsydra that flows out when it is inverted, and again when it is turned the right way up, wetting a watery heaven with its spray.
8. A hydraulic machine most skilfully representing the Primum Mobile, and violently impelling a brass snake resting on top of the water in twists and turns by water.
9. A water-vomiting hydraulic machine, at the top of which stands a figure vomiting up various liquids for guests to drink.
10. A hydraulic clock urging or carrying globes or genies up and down inside crystal tubes of five palms in height, indicating the different times.
11. A hydraulic machine, which supports a crystal goblet, from one side of which a thirsty bird drinks up water, that a snake revomits from the other side while opening its mouth.
12. A hydrotectonic machine moving armed knights from one place and a crowd returning from another by means of continual drops.
13. A two-headed Imperial Eagle, vomiting water copiously from the depths of its gullets.
14. A crowd of dancing genies driven by the silent approach of water.
15. The dove of Archytas reaching towards a crystalline rotunda and indicating the hours by its free flight.
16. The catoptric theatre, completely filled with a treasure of all sorts of delicacies, fruits, and precious ornaments.
17. An architectural perspective representing the arrangement of the rooms inside a magnificent palace.
18. A perpetual screw, the invention of Archimedes, by which it is an easy matter to lift 125 pounds with the strength of a very weak small boy.
19. A large crystalline globe full of water representing the resurrection of the Saviour in the midst of the waters.

Various thermoscopes, or thermometers which indicate the daily growth of simples, the mutations of the air, the ebb and flow of the tide, and the variation of the winds, together with experiments on the origins of springs. An extremely large concavo-convex burning mirror, with a collection of many mirrors, some of which show ghosts in the air, others show objects unchanged, others show them multiplied and others reconstitute completely undetermined species from a confused series into a beautiful form. Amongst these there is one which reconstitutes the effigy of Alexander VII.

....

A large number of mechanical clocks, one of which plays harmonious music by a concert of bells with an elaborate movement, at any hour it plays the sound, also every half-hour with a marvellous harmony of notes and sweetness of sound it plays the hymn *Ave Maris stella*. Another one indicating the time of day by the movement of a pendulum. Another, finally, giving the minutes and seconds of time. The part of the world illuminated by the sun, the increase and decrease of day and night. The current sign of the zodiac, the astronomical and Italian hours, as well as the ancient hours, or the unequal hours, which it describes along a straight line by a singular artifice. Many sundials.

...

Armillary spheres, and celestial and terrestrial globes, equipped with their meridians and pivots.

Astrolabes, Planispheres, Quadrants, a very full collection of mathematical instruments.

...

The Delphic Oracle, or speaking statue.

A Divinatory Machine for any planetary influence at the circumference of two glass spheres by genies moved uniformly by a mutually sympathetic motion. Twisting themselves to the same degree at a large distance, each of them in his sphere indicates the same point of the sign.

Various motions of solid globes bearing a resemblance to perpetual motion. A hydraulic perpetual motion by rarefaction and condensation, an Archimedean screw carrying globes up with a continual motion through helical glass channels.[11]

This list is both illuminating and opaque – while allowing us to form an idea of what some of the machines may have looked like or sounded like, it gives us little or no idea of how they were perceived by contemporaries. Let us take one of them – 'the Delphic Oracle, or speaking statue', the description of which De Sepibus leaves to the final chapter of his catalogue of the museum's contents, stating that 'we have rightly left the greatest machination of art until the final course'. What was this great 'machination'? How did it work? Why was it made? De Sepibus gives the following description of the oracle:

> Kircher has [*sic,* for 'had'] a tube in the workshop of his bedroom, arranged in such a way that the porters, in order to call him to the door when business demanded it, used not have to take the trouble to go all the way to his bedroom, but merely called him in a normal voice at the door that gave access to the open-air garden. He heard their words as clearly as if they had been present in his bedroom, and answered in the same way, through the tube [...] Later he transferred this tube to the Museum, and inserted it into a statue in such away that the statue, almost breathing life, is seen to speak with its mouth open, and its eyes moving. He named this statue the Delphic oracle, as it was in the same way, by the ingenious trick of stuffing tubes into the mouths of idols, that the ancient priests of the Egyptians and Greeks deceived the people consulting the oracle and made superstitious men give valuable offerings[12]

A manuscript draft of De Sepibus' description (in Kircher's handwriting incidentally, suggesting that he had a rather active role in the composition of the 1678 catalogue), is conserved amongst Kircher's manuscripts in the Pontifical

Figure 6.2 Speaking tubes connected to statues, from Athanasius Kircher, *Musurgia universalis*, vol. 2, p. 303

Gregorian University, in which he sometimes calls the machine the Oracle of Apollo [**See Figure 6.2**], but otherwise describes it almost identically.[13] Kircher's earlier 1673 work on sound and acoustics, the *Phonurgia nova*, gives us a more detailed account of the machine, and its changing role in the *Collegio Romano*:

> There was a repository in my Museum, between the wall and the door. At the end of the repository was an oval shaped window, looking out over the domestic garden of the *Collegio Romano*, which is about 300 palms in length and width. Inside this repository, or workshop, I adapted a conical tube to the length of the space, made from a length of 22 palms of sheet-iron, the speaking hole of which did not exceed ¼ of a palm in diameter. The tube, however, had a diameter of one palm at its aperture that then grew gradually by continuous and proportional increments in diameter so that the orifice of the part extended out of the oval window towards the garden had a diameter of three palms. We have seen how the tube was made, now we will also explain its effect.
>
> Whenever our porters had to inform me of something, either of the arrival of guests or of any other matter, so that they would not be inconvenienced by having to come to my Museum through the labyrinthine corridors of the college, while standing inside the porters' lodge they could talk to me while I remained in the remote recesses of my bedroom, and, as if they were present, they could tell me whatever they wanted clearly and distinctly. Then I too could respond in the same tone of voice according to the demands of the matter, through the orifice of the tube. Indeed nobody could say anything inside the garden in a clear voice that I could not hear inside my bedroom, and this was a thing seen as completely new and unheard of by the visitors to my museum, when they heard speech, but couldn't see who was talking. So that I would not be suspected of some prohibited Art by the astonished people, I showed them the hidden structure of the device. It is difficult to say how many people, even including many Roman Nobles, were attracted to see and hear this machine.
>
> ...
>
> It happened later that I was required to transfer my Private Museum into a more suitable, and open space in the *Collegio Romano*, that they call the Gallery. Here, the tube that I have briefly described before was also moved, and even now it is looked at and listened to under the name of the Delphic Oracle, with the following difference: the tube that previously propagated clearly spoken words plainly into a distant space, now acts secretly in ludic oracles and false consultations with a hidden and quiet voice, so that nobody present is able to perceive anything of the secret technique of the reciprocal murmured conversation. And when it is exhibited to strangers even to this day, there are not lacking those who harbour a suspicion of demons among those who do not understand the machine, for the statue opens and closes its mouth as if it was

speaking, and moves its eyes. Therefore I built this machine in order to demonstrate the impostures, fallacies and frauds of the ancient priests in the consultation of oracles. For while they gave their answers through secret tubes (described in the Oedipus), they urged the people to give offerings extravagantly, if they wanted their prayers to be answered. And consequently, by this fraud, they were able to greatly increase their wealth. In any case I would not deny that they also secretly involved demons in their works.[14]

Kircher's Delphic oracle reveals much about the role of machines in his Museum, and also much about the history of the museum itself. We are told that Kircher had a 'private museum' before he transferred his collection to the Gallery of the *Collegio Romano* after the 'official' founding of the museum with Alfonso Donnini's 1651 bequest of his collection of antiquities to the *Collegio Romano*.[15] Where was this 'private museum'? In the passage cited from the *Phonurgia Nova*, Kircher identifies it explicitly with his '*cubiculum*', or bedroom in the *Collegio Romano*. So, even before Kircher was in charge of the Gallery of the *Collegio*, his own bedroom functioned as a museum, containing within it a storage area or workshop, from which his speaking-tube originally allowed him to communicate with, or occasionally eavesdrop on, people in the College garden and the college porters, who, one imagines, must have been pleased with this labour-saving device. In England, at around the same time, another prominent mathematical magician, John Wilkins (1614–1672), made a similar speaking-tube in the gardens of Wadham College, Oxford. One day, a certain Mr Ashwell was strolling through the college, shortly after Cromwell had urged the Fellows of Oxford University to bring the Gospel to Virginia. As he passed the statue of Flora, he was astonished to hear it say to him 'Ashwell goe preach the Gospel in Virginia', in a Puritanical translation of Kircher's Jesuit machine.[16]

To return to Kircher's multi-purpose bedroom in the *Collegio Romano*, however, it may appear strange that this domestic space also functioned as a museum, and clearly attracted enough visitors to warrant the development of an intercom system. In fact, there was a long tradition in the *Collegio Romano* before Kircher's arrival of describing the bedroom of the senior mathematician of the college as the *musaeum mathematicum*. Clavius kept mathematical instruments, clocks and manuscripts in this space, a space that also served as the focus for the activities of the private mathematical academy of the *Collegio Romano* discussed in Chapter 1 above. Unlike the normal mathematics lectures that formed part of the College's public curriculum in philosophy, often taught by a junior professor, the mathematical academy was founded with the specific aim of teaching mathematics professors for the Jesuit colleges in the different provinces of the

Order. Generally, the bedrooms of Jesuits were not provided with keys, but, along with the rooms of the Superiors and the Procurator (responsible for the financial affairs of the College), the room of the senior mathematician of the College formed an exception.[17] The added security of a key meant that the mathematics professor could store valuable mathematical instruments in his domestic space.

The *musaeum mathematicum* of the *Collegio Romano* then, formed a space for advanced level mathematical teaching and for the formation of close relationships between master and disciples, relationships which generally continued through correspondence after the apprentice mathematicians left to teach the mathematical disciplines in the provinces. When Clavius died, in 1612, his correspondence, manuscripts, instruments and position as the most senior mathematician of the *Collegio Romano* were inherited by the Tyrolese Jesuit Christoph Grienberger (c. 1564–1636). After Grienberger's death in on 11 March 1636, the manuscripts collected by Clavius and Grienberger, their 'archive' of correspondence, and their instruments seem to have all passed to Kircher. So, although Kircher only occupied the position of public mathematics professor for a short time, he inherited the *musaeum mathematicum*, a space in which the building of instruments and machines was already an established tradition. Indeed, Kircher's far more modest predecessor Grienberger was rumoured to have invented a speaking statue himself. We find ample references in the works of Kircher to the documents and objects Kircher inherited. In Kircher's 1641 book on magnetism, the *Magnes*, for example, Kircher states clearly that 'I have collected together many observations concerning magnetic declination that are not to be rejected [...] partly from the Archive that I possess of mathematical letters sent from the different parts of the globe to Clavius, Grienberger and my other predecessors as Roman mathematicians of the Society of Jesus'.[18]

Emulating the private mathematical academy directed by Clavius and Grienberger before his arrival in Rome, Kircher gathered private disciples around him who were also able to avail of the instruments and documents that Kircher had inherited from his mathematical predecessors. While working as Kircher's assistant in Rome between 1652 and 1654, Kaspar Schott (1608–1666) seems to have spent much of his time leafing through the papers of Clavius and Grienberger: 'In the manuscripts of the most learned man Fr. Christoph Grienberger [...] that I found in the Clavius and Grienberger archive ', he wrote in his *Mechanica Hydraulico-Pneumatica*, 'I came across the following words about this Machine made by Bettini, and an opinion about perpetual motion'.[19] Describing a machine in which a sphere was suspended in the air and rotated about its centre, Schott wrote

> I found the following machine amongst the papers of Fr. Christoph Clavius and Fr. Christoph Grienberger, once professors of mathematics in this Roman College of ours. However it was in the handwriting of neither of them, nor was it composed by them, as it smelled of neither of their lanterns. I suspect that it was sent to Clavius by one of the disciples of Francesco Maurolico, the Abbot of Messina, for it cites a small unpublished treatise of his. But, whomsoever's manuscript it is, I have judged it fitting that it should be inserted here, since it can be applied to many things by an industrious artisan.[20]

Schott also borrowed items from the Clavius and Grienberger 'mathematical archive' that he did not acknowledge – a demonstration of how to lift a golden earth using the force of one talent, using a system of toothed wheels published in his *Magia Universalis* is lifted directly from an unpublished manuscript by Grienberger that Kircher would have possessed, as is a passage extolling the powers of mathematics and the extraordinary achievements of Archimedes in the same work.[21]

Schott and De Sepibus also inform us about instruments, experiments and machines that Kircher had inherited from Clavius and Grienberger, and subsequently transferred to the Gallery after 1651, such as a trick-lantern made by Grienberger that performed in the same way when filled with water as with oil, and a sample of water from the river Jordan that Clavius had sealed hermetically in a glass vial, perhaps the most undramatic of Kircher's museum exhibits, demonstrating the incorruptibility of water by remaining forever unchanged. A wooden astrolabe made by Grienberger was also displayed prominently in the museum, though by the time Sepibus compiled his catalogue it had been almost completely eaten away by woodworm.[22] From all these examples, it should be clear that Kircher effectively inherited a space, complete with manuscripts, instruments and experiments, that already had a well-established role in the *Collegio Romano* – the *musaeum mathematicum*, and that many of the functions of this space did not change dramatically with Kircher's arrival in Rome, when the space became his 'private museum'. Indeed, it seems that most Jesuit colleges where mathematics was taught in the mid-seventeenth century had a mathematical museum of some description, which was normally the bedroom of the senior mathematician of the college where the mathematical instruments could be locked away, though most would have been far more modest than that of the *Collegio Romano*. An example is Valentin Stansel's mathematical museum in Prague, where Jakob Johann Wenceslaus Dobrzensky de Nigro Ponte saw a hydro-magnetic fountain clock, that he described in his *Nova, et amaenior de admirando fontium . . . philosophia*.[23]

The descriptions of Kircher's Delphic oracle quoted above also reflect on other aspects of his machinic installations. Kircher claims to have built the device in order to expose the 'impostures, fallacies and frauds of the ancient priests', so the ludic machine bears a moral burden. The corruption of the good magic given by God to Adam into a tool of deception and evil-doing in the hands of the post-diluvian Egyptians is a theme that crops up frequently in the works of Kircher and Schott, and we shall return to it. In the house of a certain Francesco Serra, Kircher and Schott had seen an example of an Egyptian speaking statue designed to contain just such a speaking-tube as that hidden in Kircher's Delphic Oracle, illustrated in the *Oedipus Aegyptiacus*.[24] The section of this work dealing with Egyptian mechanics contains many examples of the tricks employed by Egyptian priests to deceive worshippers, and many of the machines in Kircher's museum relate to the debunking of Egyptian magic. A 'multimammary Goddess', for example, spraying forth liquid from her multiple breasts [**See Figure 6.3**] is described both in the *Oedipus Aegyptiacus* and in Schott's *Mechanica Hydraulico-Pneumatica*, where Schott writes: 'many thought that this work was constructed with the art of prestidigitation and of demons, but Fr Kircher clearly showed that this was a devious machination of the priests [...] and he has a small machine in his museum that he displays to this end'.[25] Describing another Egyptian device, an altar on which small gods or demons dance, Kircher writes 'A devious invention elaborately contrived by either Priests or evil demons in order to enslave the stupid and ignorant plebs in idolatrous servitude, so that nothing more effective or powerful could be devised for the cult of false gods'.[26] It is interesting that, while exposing the fraudulence of the magic of the Egyptian priests, Kircher will nonetheless not rule out their involvement with demons. One might have thought that the priests' impressive technical skills would have removed any need for traffic with real demons.

Regarding Kircher's own performances with his Delphic oracle, we are also told that he was frequently suspected of involvement with demons by his less perceptive visitors, and that he explained the functioning of the machine in order to remove suspicions of him practicing 'some prohibited Art'. Traffic with demons was no laughing matter in the mid-seventeenth century, at the height of the European witch-craze. One could well imagine that a less well-inclined audience might well view Kircher's wonders in an altogether different light. Indeed, on one of the few occasions when Kircher performed in front of a larger audience, this was precisely what happened. Kircher, in his early twenties, had recently arrived in Heiligenstadt after being stripped of his clothes and nearly killed by heretical soldiers who recognised him as a Jesuit, and a legation sent by

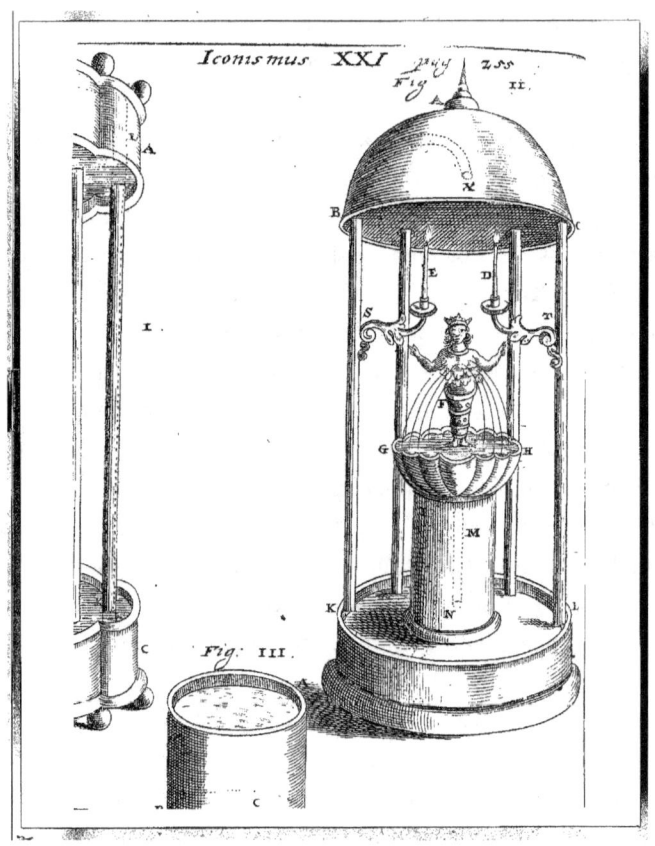

Figure 6.3 Multimammary goddess, from Kaspar Schott, *Mechanica Hydraulico-Pneumatica*, p. 255

the Archbishop-Elector of Mainz was about to be received in the town. The following excerpt is from his posthumous autobiography:

> And because it was decided to spare no magnificence to provide an appropriate welcome for the legates, I was commissioned to arrange a theatrical performance. When I exhibited this, as they saw some things that went beyond common knowledge, the legates who witnessed the performance were so excited to great admiration that some of them accused me of the crimes of Magic, with some people say other things against me. In order to free myself of such an ugly crime I was obliged to expose the mechanisms of all of the things that I had exhibited. And when this task was discharged to everybody's great satisfaction, so that they could hardly be separated from me, I also gave them a new collection of Mathematical Curiosities together with a laudatory panegyric in exotic languages composed in their honour, by which things resulted no small increase in their benevolence towards me.[27]

It is clear from this episode that Kircherian magic flirted dangerously with the boundaries between technical ingenuity and the 'prohibited art' of demonic magic. The Elizabethan magician John Dee (1527–1608), similarly came under suspicion of demonic magic in England when he constructed an automatic 'scarabeus' that flew up to Jupiter's palace during a performance of a comedy by Aristophanes, when in fact the theatrical trick was achieved by 'pneumatithmie' or by 'waights'.[28] Perhaps this very flirtation with the black arts was a source for titillation for the princely and religious audience of Kircher's wonders – an audience directly involved in the persecution of popular magic during the same period – allowing them to experience the 'armchair-thrills' of magic without being morally implicated.[29] Jesuit theatrical productions during this period were particularly famous for their stage machinery – convincing representations of hell were a speciality – and for their hard-hitting moral didacticism, both features

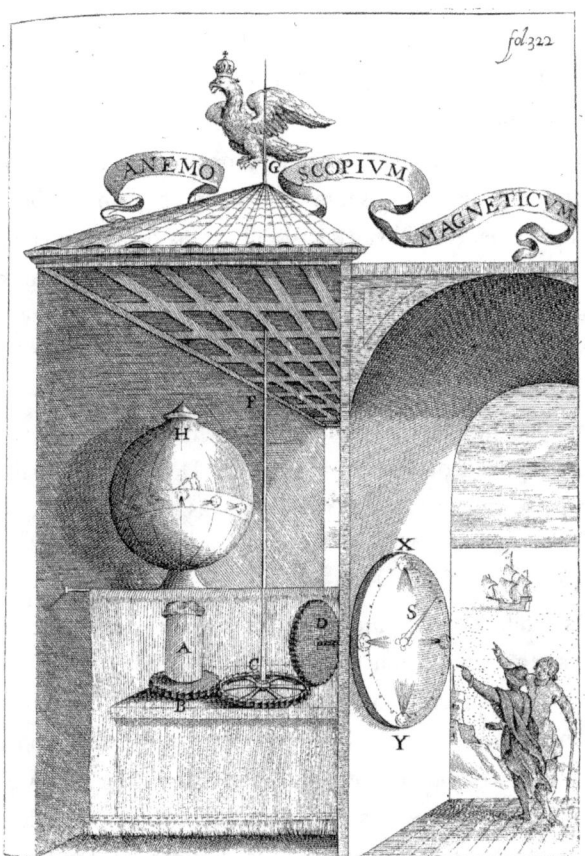

Figure 6.4 The magnetic anemoscope that Kircher built in Malta, from *Magnes, sive de arte magnetica* (1643 edn.), p. 322

that they shared with Kircher's machinic-performances, as we have seen in the case of the Delphic oracle.[30] Other inventions of Kircher's also appear to have come under suspicion of demonic magic, including the magnetic anemoscope that he built in Malta [**See Figure 6.4**] while he was supposed to be providing spiritual guidance to Landgrave Ernst of Hessen-Darmstadt, relied, like many Kircherian machines, on a hidden magnet. The magnet, rotated by a wind-vane, caused a figure of Aeolius, the god of winds, suspended in a glass sphere, to point to the direction of the wind marked on the outside of the sphere. Some of the Knights of Malta who witnessed Kircher's machine apparently suggested that it must contain a real demon, and Kircher, yet again, had to take pains to demonstrate that his brand of magic was entirely natural.[31]

Anatomies of machines and mechanical anatomies

By the time that De Sepibus' catalogue was published, the *Musaeum Kircherianum* had entered a dramatic phase of decline, only to be resurrected through the efforts of Filippo Buonanni in the early years of the eighteenth century. The famous frontispiece of De Sepibus' work, and many of its contents are misleading, as they represent Kircher's museum as occupying a space that it had long abandoned, due to General Oliva's decision to transform it into a library for the Jesuit 'scriptors', excused from teaching duties in order to devote themselves to writing works for publication. The frescoed lunettes and large windows of the space depicted and described in De Sepibus' catalogue had long been forsaken for a dark corridor, much to the dismay of the ageing Kircher. The catalogue thus presents immediate problems as a historical document of Kircher's museum. By 1678, Kircher, depicted on the frontispiece of De Sepibus' catalogue warmly welcoming a pair of visitors to his museum, was nearing death, and spending almost all of his time in the Marian shrine of the Mentorella in the hills of Lazio, where his heart was soon to be buried.[32]

De Sepibus' catalogue of the museum, then, crammed with illustrations culled from Kircher's other works, must be regarded as a monument to a dead, or at least dying and transfigured institution. In order to understand the magical nature of the machines on display in the museum, many of which had fallen into disrepair by 1678 we will have to look elsewhere. Long before De Sepibus published the catalogue, repeated attempts to publish a description of Kircher's gallery had been made by Kircher's close disciple Kaspar Schott.[33] Schott's association with Kircher had begun in 1630, when he was studying in Würzburg, a city that both Schott and his master had to abandon with the onslaught of the Swedish troops of Gustavus Adolphus in 1631. Whereas Kircher fled to the South of France, arriving in the Jesuit province of Lyon along with forty other Jesuit refugees, Schott made for Tournai, and then began a series of wanderings through Sicily, where he completed his studies and taught in a number of Jesuit colleges.³⁴ Between late 1652 and 1654, Schott was finally reunited with Kircher in Rome for an extraordinarily intense period of activity centred around the recently founded museum, a period that was to fuel his prolific output in the years that followed.³⁵ In addition to assisting Kircher in the museum, Schott performed a number of other tasks. While Kircher laboured to complete his monumental *Oedipus Aegyptiacus*, Schott patiently edited the third edition of Kircher's *Magnes*. An anonymous foreword by the 'Author's colleague in literary matters' inserted into this edition gives a graphic picture of the conscientious approach taken by Schott to this task:

> I examined and emended all of the calculations and arithmetic tables with great care. I inspected the words in Latin, Greek and Hebrew of authors who were cited in the original sources and where they had been corrupted I restored them. I compared the magnetic declinations and inclinations, and other observations sent here to the Author (who had asked for them by letters) with the autographs, and eliminated typographical errors. I inspected the diagrams even engraved on brass or wood, and emended the mistakes, restoring the missing or erroneous letters, lines and signs. For several elevations I substituted more accurate ones. From time to time I eliminated words, or added them, or changed them, when I noticed that the sense was either false, altered or unclear. In arranging the Appendices, Paradoxes, Problems, and new Experiments and Machines written by the Author, or given to me to write, I conserved an order that altered the order of the previous editions as little as possible [...] I omitted, finally, no task that I felt would contribute to the splendour of the Work.³⁶

As well as working as Kircher's editor, Schott was deeply involved with the machines of the museum, and it is to his works that we will turn to attempt to situate Kircher's machines in a magical tradition. Schott's *Mechanica Hydraulico-Pneumatica* was published in 1657, shortly after his return to Germany. Apart from the appendix, which dealt with the new 'Magdeburg' experiment carried out by Otto von Guericke to demonstrate the existence of a vacuum, Schott had composed the book while he was still in Rome with Kircher, as he explains in a 'Notice to the Reader', excusing himself for often writing as if he was still living in Rome. Schott writes that he plans 'to compose a Natural Magic, collected from the printed works and manuscripts of the most learned man Athanasius Kircher, of world-wide fame, and also from all of his notes and loose pieces of paper that are in my possession, as well as from the works of other approved authors and the inventions of ours (i.e. Jesuits), composed in all trustworthiness and as the result of much study, established through my own experiments and those of others'. His promised work, subsequently published as the *Magia Universalis Naturae et Artis*, will contain 'various, curious and exotic spectacles of admirable effects, wonders of recondite inventions, that are rightly called magic, free from all imposture and suspicion of the forbidden Art'.[37] In the meantime, Schott's *Mechanica Hydraulico-Pneumatica* consists in an exhaustive description of the hydraulic and pneumatic machines found in Kircher's museum. As he writes in the preface to the work:

> There is, in the much-visited Museum (that we will soon publish in print) of the Most learned and truly famous Author mentioned above (i.e. Kircher), a great abundance of Hydraulic and Pneumatic Machines, that are beheld and admired with enormous delight of their souls by those Princes and *literati* who rush from all cities and parts of the world to see them, and who hungrily desire to know how they are made, and so that I can satisfy their desire to know the construction of the machines, I have undertaken to show the fabric, and almost the anatomy of all of the Machines in the said Museum, or already shown elsewhere by the same author.[38]

Schott promises to give his readers detailed instructions on how to make instruments,

> for garden pleasures, for the utility of houses, for the commodities, and ornaments, particularly of Princes, who derive greater pleasure of their eyes and souls from these things than they might expect profit for their estate. Neither will we be satisfied with delighting only the eyes, we also prepare a feast for the ears, with various self-moving and self-sounding organs and instruments, that we will excite to motion and sound only by the flow of water and the stealthy approach of air, with no less ease than skill.[39]

Schott's *Mechanica Hydraulico-Pneumatica*, then, provides an eloquent 'identikit' picture of the ideal audience for Kircherian wonders, a leisured, decadent class of princes and cardinals, quite happy to turn their minds away from pressing matters of church and state in order to delight their minds, eyes and ears with the sensual pleasures provided by Kircherian machines. From the rich study of the intellectual culture of the Habsburg monarchy carried out by R.J.W. Evans, we see that this description was entirely consonant with the consuming interests of the prominent members of the Viennese courts of Ferdinand III and Leopold I.[40]

The wonders described in Schott's work give us a vivid picture of how Kircher and his disciples went about satisfying the remarkable thirst for hydraulic and pneumatic curiosities of a Catholic elite on a daily basis. In one instance, Schott describes an incident in which the two Jesuit companions came across the marvellous spectacle of a 'water-vomiting seat' in a Roman villa:

> Lately Father Kircher and I were wandering through the fields of Rome to take the air, and we went into a suburban villa, on the facade of which an elegantly made sciatheric sundial was painted. While we were looking at this curiosity, we were invited by a Noble Frenchman to inspect the building and garden more thoroughly. We entered, and first saw a most delightful pleasure-garden, filled with flowers and fruit, and ornamented with statues of all kinds. We then entered a most elegant house, ornamented with paintings, emblems, epigrams, and epigraphs in Latin, Greek and Arabic, and thoroughly filled with statues and artificious machines, so that even Pope Innocent X, as he was being carried through the same fields with the delight of his soul, entered the same house and garden, and was not reluctant to honour it with his presence. The villa belongs to Jean Laborne, a French Presbyter and Knight of the same Pope. Amongst the other things, by which I was most delighted, was a seat known as hydratic or water-vomiting because of its effect.[41]

If we are to take De Sepibus' list of machines as a guide, we are forced to conclude that the predominantly German princely audience of the productions of Kircher and Schott had a peculiar fascination with regurgitation. From the two-headed Imperial Eagle belching water copiously from its twin gullets [**See Figure 5.1**], to the 'water-vomiting hydraulic machine, at the top of which stands a figure vomiting up various liquids for guests to drink', not to mention the various birds and snakes ingesting and throwing-up water from goblets, the spectacle of retching, puking, and spewing seems to have been the very epitome of good taste and noble amusement for the visitors to Kircher's museum [**See Figure 6.5**]. Schott further confirms this impression of an 'emetophiliac' Catholic elite. One of the most endearing machines of his *Mechanica* is a '*cancer vomitor*' [**See Figure 6.6**] illustrated as a nauseous lobster, bending forlornly over the edge of a goblet in its unhappy state. One is left unsure whether sea-sickness or the drinking of the goblet's contents is responsible. Like a number of the machines illustrated in Schott's works, this device was adapted from the popular work by Daniel Schwenter (1585–1636), later expanded by Georg Philipp Harsdörffer (1607–1658), the *Deliciae Physico-Mathematicae*.[42] Perhaps the most graphic demonstration of the cult of emesis is in Schott's description of a French visitor to Rome with an unusual talent:

While I was writing this, Jean Royer, a Frenchman from Lyon, who is superior to all in the art that we have been discussing, arrived here. From his stomach he brought forth twelve or fourteen differently coloured perfumed waters, most perfect liquors, distilled wine that could be set alight, and rock oil that burned with a lamp-wick, lettuces and flowers of all kinds, with complete and fresh leaves. He also exhibits a fountain by projecting water out of his mouth into the air for the time of two *Misereres*.[43]

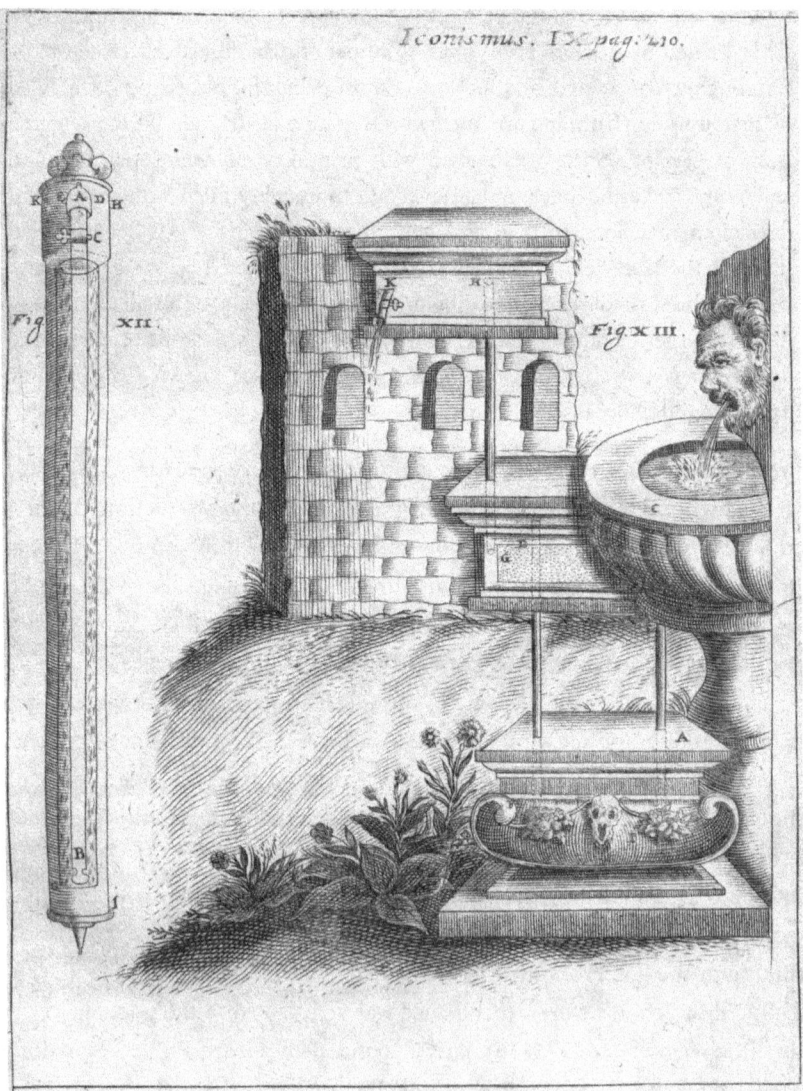

Figure 6.5 Vomiting fountain from Kaspar Schott, *Mechanica Hydraulico-Pneumatica*, p. 210

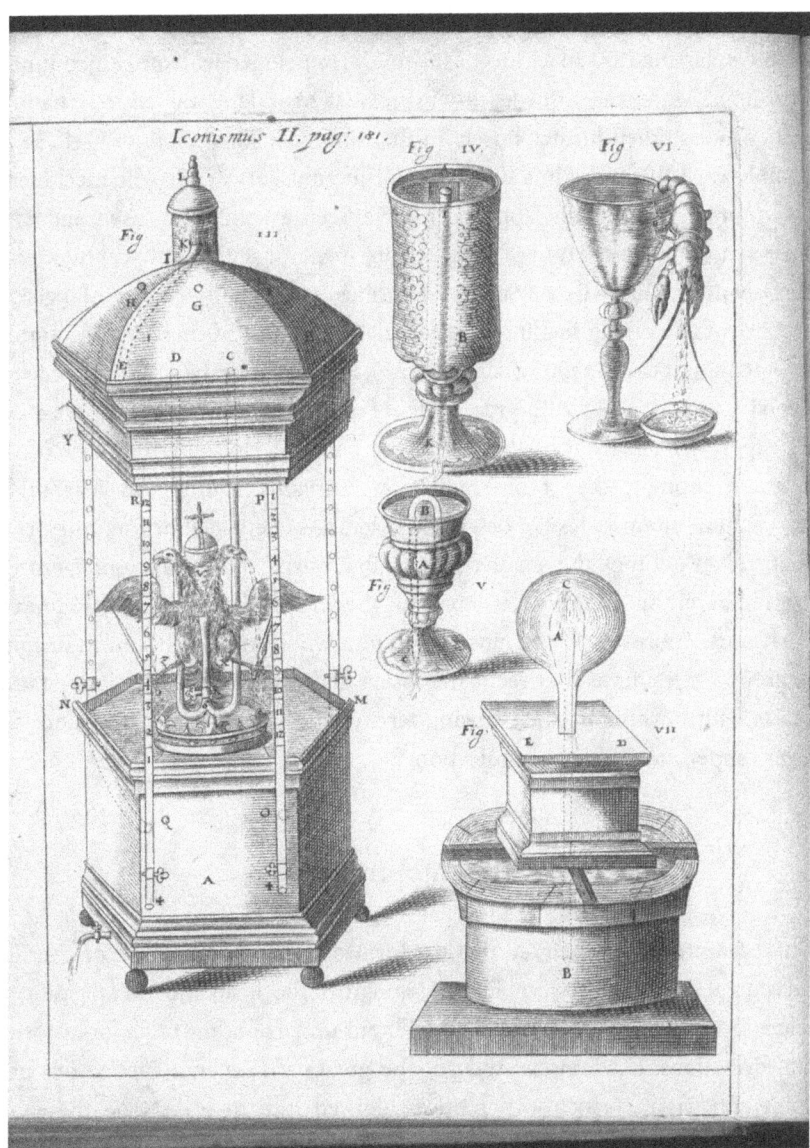

Figure 6.6 Various hydraulic machines, from Kaspar Schott, *Mechanica Hydraulico-Pneumatica*, p. 181

The description of this technicolour spectacle is followed by a letter from Kircher, in which he reassures worried readers that the digestive system of Mr Royer was entirely free of demonic interference, and that his stomach-churning feats were carried out purely through the manipulation of natural causes. Royer, it transpires, had even entertained the Emperor at Regensburg, also exhibiting his 'art' before 'five kings and many princes and learned men'. In Schott's work, Royer himself is classified as a machine – 'Machina VII', included with other incontinent 'hydropota'. Moreover, in order to ensure that his talent was entirely natural, Kircher had studied his act closely in the *Musaeum Kircherianum* itself, so he certainly earns his place in a discussion of the museum's hydraulic machines.[44] The *Miserere*, incidentally, appears to have been a commonly used and even somewhat standardised unit of time measurement for seventeenth century Jesuit experimenters. Elsewhere, Schott describes one of his more dangerous experiments involving heating a sealed glass tube full of mercury, recounting that 'after about the time in which Psalm 50, *Miserere mei Deus*, can be recited, it opened a way for itself with great violence and noise'. When Schott performed this experiment in front of the son of the Duke of Holstein, the noise of the explosion brought the young nobleman's servants running in fear of an assassination attempt. Jesuits describing Manfredo Settala's burning mirrors in Milan remarked that 'the smaller mirror, that burns at a distance of 7 *braccie*, works in barely an *Ave Maria*, whereas for the one that burns at 15 or 16 *braccie*, which works more slowly, you have to wait for a whole *Miserere*'. One can imagine the groups of Jesuits as they recite the rosary and sing hymns while incinerating objects with burning glasses, causing terrifying explosions or witnessing Jean Royer's superhuman feats of projection.[45]

The catoptric cat

Robert Darnton has remarked that the torture of cats was a source of constant amusement in early modern Europe, and that the historical investigation of arcane forms of humour has much to offer our understanding of major historical transformations. His famous study of the 'great cat massacre' carried out by a group of Parisian printer's apprentices allowed him to investigate the social tensions that formed the historical prologue to the French Revolution.[46] More recently, Thomas Hankins and Robert Silverman have used Darnton's insights in an original study of some of the more ludic machines and instruments produced by Kircher and others, in particular the sunflower clock [**See Figure 6.7**] that Kircher displayed to Nicholas Claude Fabri de Peiresc in Aix, and the 'cat piano', a grisly musical instrument, said to have been invented by Kircher, that worked by prodding the tails of cats with spikes driven by a keyboard.[47] Whereas for Darnton's Parisian apprentices, the torture of cats was a humorous means for an abused community of labourers to score a symbolic victory over their wealthy bosses, for

Kircher and his princely clients the manipulation of animals and automata was arguably a symbolic means of reinforcing the political and philosophical *status quo*. Schott recounts that one of the most 'artificious and delightful' machines in Kircher's museum was a catoptric chest, presumably identical with the 'catoptric theatre' described by De Sepibus. **[See Figure 6.8]**.

Figure 6.7 The sunflower clock, from Kircher, *Magnes, sive De Arte Magnetica* (1643 ed.), p. 644

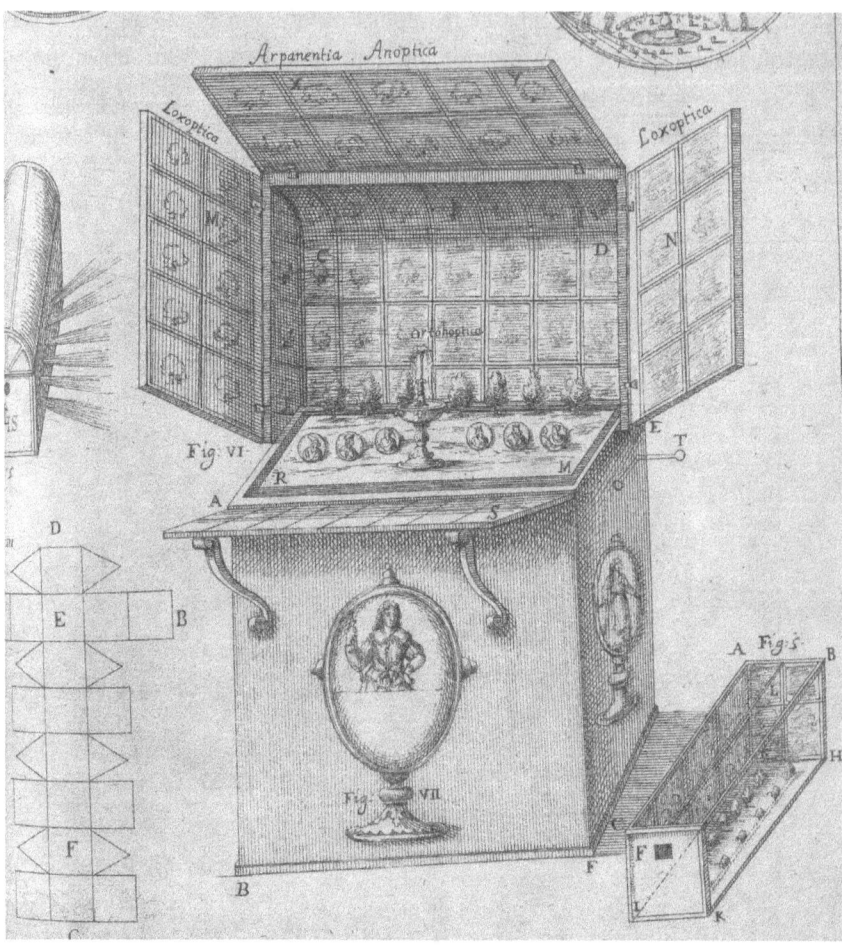

Figure 6.8 The catoptric theatre from Kircher, *Ars Magna Lucis et Umbrae* (1671 ed.), p. 776

Two other catoptric chests existed in Rome, according to Schott, one in the Villa Borghese and the other in the 'villa of some other Prince', and both exhibited wonderful spectres of objects – forests of pine trees, cities, elegantly furnished houses, treasures of gold and silver vases and pearls and infinite libraries of books, that seem so real that even those who were knowledgeable in catoptrics were sometimes fooled, and less intelligent people frequently held out their hands and attempted to take hold of the 'species of things', to the great amusement of spectators. Kircher's catoptric chest, however, far surpassed the competition, both in multiplying species and in displaying illusory scenes. It could display infinite colonnades, tables covered with all sorts of delicacies, inexhaustable treasures, to the great torment of avaricious visitors who often, according to

Schott, attempted to make off with the infinite quantities of money contained in the chest, only to be left with a handful of air. 'You will exhibit the most delightful trick', Schott informs us,

> if you impose one of these appearances on a live cat, as Fr. Kircher has done. While the cat sees himself to be surrounded by an innumerable multitude of catoptric cats, some of them standing close to him and others spread very far away from him, it can hardly be said how many capers will be exhibited in that theatre, while he sometimes tries to follow the other cats, sometimes to entice them with his tail, sometimes attempts a kiss, and indeed tries to break through the obstacles in every way with his claws so that he can be united with the other cats, until finally, with various noises, and miserable whines he declares his various affectations of indignation, rage, jealousy, love and desire. Similar spectacles can be exhibited with other animals.[48]

The catoptric chest, then, is an instrument for the manipulation and revelation of the passions. It is a theatre of social distinction, using visual illusion for the detection and display of baser human traits such as avarice and the instinctual passions of animals. An understanding of the magical art of catoptrics can allow one to trick people (and cats) into revealing their hidden natures. Kircher's emotionally confused catoptric cat is thus very different from the pampered aristocratic cats slaughtered by the Parisian artisans described by Darnton. By making a spectacle out of incivility or popular superstition, devices such as the catoptric theatre, the Delphic oracle and the various vomiting-machines shown to visitors to Kircher's museum contributed to a particular definition of early modern European civility.[49] Many of Athanasius Kircher's machines were thus civilizing machines. Descartes's *Treatise on the Passions of the Soul*, published in 1649, attempted to provide a manual to instruct his readers both to combat the effects of the passions on the soul and to dissimulate their outward manifestations.[50] The vogue for automata and machine-models of the human body in the seventeenth century was closely connected to the desire to exercise control over the body through discipline and manners. The Jesuit educational system, experienced by Descartes as a schoolboy at La Flèche, laid great emphasis on bodily comportment and behavioural discipline, epitomised by the choreographed movements of Jesuit *ballet*. The limits of the man-machine metaphor exercised a powerful fascination over Kircher's contemporaries. While Marin Mersenne (1588–1648) theorised about mechanised musical ensembles, and instruments such as the '*Archiviole*', allowing a single player to play multiple musical instruments simultaneously, and shortly after Justus Lipsius (1547–1606) had theorised about the well-disciplined army as a war-machine, Thomas

Hobbes (1588–1679) opened his *Leviathan*, published in the very year that the *Musaeum Kircherianum* was officially founded, with the famous metaphor of the commonwealth as a giant automaton, manipulated by a single monarch.[51] Peter Dear has evoked the close links between the mastery of the passions, the rise of European absolutism and the culture of automata in early modern Europe.[52]

We have frequently been led to discuss the wonders produced by Kircher and Schott in magical terms. But just what was the magic practiced by Kircher, that he took such pains to distinguish from the illicit arts that invoked the aid of demons? What were its boundaries? How did it intersect with natural philosophy, and with the mathematical arts? How did it find a home in the bosom of the Jesuit order and, especially, in Kircher's museum?

Kircherian magic: The roots of a paradigm

Kircherian machines, we have suggested, like Jesuit rhetorical devices, emblems and learned orations, helped to draw a boundary between elite and vulgar. To mount an attack on the causal knowledge at the core of the Kircherian culture of machines on physical grounds was comparable to challenging the authenticity of the *Corpus Hermeticum* and the traces of the *prisca sapientia* contained in Egyptian hieroglyphics on philological grounds. Both challenges threatened the mystical core of a structure of political power in which the Jesuit order constituted the cement linking the Counter-reformation Papacy to the Habsburg court in Vienna through a sophisticated network of intermediaries. The intellectual project of Kircher's *Oedipus Aegyptiacus*, supported by Ferdinand III, cannot be separated from Kircher's artificial magic.[53] Kircher's marvellous machines took their place alongside his wooden reconstructions of Egyptian obelisks in the *Musaeum Kircherianum*. A letter from Schott inserted into the first volume of Kircher's *Oedipus Aegyptiacus* gives us a revealing picture of the mutual legitimation that characterised Kircher's close relationship with his Habsburg-linked clients:

> In Kircher's archive, I discovered an enormous number of letters, many of which were sent by him at every moment by Princes of the Christian world, and the supreme heads of the Roman Empire, and the Most Wise Emperor FERDINAND III, the Most Serene and Most Wise Queen of Sweden Christina, many Most Eminent Cardinals of the Holy Roman Church, Most Serene Electors of the Holy Roman Empire, Most Distinguished and Illustrious Dukes, Princes, Counts, Barons and innumerable Nobles of the same Empire and other Nations, all of

whom admire and praise Kircher's learning, and thank him for the books he sent them and for his other enormous productions, they urge and solicit him to print other monuments to erudition, they offer him help and protection, they communicate secrets, and ask for arcana, and for the unravelling of arcane matters, they seek the interpretation of exotic languages, strange inscriptions, and unknown characters, and various questions. I would have appended here various long letters from Emperors, other Princes and almost all the learned Men of this century, showing singular affection and respect if the small space and the Author's modesty had permitted and if I had not reserved that for a different time and place[54]

While Kircher provided princes, young and old, with enigmas, puzzles, emblems and arcane knowledge that confirmed their social distinction, they provided him with financial support and conferred authority on his works. Elsewhere Schott tells us of a revealing dream that Kircher had in the *Collegio Romano* while suffering from a serious bout of illness. After requesting a strong sleeping-mixture of his own specification from the college pharmacy, Kircher fell into a deep sleep, and dreamt that he had been elected to the Papal throne and was overcome with joy. He received legations and congratulatory messages from all the Christian princes, applause from all peoples, and, in his dream-role as Pope, built colleges and churches in Rome for the different nations of the world, and established 'many other things for the propagation of the Catholic faith'. Schott is particularly interested in the healing capacities of Kircher's dream – the older Jesuit pronounced himself to be restored to full health the following morning. However, without too much imagination, his dream might also be seen as hinting at more than a modicum of personal ambition on Kircher's part. Although some of Kircher's other nocturnal visions were later transformed into reality, most dramatically a graphic vision of the imminent destruction of the Jesuit college in Würzburg by the Swedish armies of Gustavus Adolphus in 1631, his narcotically-induced dream of the papal tiara was never to be realized, although one is tempted to wonder what directives he might have issued in this role.[55] Despite the fact that Kircher was never elected pope, he was arguably the ruler of his own invented polity. The *Oedipus Aegyptiacus* contains no less than thirty-one separate letters of dedication for its different sections and provides us with a suggestive map of Kircher's political universe. Prominent dedicatees include: the holy Roman Emperor Ferdinand III, Pope Alexander VII, Ferdinand IV King of the Romans, the Grand Duke of Tuscany Ferdinand II de' Medici, Johann Philipp von Schönborn, Elector of Mainz; Archdukes Leopold Wilhelm and Bernhard Ignaz of Austria, Johann Friedrich Duke of Brunswick-Lüneburg, and a host of other princes, cardinals, counsellors and confessors of the Holy Roman Empire.

Kircher's *Oedipus Aegyptiacus* provides an ancient pedigree of magic that justified its revival amongst his distinguished dedicatees and their peers, a pedigree echoed in Kaspar Schott's *Magia Universalis*. In its broad lines, legitimate magic was first given by God to Adam, along with the other forms of knowledge. However, true magic was corrupted, through the 'Cainite evil', leading to the division between 'licit' and 'illicit' magic. The architect of the corruption of magic was, as Pliny recounts, Zoroaster. But which Zoroaster? A number of different Zoroasters appear in the history books. On this subject, many learned authors were in disagreement, but Kircher and Schott, aided by a manuscript of the apocryphal Book of Enoch studied by Kircher in the Greek library of Messina, are in agreement that Zoroaster is identifiable with Noah's rebellious son Cham, who learned this art from the impious Cainites before the Flood and inscribed it on stones and columns so that it would not be destroyed in the deluge, transmitting it to his followers once the waters had abated. These columns were the very columns described by St Augustine, when he wrote in the *City of God* that Cham, Noah's son, erected fourteen columns bearing the canons of the arts and the sciences, seven made of brass and seven of bricks. After propagating his magic in Egypt, where he had settled after the Flood and the linguistic confusion of the Tower of Babel, Cham left his kingdom to his son Misraim, and departed to spread the astrological and magical arts to Chaldea, Persia, Medea and Assyria, eventually obtaining the name 'Zoroaster', meaning 'living star' as he appeared to be consumed with celestial fire in his zeal to spread magical knowledge.[56]

What is magic? Schott tells us that magic is whatever is 'marvellous and goes beyond the sense and comprehension of common man'. Common men because to 'wise people or those who are more learned than the common people the causes of magical effects are normally apparent'. Natural magic, according to Schott, is 'a recondite knowledge of the secrets of nature, that applies things to things, or, to speak philosophically, actives to passives, in the correct time, place and manner, by the nature, properties, occult powers, sympathies and antipathies of individual things, bringing about some marvels in this way that appear magical or miraculous to those who are ignorant of the causes'. An example of natural magic is asbestos that resists combustion in flames, as Kircher had demonstrated very frequently in Rome. Other examples of natural magic include the magnetic marvels described by Gilbert, Cabeo and Kircher, and the effects of music on the venom of the tarantula, also described by Kircher. However, one must beware, as not all magic said to be natural is truly so, the sunflower's supposed capacity to make men invisible being an example of something that couldn't possibly happen naturally. Schott's encyclopedia of natural and artificial magic comprises four parts: Optics ('that is those things regarding sight and objects that are seen, and whatever in Optics, Catoptrics, Dioptrics, Parastatics, Chromatics, Catoptro-Dioptro-Caustics, Catoptrologics, and other similar sciences, arts, practices and secrets is rare, portentous and beyond the understanding of the common people, when they perceive rays directly, reflected or refracted at the eye'), Acoustics ('that is, whatever pertains to hearing, and the object heard, and it will explain all of hearing, sound, the human voice, harmony, the oeconomy of music, by analogy to the oeconomy of sight and vision, colours, lights, and their appearances, but only the rarer, less obvious ones that fall under praxis and operation'), Mathematics ('that is Arithmetic, Geometry, Astronomy, Statics, Hydraulics, Pneumatics, Pyrobolics, Gnomonics, Steganography, Cryptology, Hydrography, Nautical matters, and many other things, but only the rarer and more amusing and wonderful matters, and most of the practical things that come under human industry') and Physics (' whatever is wonderful, paradoxical or portentous in Nature. Of this kind are magnetism, sympathy, physiognomy, metallurgy, botany, stichiotics, medicine, meteorology, the secrets of animals, stones and innumerable other things').[57]

Natural magic has two branches in Schott's system: operative and divinatory. The latter include such arts as physiognomy, allowing a person's character to be determined by examining their features, colour and voice. Divinatory natural magic, however, cannot be used to find supernatural gifts or sins, as these don't depend on nature but on free human will. Artificial magic, or operative natural magic, is, in Schott's definition 'an art or a faculty of producing some wonder through human industry, by applying various instruments'. Schott's examples of this art, culled from an assortment of classical sources, include the glass sphere of Archimedes described by Cicero, which depicted the motions of the different planets, the flying wooden dove of Archytas, the small golden birds singing to the Byzantine emperor Leo, and the flying and singing birds and hissing serpents of Boethius [**See Figure 6.9**]. More recent pieces of artificial magic included the eagle of Regiomontanus that reportedly flew to meet Charles V when he was arriving in Nuremberg, and accompanied him to the gates of the city, and an iron fly also made by Regiomontanus that flew out of the hands of its artisan, and flew around the assembled guests, and a statue in the shape of a wolf that walked around and played a drum, that Schott had heard about from an eyewitness. The talking head reportedly made out of brass by Albertus Magnus was a further example of artificial magic for Schott. Whereas some claimed that this was a mere fable, and others suggested that it was the work of the devil, Schott disagreed, arguing that it was made by human industry alone. Kircher himself, Schott had just heard in a letter sent from Rome, was in the process of making just such a speaking statue for the visit of Queen Christina of Sweden to the *Musaeum Kircherianum*, 'a statue that will have to answer the questions that it is asked'. The Delphic Oracle, then, places Kircher's magical productions in a highly respectable historical series of artificial wonders, and rids Albertus Magnus of the suspicion of sorcery that allegedly led Thomas Aquinas to destroy his talking statue of Memnon.[58]

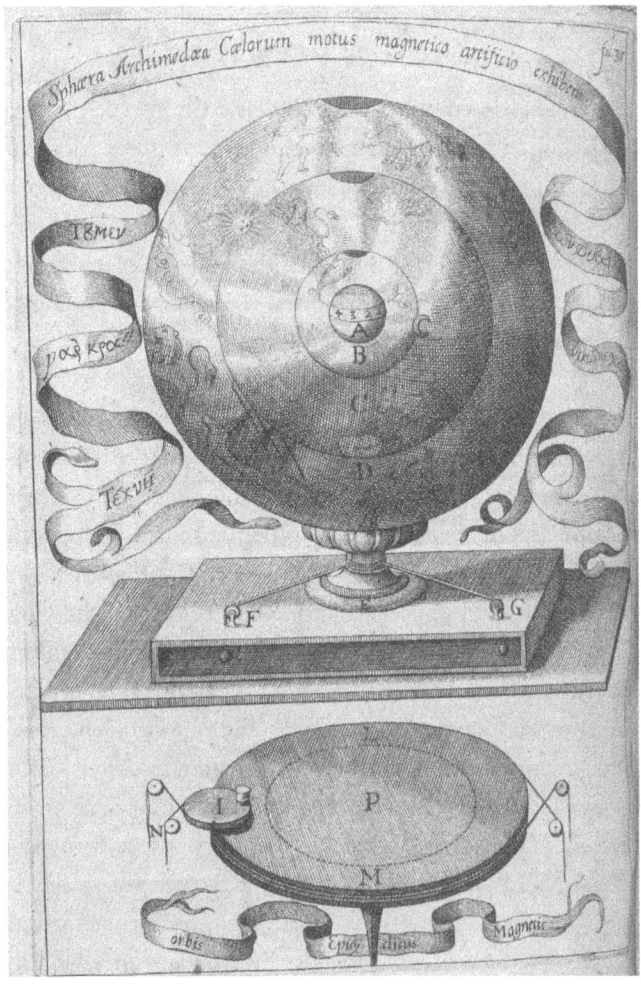

Figure 6.9 Kircher's reconstruction of the legendary sphere of Archimedes from Kircher, *Magnes, sive de Arte Magnetica* (1643 ed.) p. 305

The machines in Kircher's museum occupy a place in Schott's exhaustive account of the licit magical arts. But what exactly were the boundaries of these arts? Where is the point of transgression? Schott's answer is simple: illicit magic involves pacts with demons rather than the mere application of human industry and artifice to natural causes. Following the principal Jesuit authorities on the matter, the humanist Martin del Rio (1551–1608) and the philosopher Benito Pereira (1535–1610), Schott insists that demons are restricted to the manipulation of natural causes. Only God can affect miracles that go against the natural order. Demons are, effectively, just very good artificial magicians, manipulating natural causes with greater dexterity than even the most adroit instrumentally-enhanced human being.[59]

But what exactly is the order of nature that even demons cannot pervert? Schott's answer is unequivocal: demons are bound to obey the laws of Aristotelian natural philosophy! 'They cannot create anything, as this exceeds the power of acting naturally. Neither can they derive a substantial form immediately from a subject, without a prior alteration, because this cannot be done naturally'. Demons cannot even create a vacuum, 'as Nature abhors this and no experiment carried out until now proves that a vacuum has been made, as we have said in the *Mechanica hydraulico-pneumatica*'. If demons could not make a vacuum, what chance did Evangelista Torricelli, Valeriano Magni or Otto von Guericke stand of doing so? Schott's account of the absolute limits of artificial magic reveals its staunchly Aristotelian core. The artificial magic practiced and described by Schott and Kircher relied on an unchanging body of assumptions about the normal behaviour of the natural world. Schott's *Mechanica Hydraulico-Pneumatica* had opened with a list of the four fundamental principles underlying all hydraulic machines: the 'attractive power to avoid a vacuum', the 'power of expulsion, avoiding the penetration of bodies', the rarefactive power (i.e. the 'expulsion or attraction of water by rarefaction and condensation') and the weight of the water seeking equilibrium. The purpose of Schott's work is not to investigate the truth of these principles, which have the status of axioms. Instead, his aim is to catalogue the surprising effects that can be obtained by combining these causes in different ways.[60]

In discussing Otto von Guericke's experimental demonstration of the existence of a vacuum using his *antlia pneumatica*, Schott remarks casually that of course, the plenitude of nature is invulnerable even to an angel, and thus Guericke's device could never have produced a real vacuum. A refusal to allow the instrument to produce new natural philosophy did not put an end to Jesuit discussions of hydraulics. Instead, the device was removed from circulation in the philosophical domain and relocated within the context of the *Wunderkammer*. Schott's *Mechanica Hydraulico-Pneumatica* includes the experiments performed by Evangelista Torricelli and Gasparo Berti to demonstrate the existence of the vacuum in a section entitled *De machinis hydraulicis variis*, where they are surrounded by a ball made to spin in the air, a perforated flask for carrying wine known as the 'Sieve of the Vestal Virgin', and a 'phial for cooling tobacco smoke'. Unhealthy philosophical readings of *Machina VI* (the Torricelli and Berti tubes) are dismissed by Schott as the writings of '*Neotherici Philosophastri*' and 'insolent and unmannerly braggarts proclaiming a triumph before victory'.[61] To situate the Torricellian experiment in the context of trick fountains and water-vomiting seats was to insulate it from the Aristotelian philosophy taught in the classrooms of Jesuit colleges. In a strong sense, then, the Aristotelian physics at the basis of the artificial magic of Kircher and Schott was invulnerable, except to occasional divine intervention. Machines combined a pre-established set of causes to produce surprising effects, leaving the spectators to attempt to decipher the combination of natural causes underlying the appearances.

Schott's accounts of natural and demonic magic drew heavily on the comprehensive treatment of magic composed by the Antwerp-born Jesuit Martin del Rio, the *Disquisitionum Magicarum Libri Sex*, first published in 1599. Del Rio was a scholarly prodigy before he joined the Jesuit order. At the tender age of twenty he published a work on the Latin grammarian Gaius Solinus, later attacked by Claude Saumaise. Shortly afterwards, he published a work on Claudius Claudianus that cited more than 1,100 authors. Before he joined the Jesuit order, he occupied the important public offices of Senator of Brabant, Auditor of the army, Vice-chancellor and Procurator General. Del Rio's three-volume treatment of magic was an enormously influential work, the influence of which was felt in witch-trials as much as in the scholarly arena.[62] Chapter IV of Del Rio's work deals with artificial magic, which Del Rio divides into 'mathematical magic', deploying the principles of geometry, arithmetic and astronomy, and 'prestidigitatory magic', involving deliberate deception and sleight of hand. The former includes all the famous mechanical marvels that Schott listed. Del Rio's approach to magic is to build an impenetrable wall between supernatural phenomena, which are the prerogative of God alone, and artificial and preternatural phenomena, which can be produced by men, by demons and by angels. Preternatural phenomena are those which appear to most people to go beyond nature's capacities, but are in fact achieved through the combination of natural causes by human, demonic or angelic agents. They belong not to the 'Order of Grace', the realm of true miracles brought about by divine intervention in opposition the laws of nature, but to the Prodigious Order, reserved for phenomena that resemble miracles, but are in fact carried out through the manipulation of natural forces.[63] Kircherian thaumaturgy, then, appears to transcend what can be achieved through the human manipulation of natural powers, thus leading some to view them as being produced by demonic means. Good angels do not collaborate in magical works, according to Del Rio, so any magical feat that goes beyond human capacities, such as the production of healing effects through incantations, must be due to the 'ministry of bad angels', that is to say the companions of Lucifer, as 'no words have a natural power of healing wounds or illnesses, or driving away other injuries'.[64] Incantations employed by Catholic priests in sacraments and exorcisms did not work naturally, but through the concurrence of divine grace, and thus belonged to the Order of Grace, and are thus excluded from the natural order.

Kircher's machines ludically encouraged spectators to read them as wonders achieved through angelic or demonic concurrence. Many of the machines described in De Sepibus' list even contained small genies, angels and demons, moved by occult forces to point at letters, scales and inscriptions, a miniature automated population that positively cried out to be interpreted as preternatural, and belonging to Del Rio's prodigious order. While Descartes hypothesised a single evil genie to demolish the basis of scholastic metaphysics in the first of his *Méditations Metaphysiques*, Kircher and Schott employed an obedient mechanical army of them to uphold the core of Aristotelian physics.

Benito Pereira, Schott's other chief authority on magical matters, was one of the most influential philosophers of the Jesuit order in the late sixteenth century, despite coming under suspicion of heterodoxy for his sympathy for the philosophy of Averröes.,[65] Pereira's textbook on natural philosophy, *De Communibus omnium rerum naturalium principijs & affectionibus*, went through a great number of editions, and was widely used for teaching in Jesuit colleges. His widely read work on magic and divination, the *Adversus fallaces & superstitiosas artes, id est, De magia, de observatione somniorum, et de divinatione astrologica*, argued that demons could not pervert the natural order of the Aristotelian elements or create a vacuum, and this may have been the source for Schott's similar assertions. Pereira insists that men skilled in knowledge of nature can work great wonders by natural magic, but those who are either wicked or ignorant may only learn this art from demons, 'for scarcely any mortal or certainly very few indeed, and those men of the keenest mind who have employed diligent observation for a long time, can attain to such natural magic'.[66]

Kircher clearly considered himself to be one of the latter, and offers us his own working definition of natural magic in his *Magnes*, a definition that is pretty close to those provided by Del Rio, Pereira and Schott:

> Here I call natural magic that which produces unusual and prodigious effects through natural causes alone, excluding any commerce, implicit or explicit, with the Enemy of humankind. Of this kind are those machines that are called for this reason 'thaumatourgikai', that sometimes transmit prodigious movements to an effigy from air and water contained in siphons by a subtle art, and sometimes blow spirits into an organ arranged in a certain way to make statues burst forth in speech, and similar things, that can seem like miracles to people who are ignorant of their causes.[67]

Kircherian machines thus walked a tightrope between the demonic and the miraculous. To understand how the magical aspects of Kircher's machines were experienced by contemporaries, it may be helpful to look at how Kircher's *Musaeum* was visited.

Visiting the machines

The frontispiece of the fourth volume of the first edition of Kaspar Schott's *Magia Universalis* depicts a crowned man pointing a magic wand at a flowerbed, making a clear visual link between social status and the practice of natural magic. The opening of Schott's work provides a justification of magic that places Kircher's machines directly in the context of aristocratic visits to the Jesuit *Collegio Romano*:

> In my various long journeys through Germany, France, Italy and Sicily, and in my frequent occupation teaching mathematics both in public and in private, I have always found that almost everybody, especially Nobles and Princes, not only youths, but also men conspicuous for their learning, prudence, worldly experience and dignity displayed a propensity towards those disciplines that promise and set forward things that are marvellous, curious, hidden and beyond the comprehension of the common people. I hardly ever saw anyone, who, when he had achieved a little mastery of these matters, or had examined devices constructed from their prescription, was not thereby incited to continual study and did not surrender himself entirely to this discipline, or wish to do so if other occupations had permitted. Witnesses to this, to omit other examples, are the whole of Rome, and the most celebrated Roman College and Athenaeum of our Society, the seat and residence of Athanasius Kircher, a man of great fame in the

whole world. For, every day the inhabitants of both [city and college] look at and admire (as I myself beheld with amazement and delight of my soul when I was [Kircher's] assistant in literary matters for a few years) those works that many people hasten at every moment to behold, excited by the fame of his learning and the desire of seeing the things that he displays in his famous Museum. These works, constructed from the recondite arts and sciences, are truly deserving of wonder. The visitors are drawn from the most illustrious ranks, in doctrine and dignity, including Royalty and Cardinals, foreigners as often as natives. How many of them are instructed privately by him, even if occupied by other most grave matters, particularly the sons of Princes, recommended by very polite letters, with profit flowing into their whole nations and even into the whole Roman Church as a result![68]

Here Schott suggests that Kircher's museum in Rome functioned as a powerful magnet for a Catholic elite, attracting princely visitors to the *Collegio Romano*, and encouraging them to send their sons to be privately educated in arcane matters by Kircher. Kircher's aristocratic apprentices in magic would then return to their countries of origin, having acquired a taste for curiosity, and this would bring clear benefits both for their countries and for the Catholic church as a whole. Schott's description of the social function of the museum is consonant with the apostolic goals of the Jesuit educational system, as developed since the mid-sixteenth century. Christoph Clavius had argued, as discussed in Chapter 1, that excellence in the mathematical disciplines would aid the Jesuits to gain precious ground on the Protestant pedagogues that were enticing aristocrats away from the Catholic church.[69] We have argued above that Kircher inherited Clavius's *musaeum mathematicum*. Schott's description of the function of Kircher's museum as a magnet for a curious princely elite suggests that it had much in common with Clavius's prophetic vision of the Jesuit educational apostolate.

Courting Queen Christina

On 11 November 1651, Athanasius Kircher wrote a letter to Queen Christina in Stockholm:

> Your Majesty will know that our Society not only holds you in intimate affection, as is fitting, but also esteems and admires above all other things those rare and sublime treasures bestowed by heaven that divine bounty has hoarded up in your breast. This is especially true of this Roman College of our Society, both of

the famous men and writers and of the novices, who have come from all of the nations of the world, where we speak 35 different languages, some native to Europe, Africa and Asia, the remainder to the Indies and America. And all of them are excited by the fame of your majesty's wisdom, and attracted by some unknown sympathetic magnetism, and their only ambition is to paint the extraordinary example of all virtues that your Majesty exhibits to the world in all the colours that it deserves.[70]

Queen Christina's tour of the Collegio Romano in 1656 was the culmination of a lengthy process of rapprochement between the Queen and the Jesuit order which had begun in February 1652 when two Italian gentleman travellers, going by the names of Don Bonifacio Ponginibio and Don Lucio Bonanni, had arrived in the Royal court in Stockholm.[71] The two gentlemen, as Christina quickly divined, were in reality Jesuits, carefully disguised by long hair and beards. Paolo Casati and Francesco Malines, both highly trained in mathematics and theology, had set off from Venice on 8th December on their important mission to convert 'Don Teofilo', as Goswin Nickel, the Vicar General of the order, had instructed them to call Christina in their letters. Christina had specially asked the General for mathematically skilled Jesuits, and spent as much time with her visitors discussing Galileo's *Dialogo*, atomism, and the latest books by Bartoli and Kircher[72] as the matters of faith that were the ostensible reason for the meeting. She received a copy of Bartoli's *Dell'huomo di lettere*[73] from her Italian visitors, and probably availed of their services to send a letter to Kircher in Rome in which she expressed a desire to have a chance to talk to the famous polymath more freely in the future.[74]

Curiosity played a central role in Christina's abdication and relocation in Rome. The image of Rome which the Jesuit missionary mathematicians nurtured in the Queen's mind was one of a city in which the secrets of the natural world could be investigated under conditions of utter intellectual freedom, in stark contrast to the ascetic Lutheranism that reigned in Stockholm. Paradoxically, the very book that Kircher was to dedicate to Christina, the *Iter Exstaticum*, ran into serious difficulties on account of the atomist matter-theory which it sanctioned and which Christina also favoured.[75] The receptions of the Queen in the *Collegio Romano* were intended to continue to convey the image of the Jesuits' showpiece college as the home of cultivated Catholic curiosity.

On 18th January 1656, Queen Christina made her first visit to the *Collegio Romano*.[76] Twenty Swiss guards were placed at the door, preventing anyone from entering the building except the pupils of the lower classes, who were all meant to await the Queen in their classrooms. When the Queen arrived, the bells rang twice, and all of the Fathers, wearing cloaks, lined up inside the main door to receive her. The Queen entered the college with her entourage and the door was closed. In each class that the queen visited a pupil came forward to recite an epigram, and then presented her with a piece of printed satin brocaded with golden lace. When she had finished visiting the classes, she returned to the entrance, and went to visit the Church, where she prayed to Saint Ignatius and at the altar of Blessed Aloysius Gonzaga, while musicians sang some motets.

As she had been unable to see everything during this first visit, Christina returned to the college on 30th January. She entered by the side door, where she was received by the General, the Roman Provincial, the Rector of the College and other members of the order. Her subsequent perambulations are described meticulously in Galeazzo Gualdo Priorato's biography of the queen, and I cite from the 1658 English translation:

> She quickly went into the Library [...] Here her Majesty entertaining her self for some time, in viewing the many volumes, took pleasure too in looking on the Modell and Platforme of the City of Jerusalem, which was left by Father Villalpando, with the description of the streets, and holy places, consecrated by the journeys and passions of our Lord Jesus Christ. She then, going about the other sides, discovered some Greek and Latin Manuscripts lying open on a Table, and could judge of the Authors, shewing very great learning.
>
> She went thence into the gallery, that was near, where Father Athanasius Kircherus the great Mathematician had prepared many curious and remarkable things, as well in nature, as art, which were in so great a number, that her Majesty said, more time was required, and less company to consider them with due attention. However she stayed some time to consider the herb called Phoenix, which resembling the Phoenix grew up in the waters perpetually out of its own ashes. She saw the fountains and clocks, which, by vertue of the load-stone turn about with secret force. Then passing through the Hall, where she looked on some Pictures well done, she went through the walkes and the garden, into the Apothecaries shop, where she saw the preparation of the ingredients of herbs, plants, metalls, gemms, and other rare things, for the making of Treacle [i.e. Theriac] and balsome of life. She saw them distill with the fire of the same furnace sixty five sorts of herbs in as many distinct limbecks. She saw the philosophical calcination of ivorie, and the like. She saw extracted the spirits of Vitriol, Salt, and Aqua fortis, as likewise a jarre of pure water, which with two single drops of the quintessence of milke, was turned into true milk, the only medicine for the shortness of the breath, and affections of the breast. In fine being presented with Treacle [i.e. Theriac] and pretious oyles, she went into the sacristy, where they opened all the presses, where they keep the Plate and reliques of the Church, with the great candlesticks, and vases given them by the deceased

Cardinall Lodowick Lodowiso the founder of the Church. She honoured particularly the blood of St. Esuperantia a Virgin and Martyr, which, after a thousand and three hundred years, is as liquid as if newly shed. Then going into the Church she heard Mass, and at her departure, gave testimonies to the Fathers of her great satisfaction and content.[77]

The accounts of Christina's visits to the *Collegio Romano* resonate with the image of the College as both a *theatrum mundi* and repository of universal knowledge suggested in Kircher's letters to the Queen before her departure for Rome. Although Christina's case is conspicuous for its dramatic charge, the pattern is far from unique, and there are innumerable other examples of monarchs and aristocrats, Catholic and Protestant, being enticed into metropolitan Jesuit colleges throughout Europe rather as Chinese *literati* were initially enticed into Matteo Ricci's house, by the promise of arcane knowledge, curiosities, maps and mathematical instruments.[78] A manuscript chronicle of the *Collegio Romano* describes a large number of such ceremonial visits.[79]

The transformation of the *Collegio Romano* into a theatre of curiosity had numerous precedents throughout the century. During the festivities to mark the canonization of Saints Ignatius and Francis Xavier in 1622, the College was transformed into ancient Rome, to echo the solemn ecclesiastical rites with 'erudite allusion and ancient Apotheosis'.[80] The *Atrium* and entrances of the Collegio were decorated to represent the Roman forum, while the Aula Magna became the Campus Martius, scene of the apotheoses of the Roman Emperors. Two large globes, at the main entrance, represented the old and new worlds, divided into thirty-four Jesuit provinces, with their colleges and houses marked on *tesserae*. Plays representing important events in the lives of Xavier and Ignatius were staged by the Parthenian academicians of the College and the members of the Roman seminary. The mathematics professor Orazio Grassi staged an opera in the transformed *Aula Magna* for the occasion, the *Apotheosis of Saints Ignatius and Xavier*, set to music by Kapsberger, with elaborate stage machinery.[81] Grassi also provided geographical demonstrations (*ragioni Geografiche*) that St Francis Xavier was responsible for a larger amount of territory than any apostolic preacher, much as he had provided public mathematical demonstrations for the supra-lunary location of the comets of 1618.[82]

By the time of Christina's visit in 1656, as Gualdo Priorato's account reveals, the College could boast two further sites of courtly display: the College pharmacy and the *Museum Kircherianum*. Building of the college pharmacy commenced on 5 July 1627, shortly after the commencement of work on Orazio Grassi's church of St Ignatius,[83] but the existence of *Spetiali* is evident from the Catalogues of the College back to 1598 and beyond.[84] In 1609 the category becomes 'Aromatarius',[85] before the title of *pharmocopolae* was bestowed upon Francesco Vagioli and Francesco Savelli in the *Catalogi* of 1624–25.[86] The walls of the pharmacy were decorated with a series of (surviving) frescoed lunettes by

Figure 6.10 The spagyrical furnace of the Collegio Romano from Kircher, *Mundus Subterraneus* (1665 edn.) vol. 2, p. 392

Andrea Sarti and Emilio Savonanzi in 1629, depicting Galen, Hippocrates, Mesue, Andromachus and other authorities in medicine, botany and pharmacy. A painted panel at the centre of the ceiling depicted the patron saints of medicine, Cosmas and Damien, in the company of Saints Francis Xavier and Ignatius and the Madonna and child, a grouping lent legitimacy by the coincidence that the bull of foundation of the Jesuit order (27th September 1540) fell on the feast day of the medical saints.[87] A manuscript ground-floor plan of the Collegio[88] apparently dating from the mid-seventeenth century depicts the pharmacy as occupying at least five rooms. As well as producing the balsam of life, theriac and various other precious substances that could be distributed to potential patrons of the order,[89] the numerous books of secrets that survive suggest that the pharmacy was used for alchemical operations as well as the production of candle-wax and even substances for combatting 'carnosità', or carnality, clearly a dangerous enemy to the Jesuit way of life.[90] As a site of display, the pharmacy played a part in a visit made by Urban VIII to the Collegio Romano as early as 1631.[91] On Vincenzo Carafa's first visit to the college as General in 1646 he was shown a large parchment bearing the recipes of the theriac and other medicines produced in the Jesuit pharmacy, which had been lavishly frescoed with portraits of ancient medical writers in 1629. The enormous spagyrical furnace shown to Christina was depicted graphically [**See Figure 6.10**] in Kircher's *Mundus Subterraneus*,[92] where it bolstered Kircher's attack on alchemical charlatans.

Miracle-machines

As Clavius had gathered mathematical disciples, Kircher recruited technicians of the curious, both from the laity and from the Jesuit order. After brief periods of apprenticeship with Kircher in Rome, two of his most prolific Jesuit disciples Kaspar Schott and Francesco Lana Terzi redistributed Kircherian artificial magic to the provinces.[93]

For Kircher, as for other early modern natural magicians, art is nature's ape. Or, to turn the metaphor on its head, nature is God's work of art, and thus the natural magician bears a relationship to his technical productions that is analogous to the relationship God bears to the whole of Creation.[94] Kircherian machines can thus be compared to miniature, artificial universes, bearing encrypted messages from a playful creator. The perpetual-motion machines and emblematic clocks displayed in Kircher's museum display the microcosmic character of Kircherian machines most evidently, sometimes even bearing zodiacal and planetary symbols to make the analogy unmissable. The 'user intervention' required by machines such as Kircher's sunflower clock [See **Figure 6.7**], that so frustrated Nicholas Claude Fabri de Peiresc when the instrument was demonstrated to him in Aix-en-Provence in 1633 was not a failing in Kircher's instrument, but rather a rhetorical demonstration of the limits of the analogy between the human magus and his omnipotent forbear.[95] Other machines, as we have argued, were miniature moral universes, the catoptric chest [**See Figure 6.8**] being a striking example.

We have argued that Kircherian machines were jokes that occupied a ludic space between the demonic and the supernatural realms. What, then, are we to make of the following machine listed by De Sepibus: 'a large crystalline globe full of water representing the resurrection of the Saviour in the midst of the waters'?[96] How could Kircher dare to make a joke of the central mystery of Christianity? How could he place the resurrected Christ in a glass sphere, alongside genies, water-vomiting snakes and pagan Goddesses? Surely to place the Resurrection in this mechanical context was tantamount to reducing it to a secret combination of natural causes and denying its miraculous status?

The problem is even more striking when we look at Kircher's first published book, the *Ars Magnesia*, published in Würzburg when he was twenty-nine years old. Launching into a description of the various machines that can be constructed with the aid of the magnet, Kircher describes a device 'to exhibit Christ walking on water, and bringing help to Peter who is gradually sinking, by a magnetic trick'. 'Carve statues of Christ and Peter from the lightest material possible', Kircher's description begins,

> When a strong magnet is placed in Peter's breast, and with Christ's outstretched hands or any part of his toga turned toward Peter made of fine steel, you will have everything required to exhibit the story. With their lower limbs well propped-up on corks so that they don't totter about above the water, the statues are placed in a basin filled up to the top with water, and the iron hands of Christ soon feel the magnetic power diffused from the breast of Peter. The magnet drags the statue of Christ to it with equal motions, and insinuates itself into Peter's embrace. The artifice will be greater if the statue of Christ is flexible in its middle, for in this way it will bend itself, to the great admiration and piety of the spectators.[97]

Despite Kircher's claims, the steel-handed bending Jesus floating on a cork and drawn to a magnetic Peter does not strike us as a particularly pious artifice. Indeed, his demonstration almost seems to carry the heretical suggestion that what appeared to be miraculous was merely carried out through a clever piece of natural magic. But that can hardly be the real thrust of Kircher's demonstration. Rather, the clue to Kircher's intention can probably best be gleaned from his own definition of natural magic: feats of natural magic can resemble miracles to those who are ignorant of their true causes. Again, as in the case of the perpetual motion-machines, the analogy is limited. Real miracles by definition defy demonstration and replication. By producing wonder, fear and amusement, however, Kircher's magical machines rehearsed his visitors' reactions to the miraculous and the demonic, and trained them in civility and piety.

The *Collegio Romano* occupied a special place in the network of Jesuit colleges, and in Kircher it had a resident expert on all types of arcana. Kircher's reputed knowledge of hieroglyphics, his mechanical expertise and learned publications on an immense variety of topics encouraged numerous people to write to him 'as to the oracle' to ask for his solutions to 'difficult questions from all of the sciences'.[98] The *Musaeum Kircherianum* further legitimated the presence of the 'curious' visitor in the Collegio Romano.[99] Lorenzo Magalotti remarked in a letter to Francesco Maria de' Medici that 'a Jesuit mathematician is a rarity worthy of being put into a museum',[100] and indeed a museum was one space in which Jesuit instrumental expertise might not conflict with the institutional position of the order on controversial questions of natural philosophy.

Notes

1 An abridged version of this chapter appeared as 'Between the Demonic and the Miraculous: Athanasius Kircher and the Baroque Culture of Machines' in *The Great Art of Knowing: The Baroque Encyclopedia of Athanasius Kircher*, edited by Daniel Stolzenberg, Stanford: Stanford University Libraries, 2001. I am grateful to Stanford University Libraries for permission to include here.
2 For discussions of Kircher's machines, see particularly Thomas L. Hankins and Robert J. Silverman, *Instruments and the Imagination*, Princeton, N.J.: Princeton University Press, 1995, especially chapters 2–4, Paula Findlen, *Scientific Spectacle in Baroque Rome: Athanasius Kircher and the Roman College Museum*, in Feingold, ed., *Jesuit Science and the Republic of Letters*, pp. 225–284, Joscelyn Godwin, *Athanasius Kircher: A Renaissance man and the quest for lost knowledge*. London: Thames and Hudson; 1979, Eugenio Lo Sardo ed. *Icononismi e Mirabilia da Athanasius Kircher*. Rome: Edizioni dell'Elefante; 1999 and Adalgisa Lugli, *Naturalia et Mirabilia. Il*

collezionismo enciclopedico nelle Wunderkammern d'Europa. Milan; 1983. On Kircher's musical machines, see Jan Jaap Haspels, *Automatic musical instruments, their mechanics and their music, 1580–1820*, Niroth: Muiziekdruk C.V. Koedijk, 1987. On Kircher's magnetic devices in particular see Martha Baldwin. *Magnetism and the anti-Copernican polemic*. Journal for the History of Astronomy. 1985; 16:155–174, Jim Bennett, *Cosmology and the Magnetical Philosophy, 1640–1680*. Journal for the History of Astronomy. 1981; 12: 165–177, Silvio Bedini, *Seventeenth Century Magnetic Timepieces*. Physis. 1969; 11: 37–78. On optical and catoptric devices, see Jurgis Baltrusaitis, *Anamorphoses ou magie artificielle des effets merveilleux*. Paris: Olivier Perrin; 1969, and idem., *Le miroir*. Paris: Le Seuil 1978.

3   Filippo Buonanni, *Musaeum Kircherianum sive Musaeum a P. Athanasio Kirchero In Collegio Romano Societatis Iesu Jam pridem Incoeptum Nuper restitutum, auctum, descriptum, & Iconibus illustratum*. Rome: Typis Georgii Plachii; 1709, pp. 302–315:

4   See, for example, the classic study by Krysztof Pomian, *Collectionneurs, amateurs et curieux: Paris, Venise, XVIe–XVIIIe siècle*, Paris: Gallimard, 1987, especially chapter 1.

5   On the relationship between courtly models of behaviour and early modern science, see in particular Mario Biagioli, *Galileo Courtier: The practice of science in the culture of absolutism*. Chicago: University of Chicago Press; 1993. For *sprezzatura* see pp. 51–52.

6   On early-modern scientific 'jokes', cf Paula Findlen, *Jokes of Nature and Jokes of Knowledge: The Playfulness of Scientific Discourse in Early Modern Europe*. Renaissance Quarterly. 1990; 43:292–331.

7   Athanasius Kircher to Johann Georg Anckel (or J.M. Hirt), Rome, 16 July 1659, Herzog August Bibliothek Wolfenbüttel, Bibliotheksarchiv N° 376, quoted in John Fletcher (ed.), *Athanasius Kircher und seine Beziehungen zum gelehrten Europa seiner Zeit*. Wiesbaden: Harrassowitz; 1988, p. 105. The manuscript letters of Kircher conserved in the Herzog August Bibliothek have been made available online http://diglib.hab.de/edoc/ed000005/transcription.htm

8   See Claudio Costantini, *Baliani e i Gesuiti*. Florence: Giunti Barbèra; 1969, Ugo Baldini, *Uniformitas et Soliditas Doctrinae*: Le censure *librorum* e *opinionum*. in idem., *Legem impone subactis. Studi su filosofie e scienze dei gesuiti in Italia, 1540–1632*. Rome: Bulzoni; 1992; pp. 75–119, Michael John Gorman, *A Matter of Faith? Christoph Scheiner, Jesuit censorship and the Trial of Galileo*. Perspectives on Science. 1996; 4(3):283–320, idem., *Jesuit explorations of the Torricellian space: carp-bladders and sulphurous fumes*. Mélanges de L'Ecole Française De Rome. Italie Et Méditerranée. 1994; tome 106(fasc. 2):pp. 7–32 and Marcus Hellyer, *'Because the authority of my superiors commands': Censorship, physics and the German Jesuits*. Early Modern Science and Medicine. 1996; 1(3):319–354.

9   De Sepibus, *op. cit.* (note 6)

10 See Francesco Gizzio to Athanasius Kircher, Naples; 27 October 1668, Rome, Archives of the Pontifical Gregorian University (hereafter APUG), 564 f. 156r and, for De Sepibus' trip to Naples, Gizzio to Kircher, Naples, 28 February 1670 (APUG 559, f. 85r). For Kircher's fear that De Sepibus had died in 1674, see Gizzio to Kircher, Naples, 14 July 1674, APUG 565, f. 213rv. The manuscript correspondence of Kircher conserved in the Archives of the Pontifical Gregorian University (APUG 555–568) is now available for consultation on the Internet. See *The Athanasius Kircher Correspondence Project*, ed. Michael John Gorman and Nick Wilding, http://kircher.stanford.edu/

11 De Sepibus, *op. cit.*, pp. 2–3.

12 De Sepibus, *op. cit.*, p. 60

13 *De oraculo Delphico*, APUG 566, f. 236r, accessible via the *Athanasius Kircher Correspondence Project*, cit.

14 Athanasius Kircher, *Phonurgia nova*, Campidonae: Dreherr; 1673, p. 112.

15 On the official foundation of the *Musaeum Kircherianum* and the Donnini bequest, see Findlen, op. cit., R. Garrucci, *Origini e vicende del Museo Kircheriano dal 1651 al 1773*. La Civiltà Cattolica. 1879; Serie X Vol. XII(Quaderno 703): 727–739, Maristella Casciato, Maria Grazia Ianniello and Maria Vitale, eds., *Enciclopedismo in Roma barocca: Athanasius Kircher e il museo del Collegio Romano tra Wunderkammer e museo scientifico*. Venice: Marsilio; 1986, and R. G. Villoslada, *Storia del Collegio Romano dal suo inizio all soppressione della Compagnia di Gesù*. Rome; 1954, as well as Buonanni, *Musaeum Kircherianum*, pp. 1–3, as well as the manuscripts documenting the museum's history in APUG 35.

16 The passage, from John Evelyn's diary, is quoted in Barbara Shapiro, *John Wilkins, 1614–1672; an intellectual biography*, Berkeley: University of California Press, 1969, p. 120. See Jack Peter Zetterberg, *'Mathematical Magick' in England: 1550–1650*, Dissertation, University of Wisconsin-Madison; 1976, pp. 212 ff.

17 Archivum Romanum Societatis Iesu (hereafter ARSI) Rom. 150, I. 36r, cited in Ugo Baldini and Pier Daniele Napolitani, eds., *Christoph Clavius: Corrispondenza*, Pisa: Università di Pisa, Dipartimento di Matematica, Sezione di Didattica e Storia della Matematica; 1992, Vol. III.2, pp. 54–5, note 2.

18 Kircher, *Magnes, sive de arte magnetica opus tripartitum*, Romae: Ex Typographia Ludovici Grignani, 1641, Lib. II, Cap. II, p. 431, '[P]artim è literis ab ijs, qui iter in Indias susceperant, vel oretenus ab ijs, qui inde peregrini Romam advenerant; partim ex literarum Mathematicarum è diversis orbis terrae partibus ad *Clavium, Grimbergerum, aliosque Romanos Societatis IESU Mathematicos praedecessores meos datarum, quod penes me est, Archivio*; multas sanè, circa declinationes Magneticas haud spernendas observationes collegi'

19 Kaspar Schott, *Mechanica Hydraulica-Pneumatica*, Würzburg: Pigrin, 1657, p. 339

20 Schott, *Mechanica Hydraulica-Pneumatica*, cit., p. 300

21 See Schott, *Magia Universalis*, Pars III, Würzburg: J. G. Schönwetter, 1658, pp. 219–228 '*Machina II: Glossocomum nostrum*', discussed in Gorman, *Mathematics and Modesty*, cit., and also Schott, *Magia Universalis*, Pars I, Würzburg: J. G. Schönwetter, 1657, pp. 26–27.

22 De Sepibus, *op.cit.*, p. 13 (on Clavius's experiment) and p. 17 (on Grienberger's wooden astrolabe)

23 Jakob Johann Wenceslaus Dobrzensky de Nigro Ponte, *Nova, et amaenior de admirando fontium genio (ex abditis naturae claustris, in orbis lucem emanante) philosophia*. Ferrara: Alphonsum, & Io. Baptistam de Marestis; 1659, p. 46. On Dobrzensky de Nigro Ponte see R.J.W. Evans, *The Making of the Habsburg Monarchy: An Interpretation*. Oxford: Clarendon Press; 1979, pp. 316, 337, 339–40, 356, 369–70, 390

24 Schott, *Magia Universalis*, Pars I, Würzburg: J. G. Schönwetter, 1657, p. 42, cf. Kircher, *Oedipus Aegyptiacus hoc est universalis hieroglyphicae veterum doctrinae temporum iniuria abolitae instauratio*, Rome: Vitalis Mascardi; 1652–1654, Tom. 3, Syntag. 17, Cap. 1, p. 488

25 Schott, *Mechanica Hydraulico-Pneumatica*, cit., Pars II, Classis I, p. 255 and Kircher, *Oedipus Aegyptiacus*, cit., Tom. II^2, Classis VIII, Cap. III, Pragmatia I, p. 332.

26 Kircher, *Oedipus Aegyptiacus*, cit., Tom. II^2, Classis VIII, Cap. III, Prag. V, pp. 337–8, '*Ara deorum*'.

27 Athanasius Kircher, *Vita admodum Reverendi P. A. Kircher*, Augsburg: S. Utzschneider, 1684, pp. 30–3.

28 See Zetterberg, '*Mathematical Magick*', cit., p. 32

29 For a rich discussion of the contrast between learned and popular magic during this period see R.J.W. Evans, *The Making of the Habsburg Monarchy*, cit., Chapters 9–12.

30 The literature on Jesuit theatre is enormous, and a survey would take us beyond the scope of this article, but a classic study is Jean-Marie Valentin, *Theatre des Jésuites dans les pays de langue allemande (1554–1680) : salut des ames et ordre des cités*, Bern, Las Vegas : P. Lang, 1978 (3 vols.).

31 See Schott, *Mechanica Hydraulico-Pneumatica*, cit., p. 323 and *Iconismus XXIX*. On Kircher's time in Malta see Alberto Bartòla, *Alessandro VII e Athanasius Kircher S.I. Ricerche e appunti sulla loro corrispondenza erudita e sulla storia di alcuni codici chigiani*. Miscellanea Bibliothecae Apostolicae Vaticanae. 1989; III:7–105.

32 See the letter from Kircher to General G.P. Oliva, Rome, 5 May 1672, published in Garrucci, *Origini e vicende del Museo*, cit., also Buonanni, *Musaeum Kircherianum*, pp. 1–3, Godwin, *Athanasius Kircher*, cit., pp. 14–15.

33 See Schott to Kircher, n.p., n.d. [Würzburg, circa 1656?], APUG 567, f. 52r: 'Tutti li Padri di questa nostra Provincia stimano e amano Vostra Reverenza principalmente il nostro R. P. Provinciale, il quale vorebbe che io discrivessi e stampassi la Galeria di Vostra Reverenza', also Schott to Kircher, Würzburg, 21 October 1656: 'O[ro] se V.a

R.a volesse e potesse impiegare per mio e suo servitio, uno o due giorni, e farmi un'abbozzo, e breve descrittione della sua Galeria, significandomi brevemente le cose più riguardevole, massimamente le nuove datte doppo la mia partenza, delineandole ruditer e obiter. Vorrei descrivere a lungo ogni cosa, e farle stampare, con bellissime figure di rame, prima separatamente, e doppo nella mia *Magia Universalis Naturae et Artis*'. Apparently Valentin Stansel had been charged with composing the description for Schott, but Stansel, soon to depart for Brazil, did not send it, despite Schott's repeated pleas (e.g. 'Prego Vostra Reverenza quanto posso, e per l'amore che mi porta, e propter con humania studia, che m'impetri dal R.P. Assistente, che mi mandi la Galeria di V.a R.a descritta dal P. Stansel, o almeno le cose più principali', Schott to Kircher, Würzburg, 16 June 1657, APUG 567, f. 45r)

34 On Schott's career, see ARSI, *Lamalle: Schott*. On Kircher's arrival in Avignon, see ARSI, Lugd. 14, f. 239v, and the appendix to the Catalogue.
35 See ARSI Rom. 81 ff.64v, 88v, 114v: (Catalogue of *Collegio Romano*, 1652–4): 'P. Gaspar Sciot, socius P. Athanasii', 'P. Athanasius Chircher, scribit imprimenda'.
36 Kircher, *Magnes, sive de magnetica arte libri tres*, Rome: V. Mascardi, 1654³, sig. †† rv.
37 Schott, *Mechanica Hydraulico-Pneumatica*, cit., pp. 1–3, *Praeloquium ad Lectorem*
38 Schott, *op. cit.*, p. 3
39 Schott, *op. cit.*, p. 5
40 R.J.W. Evans, *The Making of the Habsburg Monarchy*, cit., especially ch. 9–12.
41 Schott, *Mechanica Hydraulico-Pneumatica*, cit., p. 219.
42 Daniel Schwenter and Georg Philipp Harsdörffer, *Deliciae Physico-Mathematicae, oder Mathematische und philosophische Erquickstunden*, herausgegeben und eingeleitet von Jörg Jochen Berns, Frankfurt a. M.: Keip, 1991.
43 Schott, *Mechanica Hydraulico-Pneumatica*, cit., pp. 311–12
44 ibid.
45 Schott, *Mechanica Hydraulico-Pneumatica*, cit., pp. 63–64 (on explosions) and Gioseffo Petrucci, *Prodromo apologetico alli studi Chircheriani*, Amsterdam: Janssonio-Waesbergi; 1677, p. 128 (on Settala's burning-mirrors).
46 Robert Darnton, *The Great Cat Massacre and Other Episodes in French Cultural History*, New York: Basic Books, 1984, pp. 75–104.
47 Thomas L. Hankins and Robert J. Silverman, *Instruments and the Imagination*, Princeton, N.J.: Princeton University Press, 1995, especially chapters 2–4. On the cat piano, designed to entertain a melancholy prince, see Kircher, *Musurgia universalis*, Rome: Francesco Corbelleti; 1650, Tom. I, Lib. VI, Pars IV, Caput I, p. 519 and Schott, *Magia Universalis*, cit., Pars II, pp. 372–73. Schott provides an illustration.
48 Schott, *Magia Universalis*, cit., Pars I, p. 302
49 The classic study of early modern civility remains Norbert Elias, *The civilizing process*, trans. Edmund Jephcott, Oxford: Blackwell, 1994. A contrasting view, arguing that European civility had its origins in monastic *disciplina* rather than court culture

is advanced in Dilwyn Knox, *Disciplina: The Monastic and clerical origins of European Civility* in John Monfasani and Ronald G. Musto, eds. *Renaissance society and culture: Essays in honour of Eugene F. Rice, Jr.* New York: Italica Press; 1991; pp. 107-135. Kircher would seem to demonstrate that the lines between courtly and clerical traditions are perhaps not so clear-cut as both Knox and Elias suppose. On civility see also Jacques Revel, *The Uses of Civility*, trans. Arthur Goldhammer, in Roger Chartier, ed., *A History of Private Life*, Cambridge: Harvard University Press, 1989, Vol. 3, pp. 167-205

50 René Descartes, *Les Passions de l'âme*, ed. Geneviève Rodis-Lewis, Paris: J. Vrin, 1966.

51 Marin Mersenne, *Harmonie Universelle*, Paris: S. Cramoisy, 1636 (facsimile repr. Paris: CNRS, 1963), sig. A iiij recto (on the *Archiviole*), Justus Lipsius, *De Militia Romana*, Antwerp: Plantin-Moretus, 1598), Thomas Hobbes, *Leviathan*, London: A. Crooke, 1651. For the court of Louis XIV at Versailles as a 'machine', see Apostolidès, *Le roi-machine: Spectacle et politique au temps de Louis XIV*, Paris: Editions de Minuit, 1981. On automata and political power see Otto Mayr, *Authority, Liberty and Automatic machinery in Early Modern Europe*, Baltimore: Johns Hopkins University Press, 1986. For a more dated, though entertaining, presentation of the political function of automata, see Lewis Mumford, *Authoritarian and Democratic Technics*. Technology and Culture. 1964; 5(1):1-8. On automata more generally see Derek J. de Solla Price, *Automata and the Origins of Mechanism and Mechanical Philosophy*. Technology and Culture. 1964; 5(1):9-23, who recounts the (probably apocryphal) story that Descartes constructed a 'beautiful blonde automaton named Francine, but she was discovered in her packing case on board ship and dumped over the side by the captain in his horror of apparent witchcraft', and Silvio Bedini, *The Role of Automata in the history of technology*, Technology and Culture. 1964; 5(1): 24-42.

52 Peter Dear, *A Mechanical Microcosm: Bodily Passions, Good Manners, and Cartesian Mechanism*. in Christopher Lawrence and Steven Shapin, eds. *Science Incarnate: Historical embodiments of Natural knowledge*. Chicago and London: University of Chicago Press; 1998; pp. 51-82.

53 On the context of Kircher's *Oedipus Aegyptiacus*, see especially Giovanni Cipriani, *Gli obelischi egizi: politica e cultura nella Roma barocca*. Florence: Olschki; 1993, pp. 77-167. On the question of the Corpus Hermeticum see Frances Yates, *Giordano Bruno and the Hermetic tradition*, London: Routledge & K. Paul, 1964, and Anthony Grafton, *Protestant versus Prophet: Isaac Casaubon on Hermes Trismegistus*, and idem., *The Strange Deaths of Hermes and the Sibyls*, both in idem., *Defenders of the Text: The Traditions of Scholarship in an Age of Science, 1450-1800*, on pp. 145-161 and 162-177 respectively.

54 Kircher, *Oedipus Aegyptiacus*, cit., Sig. d recto

55 For Kircher's dream of being elected Pope, see Kaspar Schott, *Physica Curiosa, sive Mirabilia Naturae et Artis*, Würzburg: Jobus Hertz; 1667 (2nd edition), Liber III

(Mirabilia Hominum), Caput XXV, pp. 455–6. Kircher's vision of the invasion of the Jesuit college in Würzburg is described in idem., Liber II (Mirabilia Spectrorum), Caput V, p. 210 and also in Kircher's posthumous autobiography, *Vita admodum Reverendi P. A. Kircher*, Augsburg: S. Utzschneider, 1684, pp. 38–41. On the use of recorded dreams as a historical source, see Peter Burke, *The Cultural History of Dreams*, in idem., *Varieties of Cultural History*, Ithica: Cornell University Press; 1997, pp. 23–42.

56 Schott, *Magia Universalis*, cit., Pars I, Prolegomena, especially pp. 8–18, cf Kircher, *Oedipus Aegyptiacus*, Tom. 2, class. 2, cap. 1 and Kircher,*Obeliscus Pamphilius*, Rome: Ludovico Grignani; 1650, bk. 1, ch. 1.

57 Schott, *Magia Universalis*, loc. cit.

58 Schott, *Magia Universalis*, Pars I, Cap. VI (p. 22 ff). In a letter to Kircher sent from Würzburg on 1 April 1656, Schott wrote 'Gaudeo vehementer, Reginam Suedice [sic] tandem visitare Museum R.ae V.ae' (APUG 561, f. 40r)

59 On the relationship between demonology and natural philosophy in seventeenth century Europe, see Stuart Clark, *Thinking with Demons: The Idea of Witchcraft in Early Modern Europe*, Oxford: Clarendon Press, 1997, especially pp. 149–311, and idem., *The rational witchfinder: conscience, demonological naturalism and popular superstitions*. in Stephen Pumfrey, Paolo Rossi and Maurice Slawinski, (eds.). *Science, Culture and Popular belief in Renaissance Europe*. Manchester and New York: Manchester University Press; 1991; pp. 222–248. On Jesuit natural magic, see Mark A. Waddell, *Jesuit Science and the End of Nature's Secrets*, Farnham: Ashgate, 2015.

60 Schott, *Magia Universalis*, cit., Pars I, Caput X, p. 39 (on demons and the vacuum) and idem., *Mechanica Hydraulico-Pneumatica*, cit., introduction, on the four fundamental principles of hydraulic machines.

61 Schott, *Mechanica Hydraulico-Pneumatica*, cit., pp. 307–8

62 Martin del Rio, *Disquisitionum Magicarum Libri Sex*, Louvain, 1599 (edition used Mainz: Henningii; 1624). Liber I, *De magia in genere, & de naturali ac artificiosa in specie*. On this work and witch-trials see Petra Nagel, *Die Bedeutung der 'Disquisitionum magicarum libri sex' von Martin Delrio für das Verfahren in Hexenprozessen*, Frankfurt am Main: Peter Lang, 1995. On Del Rio's life, see anon., [H. Langeveltius?], *M. A. del Rii. . . . Vita brevi commentariolo expressa*. Antwerp; 1609. On Del Rio's critique of Stoic drama see Roland Mayer, *Personata Stoa: Neostoicism and Senecan Tragedy*. Journal of the Warburg and Courtauld Institutes. 1994; 57:151–174.

63 On wonders and the preternatural, see especially Lorraine Daston and Katherine Park, *Wonders and the order of nature, 1150–1750*, New York: Zone Books, 1998.

64 Del Rio, *Disquisitionum Magicarum*, ed. cit., pp. 49–50.

65 On Peirera see Paul Richard Blum, "Benedictus Pererius: Renaissance Culture at the Origins of Jesuit Science", in *Science & Education* 15 (2006) 279-304.

66 Benito Pereira, *Adversus fallaces & superstitiosas artes, id est, De magia, de observatione somniorum, et de divinatione astrologica. Libri tres,* first published Ingolstadt 1591, edition used Coloniae Agrippinae, apud Ioannem Gymnicum, 1598, pp. 41, 67–8, 91.

67 Kircher, *Magnes,* 1654³, cit., Liber II, Pars 4, p. 238.

68 Schott, *Magia Universalis,* cit., Sig. ††††† recto: *Prooemium totius operis*

69 Christoph Clavius, *Discursus cuiusdam amicissimi Societatis Iesu de modo et via qua Societas ad maiorem Dei honorem et animarum profectum augere hominum de se opinionem, omnemque haereticorum in literis aestimationem, qua illi multum nituntur, convellere brevissime et facillime possit,* (c. 1594), ARSI Stud. 3, ff. 485–487 (Clavius autograph), published in *Monumenta Paedagogica Societatis Iesu, Nova editio penitus retractata,* ed. Ladislaus Lukács, Rome, Institutum Historicum Societatis Iesu, 1965–?, VII, pp. 119–122

70 Athanasius Kircher to Queen Christina of Sweden, Rome, 11 November 1651, APUG 561 ff. 50r–v (autograph draft), on 50r.

71 There is a vast bibliography on Christina, but see especially Susanna Åkerman, *Queen Christina of Sweden and her circle: The transformation of a seventeenth-century philosophical libertine,* Leiden: Brill, 1991, idem., *Cristina di Svezia: scienza ed alchimia nella Roma barocca.* Bari: Dedalo, 1990, Jeanne Bignami Odier and Anna Maria Partini, 'Cristina di Svezia e le scienze occulte', *Physis* 1983, A. 25(fasc. 2): 251–278. Georgina Masson, *Queen Christina* London: Secker & Warburg, 1968, though a popularised presentation, remains useful as an overview.

72 Kircher had arranged for a copy of his *Musurgia Universalis* to be sent to Christina in 1650. See Louys Elzevier to Athanasius Kircher, Amsterdam; 14 November 1650, APUG 568, f. 238 r–v

73 Daniello Bartoli, *Dell'huomo di lettere difeso & emendato,* Bologna: Heredi di E. Dozza, 1646.

74 See the undated letter to Kircher in APUG 556 f. 173r, in a more legibile Italian translation on f. 174r: 'Spero che hormai havremo un occasione più libera, e fedele di corrispondenza mutua, e per poter communicarmi gli più sicuramente'. Kircher eventually dedicated his 1656 *Itinerarium Exstaticum* to Christina, who mentions his plan to do so in the same letter: 'Desiderei ancor sapere, se me giudichi ancor degna a dedicarmi la sua incomparibile opera'.

75 See Carlos Ziller Camenietzki, *L'Extase interplanetaire d'Athanasius Kircher: Philosophie, Cosmologie et discipline dans la Compagnie de Jésus au XVIIe siècle,* Nuncius, 1995, X(1): 3–32. For a detailed discussion of the heliocentric controversy in Kircher's *Itinerarium Exstaticum* see especially Siebert 2006.

76 APUG 142 ff.81r–83r

77 Galeazzo Gualdo Priorato, *History of her majesty Christina Alessandra, queen of Swedland.* London: Printed for T.W., 1658, pp. 428–431.

78 See Jonathan D. Spence, *The memory palace of Matteo Ricci*, London: Faber and Faber, 1985, Pasquale M. D'Elia, *Galileo in China. Relations through the Roman College between Galileo and the Jesuit Scientist-Missionaries (1610-1640)*. Cambridge, MA: Harvard University Press, 1960, Jacques Gernet, *China and the Christian impact: a conflict of cultures*, trans. Janet Lloyd, Cambridge: Cambridge University Press, 1985, p. 22.

79 [Anon.], *Origine del Collegio Romano e suoi progressi*, APUG: 142. This manuscript forms the basis of the descriptions of ceremonial receptions given in the Collegio Romano provided in R. Garcia Villoslada, *Storia del Collegio Romano dal suo inizio all soppressione della Compagnia di Gesù*. Rome: Typis Pontificiae Universitatis Gregorianae, 1954, pp. 263-296.

80 Famiano Strada, *Saggio delle Feste che si apparecchiano nel Collegio Romano in honore de' Santi Ignatio et Francesco da N. S. Gregorio XV Canonizati All'Illustrissimo, & Eccellentissimo Signor Principe di Venosa*. Roma: Appresso Alessandro Zannetti; 1622, sig. A2 recto. On theatrical productions in the Collegio Romano during this time, see Irene Mamczarz, *La trattatistica dei Gesuiti e la pratica teatrale al Collegio Romano: Maciej Sarbiewski, Jean Dubreuil e Andrea Pozzo*. in M. Chiabò and F. Doglio, eds., *I Gesuiti e i Primordi del Teatro Barocco in Europa*. Roma: Torre d'Orfeo; 1995: 349-387 and Jean-Yves Boriaud, *La Poésie et le Théâtre latins au Collegio Romano d'après les manuscrits du Fondo Gesuitico de la Bibliothèque Nationale Vittorio Emanuele II*. Mélanges de l'École Française de Rome, Italie et Mediterranée. 1990; 102(1): 77-96.

81 See Emilio Sala and Federico Marincola, *La Musica nei Drammi Gesuitici: Il Caso dell'Apotheosis sive Consecratio Sanctorum Ignatii et Franciscii Xaverii (1622)*, in in M. Chiabò and F. Doglio, eds., *I Gesuiti e i Primordi del Teatro Barocco in Europa*, cit., pp. 389-439. For a rich contemporary Italian discussion of theatrical machinery see Nicola Sabbattini, *Pratica di fabricar scene, e machine ne' teatri* Ravenna: Per Pietro de' Paoli, e Gio. Battista Giouanelli Stampatori Camerali; 1638.

82 Strada, op. cit., p. 9, and, for the cometary presentation, [Orazio Grassi], *De tribus cometis anni MDCXVIII Disputatio astronomica publice habita in Collegio Romano Societatis Iesu ab uno ex Patribus eiusdem Societatis*. Romae: ex typographia Iacobi Mascardi; 1619, OG VI pp. 21-35, translated in Stillman Drake and C.D. O'Malley, *The Controversy on the Comets of 1618*, Philadelphia: University of Pennsylvania Press; 1960, pp. 3-19.

83 APUG 142 ff.1r-8v: *Nota delle spese fatte nella Fabrica del Collegio Romano* f. 4r :' Dal 1627 fino a tutto il 1632 furono spesi [scudi] sedicimila dugento novanta due per la fabrica della spezieria, cominciata a di 5 Luglio 1627'

84 ARSI Rom. 79 f.11v and BNR FG 1526 f.35r

85 ARSI Rom. 110 f.51v

86 Idem. f.121r

87 See *Imago Primi Saeculi Societatis Iesu A Provincia Flandro-Belgica eiusdem Societatis Repraesentata*. Antwerp: Balthasar Moretus; 1640, p. 12.
88 APUG 134, XVI, *Abbozzo iconografico del Collegio Romano*.
89 See e.g. Athanasius Kircher to Duke August of Brunswick-Lüneburg, Rome, 25 July 25, HAB BA n. 366, and the other medical gifts discussed in John Fletcher *Athanasius Kircher and Duke August of Brunswick-Lüneburg. A chronicle of friendship* in John Fletcher, John, ed., *Athanasius Kircher und seine Beziehungen zum gelehrten Europa seiner Zeit*. Wiesbaden: Harrassowitz; 1988: pp. 99–139.
90 Some manuscript books of secrets originating in the Collegio Romano are listed in *Il Fiore dell'arte di sanare*, Rome: Edizione Paracelso, 1992, pp. 565–570. The Fondo Curia of APUG also contains numerous manuscript books of secrets, including APUG: FC 2087, APUG: FC 1381, APUG: FC 562, APUG: FC 1860/2, APUG: FC 2200. The 'ceroto per la carnosità', accompanied by a crude drawing of a phallus, is described in APUG FC 2193, f. [40v]. On candlewax see APUG 134, XIV. For a study of the contents of another Jesuit pharmacy see Carmen Ravanelli Guidotti, *La Farmacia dei Gesuiti di Novellara*, Faenza: Edit Faenza, 1994. On the tradition of books of secrets during this period, see William Eamon, *Science and the secrets of nature: Books of secrets in medieval and early modern culture*. Princeton: Princeton University Press; 1994.
91 APUG 142 f. 71r, Villoslada, cit., p. 275. In 1646 Vincenzo Carafa visited the college and was shown a highly decorated parchment containing recipes for medicines produced in the pharmacy (BNR FG 1382). For the Rospigliosi family's visit to the pharmacy in 1668, see Villoslada, cit., p. 277.
92 Athanasius Kircher, *Mundus Subterraneus*, Amsterdam: Janssonius, 1665, Vol. 2 p. 392
93 On the works of Schott and Lana Terzi and their consumers, see R.J.W. Evans, *The Making of the Habsburg Monarchy: An Interpretation*, Oxford: Clarendon Press; 1979. On Lana Terzi see also Clelia Pighetti, ed., *Immagini del '600 Bresciano. L'Opera Scientifica di Francesco Lana Terzi S.I., 1631–1687*, Brescia: Comune di Brescia, 1989, idem., *Francesco Lana Terzi e la Scienza Barocca*, in Commentari dell'Ateneo di Brescia, 1985, Anno CLXXXIV, 97–117, and Andrea Battistini's critical edition of Lana's *Prodromo* (Milan: Longanesi, 1977).
94 See Kircher, *Magnes*, 1654^3, cit., pp. 22–23, *Axiomata seu pronunciata De Natura & Arte*
95 See Hankins and Silverman, *Instruments and the Imagination*, cit., pp. 14–36
96 De Sepibus, *Romani Collegii Musaeum*, cit., pp. 2–3
97 Kircher, *Ars Magnesia, Hoc est Disquisitio Bipartita-empirica seu experimentalis, Physico-Mathematica De Natura, Viribus, et Prodigiosis Effectibus Magnetis*, Würzburg: Typis Eliae Michaelis Zinck; 1631, p. 51

98 'Continentur praeterea in *Musaeo Kircheriano* Epistolarum 12 Tomi in Folio à 40 Annis ad eum datarum annuatim collecti, quos non solum Pontifices, Imperatores, Cardinales & Principes Imperii, sed & literati Philosophi, Mathematici, Physiologi ex toto orbe ad eum variis linguis, tum honoris causa, tum veluti ad oraculum, de difficillimarum quaestionum ex omnia scientia propositarum solutione exaratas miserunt', G. de Sepibus, *Romanii Collegii Musaeum Celeberrimum cuius magnae antiquariae rei* . . . Amsterdam: Ex Officina Janssonio-Waesbergiana; 1678. p. 65

99 See e.g. Kircher to General Oliva, Rome; 1672 May 5. ARSI Rom. 38 f. 172r, 'Ego sane huiusmodi legato immensa et omnigena rerum curiosarum multitudine et varietate instructo locoque constituto animatus, nihil non egi quam ut locum condigna magnificentia expensis, viribus meis etiam superioribus, qua picturis qua machinis aliisque rebus necessariis pro mea paupertate exornarem. Accidit autem ut successu temporis (Deo sit honor et gloria) Collegium Romanum per universam Europam tantam huius occasione Musaei nominis celebritatem adeptum fuerit, ut nemo exterorum, qui Collegii Romani Musaeum non vidisset, Romae se fuisse testari posse videretur.'

100 Cit. in Cochrane, Eric. *Florence in the forgotten centuries 1527–1800*. Chicago and London: The University of Chicago Press; 1973, p. 253

7

From 'The Eyes Of All' to 'Usefull Quarries in Philosophy and Good Literature': The Changing Reputation of the Jesuit *Mathematicus*[1]

> *Kircherus, Scheinerus etc. apply Mathematics to Experiments and Mechanicks etc. They are right Iesuits to make a great blaze of all things etc so as to attract more admirers and contributors to their Order.*
>
> Samuel Hartlib, Ephemerides 1648 Part 1 (January 1648)[2]

Machines, instruments and experiments occupied an ambiguous space at the boundary between theatrical and mathematical practices in the seventeenth century Jesuit college.

In this concluding chapter I will suggest that the disciplinary proceedings of the 1640s and 50s, discussed above in Chapter 3, made the role of machinery and instrumentation in producing innovative natural philosophy highly problematic. The emblematic, performative and magical capabilities of mechanical devices and mathematical instruments were thus increasingly emphasised over their roles in forging an alternative to Aristotelian natural philosophy, complicating Jesuit relations with the emerging scientific societies of mid-seventeenth century Europe.[3]

The extremely popular works of Kircher and his disciples Kaspar Schott and Francesco Lana Terzi are particularly representative of this celebration of technical *mirabilia* in Jesuit culture.[4] While the Jesuit *Revisores* were increasingly sensitive to departures from Aristotelian teachings in physics in the years after the *Ordinatio pro studiis superioribus* of 1651,[5] the same years witnessed an explosion of publications on mechanical and mathematical magic,[6] explicitly divorced from the *physica* taught in Jesuit colleges. Kircher and his disciples responded to the disciplinary crisis by reducing mathematics to pure instrumentation and manipulation, retreating from the strong cognitive claims made for mathematical knowledge by Clavius, Grienberger and Biancani in the early years of the seventeenth century.

By the 1660s and 1670s, a number of indications suggest that for Jesuits who did not enjoy the peculiarly felicitous patronage situation of Riccioli in Bologna, it was becoming extremely difficult to use the Jesuit college as the institutional locus for protracted experimentation and observation. Without the financial backing of a family like the Grimaldi, Jesuit colleges were apparently unwilling to allocate funding to the costs of building instruments and machines. Orazio Grassi, better known for his polemic with Galileo, only succeeded in having a working model made of an unsinkable ship that he designed by submitting the design to the Genoese maritime authorities. In return for the invention, Grassi pleaded for 'some help for our poor College of Savona', financial assistance that would also 'stimulate me to mature some other ideas of even greater utility in navigation'.[7] In the Roman centre, despite the vast literary productions of Kircher and his acolytes the situation for intra-collegiate experimentation was also apparently troubled in the 1650s and 1660s. There are frequent allusions in the works of Kircher and his disciples to both the barriers placed on their curiosity by religious poverty and the demands on their time made by other aspects of the Jesuit ministry.

Sending his 'mathematical organ' [**See Figure 7.1**], an instrument which allowed all types of mathematical operations to be carried out with the aid of moveable slats,[8] to the young Archduke Karl Josef of Austria, Kircher asked to be excused for 'my poverty in constructing the organ', which was made of wood, rather than the gold, silver and gems that a young prince deserved.[9] The mathematical organ reveals something of the audience and place of Jesuit mathematical productions in the changed situation. Devised to contain within a single box all of the knowledge required by a young prince in the fields of arithmetic, geometry, 'Poliorcetica, sive Fortificatoria', 'Computus [...] ecclesiasticus', 'gnomonica', 'sphericam, seu primi mobilis doctrinam', 'doctrina secundorum mobilium', 'steganographia' and music. Schott elaborated, in a series of 'preludes' on the mathematical organ that it was made in the shape of 'the pneumatic organs that are used in our [Jesuit] churches'.[10] By 'playing' on the *organum mathematicum*, arithmetic operations could be carried out (using a system based on Napier's bones), music could be composed, letters could be written in cypher and astronomical knowledge could be acquired. In the hands of Kircher, Schott and Lana Terzi, mathematics could be transformed from the essential tool of the natural philosopher, as it was represented by Clavius and Grienberger previously in the century, to the toy of a Baroque prince. Ironically,

From 'The Eyes Of All' to 'Usefull Quarries in Philosophy and Good Literature' 251

Figure 7.1 A surviving example of Kircher's *Organum Mathematicum*, conserved in the Museo Galileo, Florence

the 'manual' composed by Schott for the use of the organ ran to two volumes and over 850 pages in length, requiring the memorisation of long Latin poems for the performance of many of its operations.[11] The collegiate mathematical space inherited by Kircher from Clavius and Grienberger had changed its function dramatically. Leibniz's Prague-based interlocutor with the Jesuits, Adam Adamandus Kochanski, expressed the transformation of the insulated space of mathematical apprenticeship to a theatrical space for the display of wonders when he wrote of Kircher in 1670 that 'Mathematics cannot be treated by him, as it requires a devotion that it cannot receive from a man who is distracted by the constant visits of foreigners and the interruptions of the dignitaries of the Roman Curia.'[12]

The transformation of the Jesuit *mathematicus* into an impresario had detrimental consequences for the credibility he enjoyed in scientific circles. The *artificialia* paraded in the ceremonial displays and printed works of Kircher and his disciples Francesco Lana Terzi and Kaspar Schott were difficult to replicate in other geographical and political contexts, as witnessed by the futile efforts of the Royal Society to repeat Kircher's experimental performances, discussed below. The failure of replication brought the fragile credibility of the Jesuit mathematical practitioner under close scrutiny in the late seventeenth century, as the boundary between ludic or edifying display and natural investigation was negotiated between Jesuits and members of the emerging secular scientific societies of the 1660s.

The Museum attracted numerous foreign visitors to the College, a point that Kircher was careful to emphasize in negotiating with General Oliva in 1672 to attempt to avoid the relocation of the museum in a dark corridor due to the expansion of the college library.[13]

A few years after Queen Christina's visit to the Collegio Romano, Kircher's museum was visited by the English traveller Philip Skippon.[14] Skippon, travelling in the company of the botanists John Ray, Francis Willughby and Nathaniel Bacon as well as two servants, visited in 1664. He gives the following very detailed description of his encounter with Kircher:

> We visited father Kircher, a German Jesuit, at the *Collegium Romanum* (which is a very large and stately building belonging to the Jesuits). He shewed us his gallery, where we saw all his works, some of which are not yet printed; he hath translated an Arabick book into Latin; wherein the virtues of plants are discoursed. He said Johnston, the printer at Amsterdam, offered him 2000 for all his writings. His Roman medals were fixed within a wire grate on a turning case of shelves. This pope's picture seen in a glass that reflects it from the plaits or folds of another picture. An organ that counterfeits the chirping of birds, and at the same time a ball is kept up by a stream of air. The picture of the king of China. A picture of father Adam Schall, a German Jesuit, who is now in great favour with the king of China, being his chief counsellor; on his breast he wears the mark of his honour, which is a white bird, having a long bill, and red on the crown of its head. The picture of Deva Rex Davan Navas. The picture of Michael Rex Nepal. The rib and the tail (flat and broad) of a Syrene, which Kircher said he saw at Malta. A cross made of 300 small pieces of wood set together without glew, nails &c. Painting of Raphael Urbin on earthen dishes. A microscope discovering fine white sand to be pellucid, and of an elliptical figure; and red sand pellucid and of a globular figure. A China shoe. Two Japan razors. A Japan sword, wherewith some Jesuits had been martyr'd. A China sword, or rather a mace. Corvus Indicus, a red bird. China birds' nests like white Gum. Canada money made of little pieces of bones, and a medal of the same, which faintly represented the figure of a man. Medals of the hieroglyphical obelisks in Rome.

A cabinet door that first opened upon hinges on one side, and then upon hinges on the other. A flat and broad hoop that moved to and fro, on a declining plane, without running off; within it having a weight at A. Water put into the glass BC, and by clapping one's hand at B, without touching the water, forces the water out a good heighth out at C.

A perpetual motion attempted by this engine. D is a cistern with water, which runs down the channel E, and turns the wheel from G to F. At i the axis of this wheel is a handle that lifts up the sucker H, that forces up the water out of the cistern K K into the pipe L into the upper cistern D.

A sphere moved regularly by water that falls on the aequinoctial line which is made like a water wheel. An image that spewed out of its mouth four sorts of water, one after another. A serpent vomiting water, and a bird drinking out of the same dish. The perpetual motion we saw at Milan. The heat of a man's breath or hand, expelled water out of a glass, that afterwards turned a wheel. A brass Clepsydra made after this manner. A and B are two cisterns for water. When that in A is uppermost it falls down thro' thee four tubuli, which are the supporters into the lower cistern B, and there it springs up like a fountain, a pretty height for an hours's space; and so vice versa when B is turned up.

A notable deceptio visus in the pyramidal spire C. D. being turned one way it seemed to go up, and moved the other way it appeared as if running downwards. These and many other inventions are described in Kircher de Magnete.

Birds' nests, that are eaten by the Indians, which Wormius p. 311, calls Nidus Ichthyocollam referens.

The figure of a woman he called the oracle with a hole in her breast, which applying one's ear to, words and sentences are plainly understood, though whispered a good way off.

Flies and a lizard within amber. A paper lizard with a needle stuck in it, ran up and down a wooden pillar, being moved by a loadstone. The magnet moved several figures hanging within glass globes. One figure was moved by the loadstone, thro' wood, glass, water and lead. A cylindrical glass of water with a glass figure in it, which rises or falls as you press the air at the top of the glass with your finger; the air being pressed in the cylinder, presses that in the figure into a narrower room, and so water comes in and weighs the figure down, which rises upon lessening the pressure at the top of the cylinder. Avis Guaria, p. 308 Wormii, was seen here.[15]

Skippon's meticulous description betrays little emotion – we are not told whether the English naturalists were frightened by the Delphic oracle. Indeed, if anything Skippon even suggests a certain tedium in the face of Kircherian wonders – 'the perpetual motion machine we saw at Milan'. There is ample evidence, however, that English circles were utterly enthralled by Kircher's natural and artificial wonders, and were doomed to repeated frustration in attempting to repeat Kircher's experiments in Restoration England.

The vegetable phoenix, admired by Queen Christina in Kircher's museum, immediately the object of great interest amongst English natural philosophers, illustrates the difficulty Kircherian wonders experienced in travelling beyond the walls of his museum. In 1657 Henry Oldenburg planned a trip to Italy, hoping to bring back to England news of Kircher's 'vegetable phaenix's resurrection out of its own dust by ye warmth of ye Sun', along with other Kircherian secrets and 'remarquable things, one might have the satisfaction to be punctually informed about'[16] Oldenburg never made the trip, and the next news about Kircher's phoenix had to wait until Robert Southwell encountered an English traveller returning from Italy. Southwell reported to Oldenburg

> [H]e gives me some incouragement yt when I come to Rome I shall be able fully to satisfy you concerning Kerchers plant. He told me he was wth him and remembers to have seene in a glasse half as bigg as his head (close luted) a plant glowne up ye length of his finger with a kind of asshes at ye bottome but I found he had not beene Curious in the observation of it'[17]

On accomplishing his mission, Southwell brought disappointing news about the phoenix: 'As to the flower growing from its ashes, he had such a thing, but it is now spoiled; he made it not himself, but it was given him'.[18] Southwell nonetheless acquired 'the receipt thereof, upon a swop, wrote with his own hand; it is long and intricate, and of a nice preparation'.[19] We have no record of whether the Royal Society suceeded in reproducing the vegetable phoenix,[20] but generally attempts to replicate Kircherian wonders in London and Oxford met with little success. The trouble was not limited to England. John Bargrave recounted in graphic terms the price of failure for a Nuremberg optician:

> I bought this glass of Myn Here Westleius, an eminent man for optics at Nurenburg, and it cost me 3 pistolls, which is about 50S English. This gentleman spoke bitterly to me against Father Kercherius, a Jesuit at Rome (of my acquaintance), saying that it had cost him above a thousand pounds to put his optic speculations in practice, but he found his principles false, and showed me a great basket of glasses of his failings.[21]

Kircher's net drew in too much, according to unsympathetic English commentators in the 1650s. Robert Payne's remarks on Kircher *qua* Jesuit in 1650, while complaining about an experiment on roasted worms reported in the *Ars Magna Lucis et Umbrae* emphasize precisely this point:

> The truth is, this Jesuit, as generally the most of his order, have a great ambition to be thoughte the greate and learned men of the world; and to that end writes greate volumes, on all subjects, with gay pictures and diagrams to set them forth, for ostentation And to fill up those volumes, they draw in all things, by head and shoulders; and these too for the most part, stolen from other authors. So that if that little, which is their owne, were separated from what is borrowed from others, or impertinent to their present arguments, their swollen volumes would shrink up to the size of our Almanacks. But enough of these Mountebankes.[22]

In similar vein, on sending Descartes a copy of Kircher's *Magnes*, Constantijn Huygens had remarked that the former would find in it 'more grimaces than good material, as is normal for the Jesuits. These scribblers, however, can be useful to you in those things *quae facti sunt, non juris*'.[23]

Sir Robert Moray (1608–1673), later one of the prime movers in acquiring a charter from Charles II for the foundation of the Royal Society and its first president,[24] entered into close correspondence with Kircher in 1644, after admiring the *Magnes*.[25] While in the services of the French army in Germany, Moray consumed Kircher's books avidly and discussed their contents with Jesuits in Cologne and Ingolstadt.[26] On his return to the royal court in Whitehall, he informed Kircher of the foundation of the Royal Society, and continued to send scholars, such as the mathematician James Gregory, the naturalist Francis Willughby and others to seek Kircher's company in Rome.[27] Moray was confident that Kircher's agglomeration of information could be filtered, or threshed, to separate the wheat from the chaff:

> Whatsoever Mr. Hugens & others say of Kercher, I assure you I am one of those that think the Commonwealth of learning is much beholding to him, though there wants not chaff in his heap of stuff composted in his severall peaces, yet there is wheat to be found almost every where in them. And though he doth not handle most things fully, nor accurately, yet yt furnishes matter to others to do it. I reckon him as usefull Quarries in philosophy and good literature. Curious workmen may finish what hee but blocks and rough hewes. Hee meddles with too many things to do any exquisitely, yet in some that I can name I know none goes beyond him, at least as to grasping of variety: and even that is not onely often pleasure but usefull.[28]

Moray changed his tune in his following letter to the secretary of the Royal Society, demonstrating the increasing fragility of Jesuit scientific credibility, and linking the failure of an experiment involving the focussing of moonbeams on substances with a powerful burning-glass to Kircher's membership of the Jesuit order explicitly, writing that 'hee does but lyke other birds of his feather'.[29]

Boyle wrote to Oldenburg in 1665 to complain about the problem:

> I suppose Sr. Rob. Murry has told you, that the Expt about Salt & Nitrous water exposed to the Beames of the moone did not succeed as Kircher promises, but as I foretold. And for the same Author's Expts with Quicksilver & sea water seald up in a ring, though the want of fit glasses will, till the commerce with London be free, keepe mee unable to try: yet besides it is at most the same, but not soe probable as that wch he publishd in his *Ars Magnetica*, 20 or 30 year ago. I cannot but think it unlikely that it will succeed at least in our Climate, where by concentrating the Beames of the Moone with a large Burning-glasse, I was not able to produce any sensible Alteration, in Bodys that seeme very easily susceptible of them.[30]

Commenting to Boyle on the unhappy results of attempts to repeat Kircher's experiments, Oldenburg wrote darkly that 'Tis an ill Omen, me thinks, yt ye very first Experiment singled out by us out of Kircher, failes, and yt 'tis likely, the next will doe so too'.[31]

The replication of the wonders displayed to visitors to Kircher's museum and described in his published works was difficult. Kircher's performances and demonstrations were apparently meant to be beheld, admired and believed, but not to be repeated outside the preternatural realm of the museum of the *Collegio Romano*.

A moment of tension occurred when it seemed that the Royal Society would have to stage its own royal show for Charles II in 1663. Christopher Wren warned the President, Lord Brouncker, that 'to produce knacks only, and things to raise wonder, such as Kircher, Schottus, and even jugglers abound with, will scarce become the gravity of the occasion'.[32] The credibility of the Jesuit *mathematicus* and the credibility of his religious order in a particular local context were inextricably linked. Whereas Clavius had initially suggested that mathematical celebrity would contribute to the credibility of the order as a whole in treating spiritual matters, reactions to the work of Kircher and his disciples suggested that the reverse was also true – the credibility of the Jesuit mathematical practitioner on matters of experiment and observation might be adversely affected by his membership of the order. The global Jesuit information network exercised its complex charm over the emerging scientific societies of the mid-seventeenth century. As Mersenne had attempted to recruit the order for the measurement of global magnetic variation, John Beale, despite regarding the Jesuits as 'our most dangerous enemies'[33] suggested to Samuel Hartlib that it would be beneficial to 'provoke the Iesuites to transport the best Telescopes to their Peru, & other Southerne plantations, and from thence to make their discoveryes', in order to reap profit for the commonwealth of learning from the observations made by Jesuit astronomers.[34] A similar proposal was actually put into practice by the *Académie des Sciences* in Paris in the 1680s, when it employed Jesuits to perform astronomical observations on the China mission with a view to solving the problem of longitude.[35] Boyle's remark to Oldenburg that 'I am glad you are like to settle a correspondence with Rome, that being the chief centre of intelligence'[36] is indicative of the tension between respect and suspicion that characterised the relationship between the Royal Society and the Society of Jesus at this point.

John Dodington, secretary to the English ambassador to the Venetian republic, also endeavoured to mediate between Kircher and the Royal Society. While entrusting the education of his son to Kircher's care, even in matters of religion, Dodington asked Kircher to procure telescopes and microscopes for the use of the Society from the rival instrument makers Eustachio Divini and Giuseppe Campani.[37]

By the 1660s, experiments and claims about nature were finding it ever more difficult to travel beyond the walls of the *Collegio Romano* to gain currency in wider scientific circles, especially in the economic centres of England, France and the Low Countries. The environments in which the Jesuit *mathematicus* worked – including classroom, court and Curia – demanded different modes of self-presentation, and required the Jesuit scientist to acquire fluency in different 'languages' and the facility with which Jesuit scientists adapted their work to these many different environments had serious consequences for the way in which Jesuit statements about the natural world were evaluated by natural philosophers outside the order. During a period of disciplinary formation, when institutions were arising around Europe that ostensibly wish to insulate the investigation of the natural world from other forms of human activity, the cultural ferment characteristic of the Jesuit collegiate network was perceived by many members of the new societies and academies of the seventeenth century as a polluting force, notwithstanding the considerable material and social resources wielded by the order. The mixing of clerical, courtly, theatrical and scientific forms of life was particularly frowned upon by the members of the early Royal Society, where Jesuit natural philosophers were perceived as 'jugglers', using sleight of hand to deceive and impress courtly patrons, whether during ceremonial visits or in folio volumes of mathematical and experimental wonders. When the 'chaff' was removed from their experimental reports, the practices they described had to be sanitised and deprived of the clerical and emblematic overtones that supposedly coloured the motives of the Jesuit experimenter, rendering him a passable reader, but not a reliable interpreter, of the book of nature.

The relationships between the emerging scientific societies and academies of the 1650s and 1660s were frequently characterised by a desire to harness the information-gathering capacities of the Jesuits while avoiding the pollution entailed by admitting Jesuits as members or surrendering information to the services of Jesuit print culture. In the case of the Parisian *Académie des Sciences*, John Milton Hirschfield has suggested that strong Jansenist influences on the early *académie* were responsible for Jesuit exclusion, although Florence Hsia has demonstrated that the relationship of the early *académie* with the Jesuits requires a more nuanced treatment.[38] In the case of the early Royal Society, Kaspar Schott's attempt to enter into an epistolary exchange of unpublished experimental and technical secrets with Robert Boyle was carefully deflected by the latter, who replied to Schott to say 'That I long to see his works increased by the accession of his *Technica Curiosa*; towards which, I fear, I shall not contribute much, both because of my being no better stocked with rarities than I am, and because I know not what particular subjects he treats of in it'.[39] Boyle's cautiousness is to be contrasted with the exuberance with which Kircher showered visiting virtuosi with curiosities, as witnessed by Robert Southwell's surprised exclamation that Kircher was 'very easy to communicate whatever he knows; doing it, as it were, by a maxim he has'.[40]

This book has been guided by the view, shared by a number of studies of Jesuit science over the past two decades, that it is high time to abandon a conflictual approach to the relationship between early modern science and Catholicism. Rather than replace conflict with an equally inappropriate image of harmony, my aim has been to look at the ways in which a religious order, the Society of Jesus, developed an institutional framework within which certain types of activities – mathematical study, experimentation, correspondence concerning natural phenomena – could be given a specific meaning. Avoiding analysing the 'influence' of Catholicism on science, an astrological form of discourse that Michael Baxandall would have banished from historical analysis,[41] I have attempted to look at the way Jesuit mathematical practices and Jesuit spirituality, discipline, bureaucracy, theatre, pedagogy and emblematics fed into one another, to adopt a more organic metaphor, focussed primarily on the Jesuit mathematicians of the *Collegio Romano*. Book censorship, the movements of people, architectural projects for Jesuit churches and colleges are just some of the practices which were subject to the centralised accumulation of reports, plans and manuscripts.

For all the importance of the *Collegio Romano* in the Jesuit system, it is important to emphasize that other key centres emerged in the network of Jesuit colleges of the sixteenth and seventeenth centuries, and the Jesuit network emerged as a network of multiple hubs, with locally varying practices rather than a centralised 'hub and spokes' network radiating from Rome. The tension between an enormously sophisticated centralised bureaucracy and a flexibility in adapting to local circumstances is perhaps the most distinguishing feature of early modern Jesuit culture in all of its aspects.

Membership of the Jesuit order offered a number of different types of tools and resources to the mathematical practitioner. Perhaps Christoph Clavius puts it best when he points out that

> It is only the Society [of Jesus] that can pursue [eminence in different disciplines] both very quickly and very easily. For it possesses diverse and most beautiful minds of youths, it has free time, it has masters [and] it has the authority to direct its subjects to whichever kind of study suits them the most.[42]

This book began with a satirical description of the strange political geography of the Jesuit 'monarchy'. The scientific practices, authority, printed output, and ethical codes of Jesuit mathematical practitioners were influenced by this geography in complex ways. Combining a flexible, accommodative approach to local geographical contexts with a highly globalist apostolate, the Jesuit order formed a laboratory for the exploration of different solutions to the problem of making locally produced knowledge universal. The creation of a globally distributed network of trained mathematical practitioners, obliged to engage with widely different local conditions but bound by a paternal bond to a Roman mathematical authority, generated a collective approach to the investigation of the natural world with no obvious precedent. The codes of discipline and deportment promoted by the Jesuit order, emphasizing the sacrifice of individual pride in the interests of the collectivity, were ideally suited to the surrendering of information collected at remote stations for central processing in Rome.

The ultimate justification for mathematical expertise provided by Clavius was its importance in furthering Jesuit apostolic goals, by involving Jesuit mathematicians in conversations with curious aristocrats, and also as leading educators of catholic Europe, schooling figures including Descartes, Mersenne, Helvetius and Torricelli in mathematics and natural philosophy. Mathematical practices were also tied to other areas of the Jesuit apostolate, including architecture, music and the design of stage machinery for Jesuit dramatic productions. The epistemological confidence displayed by Clavius and his disciples, which relaxed the borders separating mathematical knowledge from natural philosophy, was later curbed by the renewed disciplinary vigour that marked the crisis after the death of General Muzio Vitelleschi. This renewed demarcation of the boundary between mathematics and natural philosophy, led to a greater emphasis on the rhetorical, emblematic and magical facets of mathematical practice, culminating in the flamboyant works of Athanasius Kircher, Kaspar Schott and Francesco Lana Terzi in the second half of the seventeenth century.

This book has argued that many of the practices associated with the emergence of 'modern science' were shaped by Jesuit institutions, from expert book censorship to information-gathering correspondence networks, to public experimental performances and instrumental observations. The Society of Jesus was no monolithic entity in the seventeenth century, but a network rife with internal disagreements and tensions surrounding the appropriate ways to investigate the natural world, and ongoing negotiations for status between the mathematical arts and the 'superior' sciences of philosophy and theology. Precisely the eclecticism, and legendary accommodationism, of Jesuit practices, adapting to radically different contexts and cultures, from the German courts of the Thirty Years' War to the Ming court in Beijing, required repeated earnest, but ultimately unsuccessful, efforts to enforce centralised authority on the provinces, and to clamp down on divergences from approved approaches. Anti-Jesuit propaganda, particularly in Protestant countries, such as England and the Low Countries but also for example under Jansenist influence in France, encouraged a discrediting of the Jesuit *mathematicus*, based on perceived dogmatic obedience to Rome, preventing Jesuits from truly embracing the 'new philosophy', but being gradually reduced to mere obedient data-gatherers, due to their global reach, and access to training and equipment. In a sense, rather than natural philosophers, the Jesuit missionaries themselves became reduced to the status of observational instruments, rather like the gnomons of Athanasius Kircher's *Horoscopium Catholicum* [**See Figure 5.5**]. The flourishing of Jesuit magic in the Catholic courts of Europe was a realm where the dangers of experimentation could be insulated from doctrinal consequence. In this context, the complex, collective role of the Jesuits as pioneers of so many of the systems and processes that were so readily adopted by the emerging scientific societies of the seventeenth century could be easily effaced, and ultimately replaced by a more simplistic, seductive narrative of heroic individual discoveries against a backdrop of monolithic opposition from the Catholic Church.

Notes

1. A version of this chapter appeared previously as Michael John Gorman 'From "The Eyes of All" to "Usefull Quarries in philosophy and good literature": Consuming Jesuit Science, 1600–1665', in *The Jesuits: Culture, Sciences and the Arts, 1540–1773*, edited by John W. O'Malley, S.J., Gauvin Alexander Bailey, Steven J. Harris, and T. Frank Kennedy, S.J*,* Toronto: University of Toronto Press, 1999, pp. 170–189.
2. Hartlib Papers, Sheffield, 31/22/1A-B, digital edition, ed. Mark Greengrass et al., Ann Arbor: UMI, 1995.
3. For a good survey of the changing intellectual reputation of the Jesuits in the later seventeenth and eighteenth centuries, see especially Mordechai Feingold, 'Jesuits: Savants', in Feingold, Mordechai, ed., *Jesuit Science and the Republic of Letters*, Cambridge, MA., London: The MIT Press, 2003.
4. See especially Athanasius Kircher, *Magnes, sive de arte magnetica opus tripartitum*, Rome: H. Scheus, 1641 (further editions: Cologne: J. Kalcoven, 1643^2; Rome, B. Deversin and Z. Masotti, 1654^3), idem., *Ars Magna lucis et umbrae*. Romae: H. Scheus; 1646 (further edition: Amsterdam: J. Jansson van Waesberghe, 1671^2), idem., *Musurgia universalis sive ars magna consoni et dissoni*, Rome: heirs of F. Corbelletti, 1650, idem., *Mundus subterraneus*, Amsterdam: J. Jansson van Waesberghe, 1664-5 (further edition: Amsterdam: J. Jansson van Waesberghe, 1678), idem., *Magneticum naturae regnum*, Rome: I. de Lazaris, 1667 (further edition: Amsterdam: J. Jansson van Waesberghe, 1667), idem., *Phonurgia nova*, Kempten: R. Dreher, 1673 (for a full bibliography of Kircher's works, see John Fletcher, ed., *Athanasius Kircher und seine Beziehungen zum gelehrten Europa seiner Zeit*. Wiesbaden: Harrassowitz; 1988, pp. 179–190), [Kaspar Schott], *Joco-seriorum naturae et artis, sive Magicae naturalis centuriae tres*. Frankfurt: Apud Ioannem Arnoldum Cholinum; 1667, idem., *Magia Universalis naturae et artis*. Herbipoli: Excudebat Henricus Pigrin; 1657–1659 (4 vols.), idem., *Mechanica Hydraulico-Pneumatica.* . Würzburg: H. Pigrin; 1657, and idem., *Technica Curiosa sive mirabilia artis, Libris XII. comprehensa*. Würzburg 1664 (further edition: Würzburg: Jobus Hertz; 1687^2), Francesco Lana Terzi, *Prodromo overo saggio di alcuni inventioni nuove premesso all'arte Maestra*. Brescia: Rizzardi, 1670, idem., *Magisterium naturae et artis, opus physico-mathematicum*. Brescia: Rizzardi; 1684. On Lana Terzi, see M.J. Gorman, 'L'académie invisible De Francesco Lana Terzi: Les Jésuites, L'expérimentation et la Sociabilité Scientifique au Dix-Septième siècle,' in Daniel Odon-Hurel, ed., *Académies et sociétés savantes en Europe: 1650–1800: actes du colloque de Rouen, 1995*, Paris: Champion Verlag, 2001, pp. 409–432
5. See Claudio Costantini, *Baliani e i Gesuiti*. Florence: Giunti Barbèra; 1969.
6. On the background of mathematical magic in England in this period see Jack Peter Zetterberg, *'Mathematical Magick' in England: 1550–1650*. PhD dissertation,

University of Wisconsin-Madison; 1976. On the European context, see especially William Eamon, *Science and the secrets of nature: Books of secrets in medieval and early modern culture*. Princeton: Princeton University Press; 1994 and R. J. W. Evans, *The Making of the Habsburg Monarchy: An Interpretation*. Oxford: Clarendon Press; 1979, especially chapters 9–12. On natural magic and the Jesuits, see especially Mark A. Waddell, *Jesuit Science and the End of Nature's Secrets*, Farnham: Ashgate, 2015.

7 Grassi to the Genoese Senate, Savona, 25 August 1652, Archivio di Stato di Genova, Archivio Segreto, Litterarum, 1988, in Claudio Costantini, *Un Batello Insommergibile Ideato da Orazio Grassi*. Nuova Rivista Storica. 1966; 50: 731–737, on p. 734.

8 See Mara Miniati, *Les* cistae mathematicae *et l'organisation des connaissances au XVIIe siècle*, in *Studies in the History of scientific instruments*, London: Turner books, 1989, pp. 43–51. In the years from 1649 to 1663 Kircher produced a number of related instruments, commencing with the *Arca Musurgica* (1649–50) and continuing with the *Organum mathematicum* (1661), the *Cista steganographica* (1663) and the *Arca Glottotatica* (1663). On the *Arca Musurgica* see Kircher, *Musurgia Universalis*, Rome: Ex Typographia Haeredum Francisci Corbelletti; 1650, II, 184/5, Iconism. XIV *Arcae Musurgicae novum inventum*, and Johann Gans to Kircher, Vienna, 6 February 1649, APUG 561, f. 133. On the *arca glottotactica* and *cista steganographica* see Kircher, *Polygraphia nova*, Rome: Ex typographia Varesij, 1663, pp. 85, 128 respectively.

9 Kircher to Archduke Karl Josef, Rome, 7 August 1661, APUG 555, f. 98r (draft).

10 Kaspar Schott, *Organum Mathematicum Libris IX. explicatum a P. Gaspare Schotto e Societate Jesu, quo per paucas ac facillime paraboles Tabellas, intra cistulam ad modum Organi pneumatici constructam reconditas, pleraeque Mathematicae Disciplinae, modo novo ac facili traduntur . . . Opus posthumum*. Herbipoli: Jobus Hertz; 1668.

11 Schott, op. cit., p. 54, *Praeludium I*.

12 'Mathematica enim ab eo tractari non possunt, ut quae applicationem aliquam requirunt, quam ille obtinere non potest, quotidianis visitationibus Exterorum, et interpellationibus Magnatum Curiae Romanae distractus', Adam Adamandus Kochanski to Leibniz, 7 June 1670, in Leibniz, *Philosophischer Briefwechsel herausgeben von der Akademie der Wissenschaften der DDR*, Erster Band, 1663–1685. Akademie-Verlag Berlin, 1987, pp. 46–48.

13 'Accidit autem ut successu temporis (Deo sit honor et gloria) Collegium Romanum per universam Europam tantam huius occasione Musaei nominis celebritatem adeptum fuerit, ut nemo exterorum, qui Collegii Romani Museum non vidisset, Romae se fuisse testari posse videretur', Kircher to Oliva, Rome, 5 May 1672, published in R. Garrucci, *Origini e vicende del Museo Kircheriano dal 1651 al 1773*. La Civiltà Cattolica. 1879; Serie X Vol. XII (Quaderno 703): pp. 727–739.

14 On Skippon see Peter Burke, *The discreet charm of Milan: English travellers in the seventeenth century*, in idem., *Varieties of cultural history*, Oxford: Polity Press, 1997, pp. 94–110.
15 Philip Skippon, *An Account of A Journey made Thro' Part of the Low-Countries, Germany, Italy and France*. in A. and J. Churchill, *A Collection of Voyages and Travels*. London: J. Walthoe; 1732; pp. 359–736, on pp. 672–4.
16 Oldenburg to Boyle, Saumur, 19 March 1657, OC I pp.155–156.
17 Southwell to Oldenburg, Montpellier; 20 October 1659, OC I, pp. 323–325.
18 Southwell to Boyle, n.p., 30 March 1661, in *The works of the honourable Robert Boyle*, ed. Thomas Birch, London: J. & F. Rivington, 1772 (2nd edition), VI, pp. 297–300.
19 ibid.
20 Boyle did however allude to the palingenetic experiment in *A Discourse about the possibility of the resurrection* (1675) in Boyle, *Works*, cit., 4, p. 194.
21 Quoted in John Bargrave, *Pope Alexander the Seventh and the College of Cardinals, with a Catalogue of Dr. Bargrave's Museum*, ed. J.C. Robertson. London; 1867.
22 R[obert] P[ayne] to Gilbert Sheldon, Oxford, 16 December 1650, British Library Ms. Lansdowne 841 ff. 33r-v, on 33v.
23 Constantyn Huygens to Descartes, n.p., 7 January 1643, published in Leon Roth, ed., *Correspondence of Descartes and Constantyn Huygens 1635–1647*, Oxford, Clarendon Press, 1926, pp. 185–6, cited in John L. Heilbron, *Electricity in the 17th and 18th centuries. A study in early modern physics*, Berkeley, California: University of California Press; 1979, p. 106.
24 On Moray see Alexander Robertson, *The Life of Sir Robert Moray. Soldier, Statesman and Man of Science (1608–1673)*, London, 1922.
25 Moray to Kircher, Ingolstadt, 1 June 1644, APUG 557 363r-v.
26 Moray to Kircher, Ingolstadt, 7 September 1644, APUG 557 323ar-av, Moray to Kircher Ingolstadt, 24 January 1645; APUG 568 ff. 74r-75v, Moray to Kircher, Paris, 12 March 1645, APUG 557 ff. 271r-v, Moray to Kircher, Cologne, 21 November 1655; APUG 568 ff. 39r-v, Moray to Kircher, Cologne, 28 January 1656; APUG 568 ff. 20r-21v, Moray to Kircher, Rotterdam, 6 August 1657; APUG 568 ff. 196r-197v.
27 Moray to Kircher, Whitehall, 25 July 1663, APUG 563 ff. 212 r-v
28 Moray to Oldenburg, Oxford; 19 October 1665; OC II: 574–576.
29 Moray to Oldenburg, Oxford, 16 November 1665 in OC II: 608–611
30 Boyle to Oldenburg, Oxford [?]; 18 November 1665, OC II: 613–614.
31 Oldenburg to Boyle, London, 21 November 1665, OC II: 615–617
32 Wren to Lord Brouncker, Oxford, 30 July 1663 in Thomas Birch, *The history of the Royal Society of London*, 4 vols., London, 1756, Vol. 1, p. 288. On the popular topos of Catholics (and especially Jesuits) as jugglers, see Rob Iliffe, 'Lying wonders and juggling tricks: Religion, Nature and Imposture in Early Modern England',

forthcoming in D. Katz and J. Force, eds., *'Everything Connects': Essays in Honor of Richard H. Popkin*, Dordrecht: Kluwer, 1998.
33 Beale to Boyle, Yeovell, 30 July 1666, in *Works*, VI pp. 408–410. On Beale's involvement with the early Royal Society, see Mayling Stubbs, 'John Beale, Philosophical Gardener of Herefordshire Part II', *Annals of Science*, 1989, 46: 323–363, although Stubbs does not discuss Beale's interest in the Jesuits.
34 Beale to Hartlib, 18 January 1658, Hartlib Papers, Sheffield, 51/55A
35 See Florence Hsia, 'Jesuits, Jupiter's Satellites, and the Académie Royale des Sciences', in *The Jesuits: Culture, Sciences and the Arts, 1540–1773*, edited by John W. O'Malley, S.J., Gauvin Alexander Bailey, Steven J. Harris, and T. Frank Kennedy, S.J, Toronto: University of Toronto Press, 1999, 241–257.
36 Boyle to Oldenburg, n.p., 3 April 1668, OC IV: 299, cit. in John L. Heilbron, *Electricity in the 17th and 18th centuries. A study in early modern physics*, Berkeley, California: University of California Press; 1979, p. 108.
37 Dodington to Kircher, Venice, 6 December 1670, APUG 559 ff. 37r–38v: 'Et essendomi abastanza noto, la sua rara e vertuosa qualità sono importuno a pregarla che si degni a farmi noto se havesse la P. S. A. qualche Istrumento mechanico curioso, o qualche isperienza degna di consideratione, o qualche libro, desiderando restarmi favorito senza sua spesa, per poterli inviare al Sigr. Cav. Roberto Murry overo alla Società Reale, in Inghilterra, venendomi da questi S[ignor]i fatta frequente instanza', Dodington to Kircher, Venice, 17 January 1671, APUG 559, ff. 21r–22v 'Prego in oltre la P. S. d'informarsi quale delli Doi Artefici Eustachio Divini o Gioseppe Campani sia più perfetto nell'edificio de Telescopij e Microscopi, e con l'informatione inviarmiene due per sorte presi d'ambedue l'Artefici, e perche devono servire per l'uso della Società Reale in Londra sono a supplicarla de più perfetti'. See also Dodington to Kircher, Venice, 30 August 1670, APUG 559 ff. 19r–20v, Dodington to Kircher, Venice, 3 January 1671; APUG 560 ff. 23r, 24v, Dodington to Kircher, Venice, 21 March 1671; APUG 560 ff. 97rv, 98v, Dodington to Kircher, Venice, 24 October 1671; APUG 560 ff. 104r–105r. On Divini and Campani see Maria Luisa Righini Bonelli and Albert van Helden. *Divini and Campani: A forgotten chapter in the history of the Accademia del Cimento*. Annali dell'Istituto e Museo di Storia della Scienza (Supplemento). 1981; Monografia N.5 (Fasc. 1).
38 John Milton Hirschfield, *The Académie Royale des Sciences 1666–1683*, New York: Arno Press; 1981, Chapter 2, Hsia, op. cit.
39 Boyle to Schott, undated draft, in Boyle, *Works*, VI: 62–3.
40 Southwell to Boyle, n.p., 30 March 1661, in Boyle, *Works*, VI: 297–300.
41 Michael Baxandall, *Patterns of Intention: On the Historical Explanation of Pictures*. New Haven and London: Yale University Press; 1985, pp. 58–62.
42 'Sola Societas est, quae hanc consequi cum brevissime, tum facillime possit. Habet enim varia, pulcherrimaque iuvenum ingenia, habet otium, habet magistros, habet

auctoritatem cogendi subditos ad id genus studii, cui erunt aptissimi', [Christoph Clavius], *Discursus [. . .] de modo et via qua Societas ad maiorem Dei honorem et animarum profectum augere hominum de se opinionem, omnemque haereticorum in literis aestimationem, qua illi multum nituntur, convellere brevissime et facillime possit,* (c. 1594) in MP VII, pp. 119–122.

Index

Académie des Sciences 257, 258
acoustics 225
Acquaviva, Claudio 66
Aerarium (Bettini) 45, 56
Agni Dei 97
Allacci, Leone 94, 101
anamorphosis 61
Anastagi, Paolo 140
Apiaria (Bettini) 44–5, 45–8, 49
Archimedes 19, 45, 55, 61
Aristotelian teachings 101–2, 147, 201
    Grienberger 65, 66, 70–1
    magic 227, 228, 229
    Magni 128
Aristotelian terminology 135
"Aristotle's wheel" 127
Arriaga, Roderigo de 103, 139, 140, 141
artificial magic 225–6, 227, 228
astronomy 95–6
authorship 43. *See also* self-abnegation

Baldini, Ugo 23
Baliani, Giovanni Battista 126–7
Bardi, Giovanni 64, 65–8, 69–70
Bargrave, John 254
Baroque culture 199
Bartoli, Daniello 132
Beale, John 256–7
Bellarmine, Robert 89, 98
Berti, Gasparo 127–8, 132, 142, 228
Bettini, Mario 44–9, 56, 207
Biagioli, Mario 13, 43
Biancani, Guiseppe 25, 53, 66, 76–7n45, 173
Bibliotheca Selecta 18–19
Bibliotheca Universalis 18, 19
Bidermann, Jakob 41–2, 43
Bireley, Robert 140
Borri, Cristoforo 175
Bosgrave, James 23
Boyle, Robert 43, 147, 256, 258
Buonanni, Filippo 197, 212

Calandrino, Nicolò 27
calendar reform 14–17
Carafa, Vincenzo 100, 102, 236
Carolinum 139–40
Cartesianism 102–3, 139
Casati, Paolo 133–5, 138
"cat piano" 218
catoptric chests 219–21
cats 218–19, 221
Cazré, Pierre 102–3
celestial fluidity 90
Cenodoxus (Bidermann) 41–2
censorship 1, 2, 44–5, 89–91, 184–5, 194n63. *See also Ordinatio pro studiis superioribus*
Cesi, Federico 55, 56, 67, 70
Chigi, Fabio 132–3, 138, 143, 146
Christina, Queen of Sweden 231–4
Ciermans, Joannes 178–9
civility 221–2
Clavius, Christoph 3
    Bibliotheca Selecta 19
    calendar reform 16–17
    correspondence 207
    death 63
    disciples 23–8
    educational policy 20–2
    engraving by Villamena 28–30
    mathematical academy 168
    mathematical authority 9–10
    mathematical celebrity 256
    mathematical practice 13–14, 22, 25
    musaeum mathematicum 206
    Opera Omnia 28–30
    relationship to Grienberger 52–3
    Society of Jesus 259
    water experiment 208
Codina Mir, Gabriel 11–12
Collège de Clermont 135
Collegio Romano 3, 9, 23, 231–2, 238
    college pharmacy 234–6
    mathematical academy 168, 206–7

mathematical culture 53–5
mathematical museum *(musaeum mathematicum)* 6, 53, 168, 206, 207, 208
   public mathematics 55–6, 61–2
   Problemata 56–61, 62–3, 65
   Queen Christina 231–4
Collegium Vilnensis 137–8
continuum 127, 138, 139
Copernicus, Nicolaus 15, 95
Cornaeus, Melchior 141, 145
correspondence 169–71
   magnetic geography 173–81

Darnton, Robert 218
De Lugo, Giovanni 139
De Revolutionibus (Copernicus) 15, 95
De Sepibus, Giorgio 201, 204, 208
Dear, Peter 222
Dee, John 211
Del Rio, Martin 227, 228–9
Delphic Oracle (speaking statue) 71, 203, 204–6, 209, 226
demonic magic 209–11, 212, 227, 228, 229
Demonstratio ocularis (Magni) 128–31, 137
deportment 43. *See also* self-abnegation
Descartes, René 87, 139, 221, 229
di Grazia, Vincenzo 64
Discorsi (Galileo) 127, 128
Dobrzensky, Jakob 142
Dodington, John 257
Donne, John 14, 15

earth-moving machine 60
eclipse observations 175–6, 184
education 11–12
educational policy 20–2
Egyptian mechanics 209
elite 199, 200, 214, 215
epistemological modesty 95
Eucharist 134, 136, 139, 143
extrinsic values 98

Fabri, Honoré 102, 104, 125
Falckestein, Johann 23–4
Favaro, Antonio 88
Ferdinand II, Emperor 94, 96, 139, 141

Ferdinand III, Emperor 94, 141
Finé, Oronce 173
floating bodies debate 64–70, 81n99
Florent, Michael 173–4
Fortescue, George 71
Fumaroli, Marc 57

Galileo Galilei 43, 44, 52
   attack on Scheiner 99
   Discorsi 127, 128
   exegesis 116n72
   extrinsic v. intrinsic values 98–9
   floating bodies debate 64–5, 67, 69, 81n99
   longitude at sea 173
   relationship with Grienberger 63–4, 76n38, 87
   vacuum experiments 126–7
Galileo Trial 86–8, 91–6
Gassendi, Pierre 175
"Geographical Plan" (Kircher) 169–70, 173, 182–4, 191n17
geographical reform 171–81, 184–7
geometry 63
Germann, Georg 17
Gesner, Conrad 18, 19
Ghetaldi, Marino 27–8
glass 130, 135
Grandamy, Jacques 177, 180
Grassi, Orazio 52, 91, 104, 146, 234, 250
Gregorian calendar 14–17
Grienberger, Christoph 24, 27–8, 43–5
   Aristotelian teachings 65, 66, 70–1
   Collegio Romano 53–5
   correpondence with Clavius 52, 53, 54–5
   death 207
   floating bodies debate 64–6, 67, 68–9
   instruments 45–52
   literary portrait 71
   public mathematics 55–6, 61–2
   Problemata 56–61, 62–3, 65
   relationship with Galileo 63–4, 76n38, 87
   trick-lantern 208
Grimaldi, Francesco Maria 186
Guericke, Otto von 141, 143–6, 227
Guiducci, Mario 99
Guldin, Paul 24, 85

Habsburg dynasty 96–7, 99–100, 222–4
Hartlib, Samuel 135, 249
Hay, John 23
Heerbrandus, Jakobus 15
"heliotropic telescope" 49, 51–2
Hobbes, Thomas 221–2
Horoscopium Catholicum 181–2, *183*, 188
Huygens, Constantijn 255
hydraulics 141–2, 143–6
    Mechanica Hydraulico-Pneumatica
        (Schott) 213–18, 227, 228

Ignatius Loyola 21, 42, 72n9, 100–1, 170
Inchofer, Melchior 85, 87, 109–10n29
    "Monarchy of the Solipsists"
        (*Monarchia solipsorum*) 100–1
    Tractatus Syllepticus 88–9, 91, 93, 95,
        110n31
Indepetae 24
Ingoli, Francesco 15
instruments 45–52, 184–7. *See also*
    machines
intrinsic values 98

jactantia / jactatores 43
Jesuit correspondence 169–71
    magnetic geography 173–81
Jesuit infrastructure 188–9
Jesuit mathematical community 9–10
Jesuit mathematical culture 53–5. *See also*
    public mathematics
Jesuit mathematical practice 13–14, 22,
    259–60
    Clavius' disciples 23–8
    discredit 260
    loss of credibility 256–8
Jesuit missionaries 167–8
    global network 169–70
Jesuit natural philosophy 2, 3
Jesuit network 259
Jesuit "Zenonism" 139
Jesuits 1–2. *See also* Society of Jesus

Kaczorononski, Simon 26–7
Karl, Archduke 96
Kircher, Athanasius 24
    encounter with Philip Skippon 252–3
    encounter with Queen Christina 231–2
    Galileo Trial 86–7

"Geographical Plan" 169–70, 173,
    182–4, 191n17
Horoscopium Catholicum (sundial)
    181–2, *183*, 188
Jesuit infrastructure 188–9
loss of credibility 252–6, 257
machines 197, 199, 200, 201–6, 213–18,
    229, 230–1, 236–8
    catoptric chests 219–21
    magnetic anemoscope *211*, 212
    mathematical organ 250–1
    speaking statue (Delphic Oracle)
        71, 203, 204–6, 209, 226
    sunflower clock *219*, 237
magic 209–11, 224, 229, 230, 238
magnetic geography 171–81
Martini 168–9
Musaeum Kircherianum 197, *198*, 200,
    212, 231, 238
musaeum mathematicum 206, *207*,
    208
Phonurgia nova 205
relationship with Habsburg dynasty
    222–4
Riccioli 185, 194n63
vacuum experiments 127, *128*, 145, 146
vegetable phoenix 254
Zucchi 131, 132–3
Kojalowicz-Wijuk, Albertus 137, 138, 142
Krüger, Oswald 137, 138, 179–80

Lamormaini, William 139, 140, 141
Leopold, Archduke 96, 97, 99–100
letter writing 170–1
light 135
Lipsius, Justus 221
longitude at sea 76–7n45, 169, 173–5, 176
lunar eclipses 175–6

machines 197, 199, 200, 201–6, 207–8, 229,
    230–1, 236–8. *See also*
    instruments
    catoptric chests 219–21
    and civility 221–2
    earth-moving machine *60*
    magnetic anemoscope *211*, 212
    mathematical organ 250–1
    Mechanica Hydraulico-Pneumatica
        (Schott) 213–18

speaking-statue (Delphic Oracle) 71, 204–6, 209, 226
sunflower clock 219, 237
Maestlin, Michael 15–16, 17
Magalotti, Lorenzo 125, 238
magic 209–11, 212, 213–14, 224–30, 236, 238
Magiotti, Rafaello 127–8
magnetic anemoscope *211*, 212
magnetic declination 167–9, 175, 176–7, *178*
magnetic geography 171–81
Magni, Valeriano
    conflicts with Jesuits 140, 141
    Galileo Trial 92
    incarceration 143
    vacuum experiments 128–31, 133–5, 137, 137–8, 142–3, 144
Malaspina, Pietro Francesco 25
Malcolm, Noel 169
Martini, Martino 167–9, 180–1
mathematical academy 168, 206–7
mathematical community 9–10
mathematical continuum 127, 138, 139
mathematical culture 53–5
mathematical museum (*musaeum mathematicum*) 6, 53, 168, 206, 207, 208
mathematical organ 250–1
"mathematical phenomenism" 95
mathematical practice. *See* Jesuit mathematical practice
mathematics 19, 225. *See also* physico-mathematics; public mathematics
Mechanica Hydraulico-Pneumatica (Schott) 142, 143–4, 146, 207–8, 213–18, 227, 228
    antlia pneumatica *145*
    multimammary goddess 209, *210*
mechanical clocks 85
Mercurian, Everard 13, 18
Mersenne, Marin 139, 175, 221
missionaries 167–8
    global network 169–70
modesty 42, 57, 95
"Monarchy of the Solipsists" (*Monarchia solipsorum*) 2–3, 100–1
Moray, Robert 255

Mordente, Fabricio 54
motion 62–3
Musaeum Kircherianum 197, *198*, 200, 212, 231, 238
musaeum mathematicum (mathematical museum) 6, 53, 168, 206, 207, 208

Nadal, Jerónimo 10–11, 12
natural magic 213–14, 224–5, 228, 229, 230, 236, 238
Naudé, Gabriel 86
navigation 167–9. *See also* longitude at sea
New Almagest (Riccioli) 187
Noël, Étienne 135–7

observational instruments 45–52, 184–7
Oedipus Aegyptiacus (Kircher) 209, 222–4
Oldenburg, Henry 254, 256
Opera Omnia (Clavius) 28–30
operative magic. *See* artificial magic
optics 225
Ordinatio pro studiis superioribus 90, 103–4
Ossiander, Lucas 15

Pallavicino, Sforza 95, 102, 125, 147
pantograph *48*, 96
Pascal, Blaise 135–7, 142
Paul II, Pope 97
Payne, Robert 254–5
Peace of Prague 141
Peace of Westphalia 141
pedagogy 11
peer review 1, 2. *See also* censorship
Peiresc, Nicholas Fabri de 86, 87–8, 107–8n, 175, 176, 218, 237
Pereira, Benito 19, 62, 227, 229
Philippus, Henricus 85
philosophy 147
Phonurgia nova (Kircher) 205
physico-mathematics 63–70
physics 225
Piccolomini, Francesco 94
Polanco, Juan de 170–1
Possevino, Antonio 10, 16, 17, 18–19
print culture 10
Problemata 56–61, 62–3, 65
    floating bodies debate 65–8, 69–70

Prodromus (Scheiner) 93–4
public mathematics 55–6, 61–2
   Problemata 56–61, 62–3, 65
      floating bodies debate 65–8, 69–70

Ratio Studorium 11, 12, 20, 56, 57
Reformed Geography (Riccioli) 184
Revisores 89, 102, 136, 249
Riccardi, Niccolò 98
Ricci, Matteo 23, 168
Riccioli, Giambattista 182, 184–7
Rosa Ursina (Scheiner) 49–52
Royal Society 255, 256, 257, 258
Rubino, Giovanni Antonio 24
Rules of the Society of Jesus 42

Salino, Bernardino 24, 25–6
Santi, Leone 102
Scaliger, Joseph 17
scenographic instrument 45–8, 49
Scheiner, Christoph 43
   censura of Inchofer's *Tractatus Syllepticus* 91, 93, 95
   Galileo Trial 86–8, 91–6
   Galileo's attack on 99
   pantograph *48*, 96
   Prodromus 93–4
   relationship with Habsburg dynasty 96–7, 99–100
   Rosa Ursina 49–52
Schönborn, Johann Philipp von 143, 144
Schott, Kaspar
   association with Kircher 212–13
   catropic chest 221
   Habsburg dynasty 222–3
   magic 224–8
   mathematical organ 250–1
   Mechanica Hydraulico-Pneumatica 142, 143–4, 146, 207–8, 213–18, 227, 228
      antlia pneumatica *145*
      multimammary goddess 209, *210*
   Musaeum Kircherianum 231–2
scientific counter-revolution 1

scientific practice 42–4
Scotti, Giulio Clemente 100
scriptors 43
self-abnegation 43, 57, 188
Shapin, Steven 43
Skippon, Philip 252–3
Society of Jesus 3, 57, 260
solar eclipses 175–6
Southwell, Robert 254, 258
speaking statue (Delphic Oracle) 71, 203, 204–6, 209, 226
Stansel, Valentin 208
Stelluti, Francesco 67, 70
Straet, Jan van der (Strandanus) 174
sundial (*Horoscopicum Catholicum*) 181–2, *183*, 188
sunflower clock *219*, 237

"telescopic heliotrope" 49, 51–2
Torricelli, Evangelista 125, 126, 128, 141, 142, 228
Tractatus Syllepticus (Inchofer) 88–9, 91, 93, 95, 110n31

University of Prague 139–40

vacuum experiments 125–31, 227–8
   Jesuit reactions 131–9, 141–7
vegetable phoenix 254
Villalpando, Juan Bautista 27
Vitelleschi, Muzio 100, 111n40
   censorship 90, 91
   continuum of indivisibles 138
   Galileo 92
   Galileo Trial 88
   "Geographical Plan" (Kircher) 173
   mechanical clocks 85

Wilding, Nick 175, 176
Wren, Christopher 256
Wyclif, John 139

"Zenonism" 139
Ziegler, Johann Reinhard 17, 25, 28, 30
Zucchi, Niccolò 131–3, 138, 145, 146

www.ingramcontent.com/pod-product-compliance
Lightning Source LLC
Chambersburg PA
CBHW072128290426
44111CB00012B/1823